Lecture Notes in Computer Science   9927

Commenced Publication in 1973
Founding and Former Series Editors:
Gerhard Goos, Juris Hartmanis, and Jan van Leeuwen

Editorial Board

David Hutchison
  Lancaster University, Lancaster, UK
Takeo Kanade
  Carnegie Mellon University, Pittsburgh, PA, USA
Josef Kittler
  University of Surrey, Guildford, UK
Jon M. Kleinberg
  Cornell University, Ithaca, NY, USA
Friedemann Mattern
  ETH Zurich, Zurich, Switzerland
John C. Mitchell
  Stanford University, Stanford, CA, USA
Moni Naor
  Weizmann Institute of Science, Rehovot, Israel
C. Pandu Rangan
  Indian Institute of Technology, Madras, India
Bernhard Steffen
  TU Dortmund University, Dortmund, Germany
Demetri Terzopoulos
  University of California, Los Angeles, CA, USA
Doug Tygar
  University of California, Berkeley, CA, USA
Gerhard Weikum
  Max Planck Institute for Informatics, Saarbrücken, Germany

More information about this series at http://www.springer.com/series/7409

Jennifer A. Miller · David O'Sullivan
Nancy Wiegand (Eds.)

Geographic Information Science

9th International Conference, GIScience 2016
Montreal, QC, Canada, September 27–30, 2016
Proceedings

Springer

Editors
Jennifer A. Miller
University of Texas at Austin
Austin, TX
USA

Nancy Wiegand
University of Wisconsin-Madison
Madison, WI
USA

David O'Sullivan
University of California at Berkeley
Berkeley, CA
USA

ISSN 0302-9743   ISSN 1611-3349   (electronic)
Lecture Notes in Computer Science
ISBN 978-3-319-45737-6   ISBN 978-3-319-45738-3   (eBook)
DOI 10.1007/978-3-319-45738-3

Library of Congress Control Number: 2016949627

LNCS Sublibrary: SL3 – Information Systems and Applications, incl. Internet/Web, and HCI

Printed on acid-free paper

This Springer imprint is published by Springer Nature
The registered company is Springer International Publishing AG Switzerland

Preface

The 9th International Conference on Geographic Information Science, held during September 27–30, 2016, continued the highly successful GIScience series of conferences. GIScience regularly brings together more than 200 international participants from academia, industry, and government organizations to discuss and advance the state of the art in the field of geographic information science. Since 2004, the biennial conference has alternated between locations in North America and Europe. For example, GIScience 2012 was held in the American Midwest, in Columbus, Ohio, and GIScience 2014 was located in the heart of Europe, hosted by the Vienna University of Technology in Austria. GIScience 2016 took place in Montreal, Canada.

Since its inception in 2000, the biennial GIScience conferences have adopted a two-track program soliciting submissions that describe the highest quality of completed research as well as those that describe the latest work in progress. There is a full-paper track with papers up to 15 pages and a short-paper track (formerly referred to as the extended abstract track) with papers of 4 pages and up to 1500 words. The full papers are contained in these proceedings, and they are complemented by additional and separate short-paper proceedings that are distributed digitally at the conference.

The two-track program is ideally suited to the diversity of disciplines that converge on GIScience, which include (but are not limited to) geography, cognitive science, computer science, engineering, information science, linguistics, mathematics, philosophy, psychology, social science, and (geo)statistics. The combination of full papers and short papers presented through talks and poster sessions has a proven record of delivering an exciting conference program that is both fast-moving and high quality.

The accepted papers provide a snapshot of the breadth of active research topics in the vibrant field of GIScience. The full and short papers showcase a mix of advanced research in topics of long-standing relevance to GIScience (e.g., spatial analysis, spatial cognition, geovisualization, and geo-ontologies) as well as topics of more recent research interest such as user-generated content, linked data, computational movement analysis, and text-based navigation systems.

For GIScience 2016, there were 63 full-paper submissions. Each paper was thoroughly reviewed by two to four independent members of the international Program Committee. Based on these reviews, supplemented by careful meta-reviews from the program chairs, 21 papers were selected for presentation, corresponding to an acceptance rate of 33 %. A total of 152 short papers were submitted, 51 of which were accepted for oral presentation and 54 of which were accepted as poster presentations (69 % acceptance rate).

The conference could not have happened without the work of the general chairs, Renee Sieber and Scott Bell, and the local organizer Raja Sengupta. We would also like to thank Steven Farber and Michael Widener as the workshop and tutorial chairs. In addition we are appreciative of the work done by Mir Abolfazl Mostafavi and Gaurav Sinha as sponsorship co-chairs and of Jing Hoon Teo for logistical support. We are also deeply grateful to the GIScience Program Committee for their considered and thorough reviews, as well as those additional reviewers who also contributed their expertise. We would like to thank all the authors who contributed to the conference by submitting full and short papers. Most importantly, we would like to thank all those who came to GIScience as presenters and participants, without whose contributions there would be no conference. The submitters, reviewers, and participants are critical to keeping the field of GIScience fresh, cutting-edge, and alive.

August 2016

Jennifer A. Miller
David O'Sullivan
Nancy Wiegand

Organization

General Chairs

Renee Sieber   McGill University, Canada
Scott Bell   University of Saskatchewan, Canada

Program Co-chairs

Jennifer A. Miller   University of Texas-Austin, USA
David O'Sullivan   University of California-Berkeley, USA
Nancy Wiegand   University of Wisconsin-Madison, USA

Workshop and Tutorial Chairs

Steven Farber   University of Toronto, Canada
Michael Widener   University of Toronto, Canada

Local Organizing Chair

Raja Sengupta   McGill University, Canada

Logistical Support

Jing Hoon Teo   McGill University, Canada

Program Committee

Ola Ahlqvist   The Ohio State University, USA
Jared Aldstadt   University at Buffalo, USA
Kate Beard-Tisdale   University of Maine, USA
Itzhak Benenson   Tel Aviv University, Israel
David Bennett   The University of Iowa, USA
Luke Bergmann   University of Washington, USA
Ling Bian   University at Buffalo, USA
Boyan Brodaric   Geological Survey of Canada, Canada
Dan Brown   University of Michigan, USA
Chris Brunsdon   National University of Ireland, Maynooth, Ireland
Barbara Buttenfield   University of Colorado Boulder, USA
Nicholas Chrisman   CaGIS
Jonathan Cinnamon   University of Exeter, UK
Christophe Claramunt   Naval Academy Research Institute, France

Atsuyuki Okabe	The University of Tokyo, Japan
Edzer Pebesma	University of Muenster, Germany
Karin Pfeffer	University of Amsterdam, The Netherlands
Ross Purves	University of Zürich-Irchel, Switzerland
Martin Raubal	ETH Zurich, Switzerland
Tarmo Remmel	York University, Canada
Britta Ricker	University of Washington Tacoma, USA
Simon Scheider	ETH Zürich, Switzerland
Raja Sengupta	McGill University, Canada
Shih-Lung Shaw	University of Tennessee, USA
Takeshi Shirabe	Royal Institute of Technology (KTH), Sweden
Alex Singleton	University of Liverpool, UK
Gaurav Sinha	The Ohio University, USA
Seth Spielman	University of Colorado Boulder, USA
Emmanuel Stefanakis	University of New Brunswick, Canada
Kathleen Stewart	University of Maryland, USA
Martin Swobodzinski	Portland State University, USA
Jim Thatcher	University of Washington Tacoma, USA
Jean-Claude Thill	University of North Carolina Charlotte, USA
Sabine Timpf	University of Augsburg, Germany
Ming-Hsiang Tsou	San Diego State University, USA
Marc Van Kreveld	Utrecht University, The Netherlands
Monica Wachowicz	University of New Brunswick, Canada
Shaowen Wang	University of Illinois at Urbana-Champaign, USA
Robert Weibel	University of Zurich, Switzerland
John Wilson	University of Southern California, USA
Matthew Wilson	University of Kentucky, USA
Stephan Winter	The University of Melbourne, Australia
Michael Worboys	University of Greenwich, UK
Dawn Wright	Environmental Systems Research Institute and Oregon State University, USA
Ningchuan Xiao	The Ohio State University, USA
Phil Yang	George Mason University, USA
Eunhye Yoo	University at Buffalo, USA
May Yuan	University of Texas at Dallas, USA
Naijun Zhou	University of Maryland at College Park, USA

Platinum Sponsor

University Sponsors

Contents

Spatial Algorithms

Computing River Floods
Using Massive Terrain Data

Cici Alexander[1], Lars Arge[2], Peder Klith Bøcher[3], Morten Revsbæk[4],
Brody Sandel[2,3], Jens-Christian Svenning[3], Constantinos Tsirogiannis[2,3(✉)],
and Jungwoo Yang[4]

[1] Dynamiques de l'Environnement Côtier IFREMER, Issy-les-Moulineaux, France
calexand@ifremer.fr
[2] MADALGO, Aarhus University, Aarhus, Denmark
{large,constant}@madalgo.au.dk
[3] Department of Bioscience, Aarhus University, Aarhus, Denmark
{peder.bocher,brody.sandel,svenning}@bios.au.dk
[4] SCALGO, Aarhus, Denmark
{morten,jungwoo}@scalgo.com

Abstract. Many times in history, river floods have resulted in huge
catastrophes. To reduce the negative outcome of such floods, it is important to predict their extent before they happen. For this reason, special-
ists use algorithms that model river floods on digital terrains datasets.
Nowadays, massive terrain datasets have become widely available. As
flood modeling is an important part for a wide range of applications, it
is crucial to process such datasets fast even with standard computers.
Yet, these datasets can be several times larger than the main memory of
a standard computer. Unfortunately, existing flood-modeling algorithms
cannot handle this situation efficiently. Hence they have to sacrifice output quality for time performance, or vice versa.

In this paper, we present a novel algorithm that, unlike any previous
approach, can both provide high-quality river flood modeling and handle
massive terrain data efficiently. More than that, we redesigned an existing popular flood-modeling method (approved by European Union and
used by authorities in Denmark) so that it can efficiently process huge
terrain datasets. Given a raster terrain $\mathcal{G}$ and a subset of its cells representing a river network, both algorithms estimate for each cell in $\mathcal{G}$ the
height that the river should rise for the cell to get flooded. Based on our
design, both algorithms can process terrain datasets that are much larger
than the main memory of a computer. For an input raster that consists
of N cells, and which is so large that it can only be stored in the hard
disk, each of the described algorithms can produce its output with only
$O(\text{sort}(N))$ transfers of data blocks between the disk and the main memory. Here $\text{sort}(N)$ denotes the minimum number of data transfers needed
for sorting N elements stored on disk. We implemented both algorithms,
and compared their output with data acquired from a real flood event.
We show that our new algorithm models the real event quite accurately,

MADALGO—Center for Massive Data Algorithmics, a Center of the Danish
National Research Foundation.

© Springer International Publishing Switzerland 2016
J.A. Miller et al. (Eds.): GIScience 2016, LNCS 9927, pp. 3–17, 2016.
DOI: 10.1007/978-3-319-45738-3_1

more accurately than the existing popular method. We evaluated the efficiency of the algorithms in practice by conducting experiments on massive datasets. Each algorithm could process a dataset of 268 GB size on a computer with only 22 GB working main memory (twelve times smaller than the dataset itself) in at most 31 h.

1   Introduction

Throughout history, river floods have caused large disasters. Usually induced by heavy rainfall, such floods can lead to casualties and huge financial damage for the local communities. A recent example is the catastrophic flood of the Indus river in Pakistan that took place in 2010 [9]. This flood claimed approximately two thousand lives, and about one fifth of the total area of the country ended up covered by water. Society wants to predict such floods, so that measures can be taken in advance to reduce the harm done. Therefore, it is important for people to know which regions around a river have the highest risk of getting flooded when the level of the river rises.

Today, hydrologists use computers to model river floods; they use specialised software to simulate flood events based on digital representations of terrains and rivers. Such terrain representations are widely known as Digital Elevation Models (DEMs). The most popular type of DEMs is the so-called *grid* or *raster* DEMs. In a raster DEM the domain of the terrain is divided into square cells of equal size, and each cell is associated with an elevation value.

One method for modeling river floods on DEMs is the method introduced by Berg Sonne [12]; let $\mathcal{G}$ be a raster terrain and let $R(\mathcal{G})$ be the set of cells in $\mathcal{G}$ that represents the region covered by a river network in this terrain. Also, let x be a positive real. Given $\mathcal{G}$, $R(\mathcal{G})$ and x, the method estimates which cells in $\mathcal{G}$ will get flooded if the level of the river $R(\mathcal{G})$ rises uniformly by x meters. Of course, a flood is a very complex phenomenon and is influenced by many factors, some of which are difficult to determine. Therefore, we cannot expect that a flood can be modeled precisely by the output of any method, no matter how involved this method is. Yet, the method proposed by Berg Sonne is today considered a quite accurate tool for modeling river floods. Hence, after approval by the European Union it is used by the state authorities of Denmark [12].

However, Berg Sonne's method has a major drawback; it cannot process massive DEMs. Recent advances in Lidar technology have made it possible to produce detailed and huge DEM datasets. In many cases, such a dataset is so large that it cannot fit in the main memory of a standard computer. Hence, the dataset has to be stored mainly on disk. Since the computer's processor can only handle data that appear in the main memory, blocks of data have to be transferred between the disk and the memory in order to process the dataset. We call a transfer of a single block of data between the disk and the memory an *I/O-operation*, or an I/O for short. The problem here is that a single I/O is an extremely slow operation; it can take about the same time as a million CPU operations. Therefore, when it comes to processing huge amounts of data, it is important to process the dataset in a way that we minimize the number of data transfers between the disk and the memory. Otherwise, the whole process becomes practically infeasible.

Standard algorithms are often designed based on the assumption that all input data fit in the main memory. Hence, usually they cannot handle massive datasets. This is also the case for algorithms that are used to model river floods; to the best of our knowledge, there does not exist any such algorithm that can deal with this problem[1]. Nowadays flood modeling has become a crucial part for a wide range of professionals, from researchers to freelance civil engineers. Therefore, it is important that this modeling can be performed efficiently even on standard computers. Yet, even moderately large terrain datasets can be several times larger than the main memory of a standard computer. Therefore, the users of up-to-date hydrological software are forced to choose between two approaches. In the first approach, the resolution of the input DEM is reduced (so it fits entirely in the main memory). Thus, a large amount of detail in terrain data is thrown away. Important features on the landscape, such as ditches and levees, may not be depicted anymore on the resulting terrain. When it comes to modeling a river flood, this results in incorrect estimations. In the other approach, users divide the massive DEM into smaller tiles and each tile is processed independently; in this way, when processing a single tile, we do not take into account how the rest of the landscape affects the flood in that region. Therefore, there is a need for developing algorithms that, on one hand model river floods accurately, and on the other hand efficiently handle massive terrain datasets.

Our Results. Inspired by the above, we designed two I/O-efficient algorithms that can be used for modeling river floods. The first algorithm is an adaptation of Berg Sonne's method that can handle massive raster terrains. The second algorithm is a novel method that we introduce for modeling river floods. For each of these algorithms, the input is a raster $\mathcal{G}$, and a subset $R(\mathcal{G})$ of cells in $\mathcal{G}$ representing the area covered by a river network. Each of our algorithms returns for each cell $c \in \mathcal{G}$ a value $f(c)$, indicating the minimum number of meters the river level should rise before c gets flooded. We call this value the *resistance* value of c. Given the resistance values $f(c)$ for every $c \in \mathcal{G}$, and a positive integer x, we can then easily extract the part of the terrain that is flooded if the river level rises uniformly by x meters. As we describe later in detail, each of the algorithms that we present uses different criteria for computing resistance values, hence they produce different outputs.

To process massive datasets efficiently, we have designed our algorithms based on the I/O-model of Agarwal and Vitter [3]. The performance of an algorithm in the I/O-model is measured as the number of I/Os (transfers of data blocks) that take place during its execution between the disk and the main memory. This measure of performance is called the *I/O-efficiency* of the algorithm. To describe the I/O-efficiency we need three parameters; the size N of the input data, the size of the internal memory M, and the size B of a single block of data that can

---

[1] This is not the case for other types of floods, which have received ample attention in this context [7].

be transferred from and to the disk. Two basic processes that take place during the execution of most algorithms is scanning and sorting. We can scan a set of N elements stored in the disk with $O(\text{scan}(N))$ I/Os, where $\text{scan}(N) = N/B$. We can also sort a set of N records in an I/O-efficient manner with $O(\text{sort}(N))$ I/Os, where $\text{sort}(N) = N/B \log_{M/B} N/B$ [3]. To compute the output for a raster that has N cells, each of our algorithms require $O(\text{sort}(N))$ I/Os in the worst case.

We implemented both algorithms and measured their efficiency in practice. For our measurements we used a terrain dataset of 268 GB, which we processed on a computer with 22 GB of working main memory (roughly twelve times smaller than the size of the described dataset). For this setting, to process the entire 268 GB dataset, our adaptation for Berg Sonne's method required roughly 24 h, and our new algorithm roughly 31 h.

We also conducted experiments to evaluate whether our algorithms can model adequately real flood events. To do this, we used as reference a vector dataset which outlines the river flood that took place in Pakistan in 2010 [9]. Using a variety of experiment settings, we showed that our algorithm provided on average more accurate results than Berg Sonne's method. To understand the reasons behind this, we used both algorithms to model river floods on a massive raster that represents the terrain in Denmark. Among other artifacts, the method by Berg Sonne produced flooded regions that had larger size than the ones calculated by our algorithm. As we explain, one reason for this is that Berg Sonne's method produces very small resistance values for areas along the entire coastline of the terrain.

2   Description of the Algorithms

Problem Definition and Notation. Let $\mathcal{G}$ be a grid terrain that consists of N cells. For every cell $c \in \mathcal{G}$ we use $h(c)$ to indicate the elevation of the terrain at this cell. We denote the cell that appears at the i-th row and j-th column of $\mathcal{G}$ by $\mathcal{G}(i,j)$. We assume without loss of generality that the center of grid cell $\mathcal{G}(i,j)$ has xy-coordinates (j,i). For any cell $c \in G$, we denote this center point by $p(c)$. We call the xy-*distance*, or simply the *distance*, between two cells in $\mathcal{G}$ the 2D Euclidean distance between their cell centers on the xy-domain of $\mathcal{G}$. Let C be a set of cells in $\mathcal{G}$ and let c be a cell that belongs to this set. We say that c is the *closest cell* in C to another cell c' if c has the smallest xy-distance to c' compared to any other cell in C.

We use $R(\mathcal{G})$ to denote a subset of the cells in $\mathcal{G}$ that belong to a river network of the terrain. We call these cells the *river cells* of $\mathcal{G}$. The cells in $R(\mathcal{G})$ represent the river network in $\mathcal{G}$ when there is no flood. This implies that the elevation value of each cell in $R(\mathcal{G})$ approximates the average height of the river level at this location when no flood occurs. For the algorithms that we present, we assume that $R(\mathcal{G})$ is provided as part of the input.

Let h_{rise} be a positive real. We say that there is a *river rise* of h_{rise} meters, or that *the river rises* by h_{rise} meters, when for each cell $c \in R(\mathcal{G})$ the river level rises to elevation $h(c) + h_{\text{rise}}$. We call h_{rise} the *rise value*. Thus, when a river rises we assume that its level increases by the same amount at all river cells.

We study the following problem. Given a terrain $\mathcal{G}$ and its river network $R(\mathcal{G})$, we want to compute for every cell $c \in \mathcal{G}$ a value f(c) that estimates the minimum value h_{rise} such that c gets flooded when the river rises by h_{rise} meters. We call this value the *resistance* value of c. Each of the two algorithms that we present in this paper defines these resistance values in a different way; hence, for the same input grid the output between the two algorithms may differ substantially. In the description of each algorithm we provide a detailed definition for the resistance of a grid cell for this algorithm. For both algorithms, it is assumed that all river cells are flooded by default. Therefore, for both approaches we imply that the resistance value of every river cell is set to zero.

2.1   Adaptation of Berg Sonne's Method

The first algorithm that we describe is based on the flood modeling method introduced by Berg Sonne [12]. Originally, this method was designed to solve a more simple problem than the one that we examine. In particular, the input of the original method is a raster $\mathcal{G}$, the river network $R(\mathcal{G})$, and a rise value h_{rise}. Instead of computing flood resistance values, the method outputs the cells in $\mathcal{G}$ that are considered to get flooded when $R(\mathcal{G})$ rises by h_{rise} meters. We call this version of the method *ProximityFlood*. Below, we first explain how *ProximityFlood* calculates the flooded cells in $\mathcal{G}$ for a given rise value h_{rise}. Then, we show how we can use this method to design an I/O-efficient algorithm that computes a flood resistance value for each input cell[2].

ProximityFlood consists of two steps. In the first step, every cell $c \in \mathcal{G} \setminus R(\mathcal{G})$ gets associated with a single river cell in $R(\mathcal{G})$; this is the river cell from which we consider that c can potentially get flooded. We call this cell the *source* cell of c, and we denote this by source(c). The source cell for every $c \in \mathcal{G} \setminus R(\mathcal{G})$ is defined as the river cell $c' \in R(\mathcal{G})$ that has the smallest xy-distance from c. After calculating source(c) for every non-river cell c, the height difference between source(c) and c is computed and stored together with c. We call this value the *obstruction* value obst(c) of c.

In the second step, we extract the cells in $\mathcal{G}$ that are considered to get flooded when the river rises by h_{rise} meters. More specifically, we extract any cell c that (a) has an obstruction value obst[c] $\leqslant h_{\text{rise}}$ *and* (b) there exists a path of cells between c and a river cell c_R such that any non-river cell c' in this path has an obstruction value obst(c') $\leqslant h_{\text{rise}}$. Notice that, in this way, not all cells with obstruction $\leqslant h_{\text{rise}}$ are flooded.

---

[2] Some implementations of *ProximityFlood* include an extra preprocessing step where the heights of the river cells are adjusted to make it consistent with the rest of the terrain data. This prevents artifacts (e.g. rivers that flow upstream) that may appear when river data are combined with DEMs acquired from a different sources. In our description of *ProximityFlood* we do not include this preprocessing step; we consider that this has to do more with configuring the datasets rather than with the method itself. Yet, this step can be also handled I/O-efficiently, given realistic assumptions on the memory size.

Method *ProximityFlood* can be used to model a single flood event at a time. On the other hand, if we want to study which regions get flooded for different rise values then we have to run this method many times, once for each distinct rise value h_{rise}. To avoid this, we instead choose to compute for each cell c the minimum rise value h_{rise} for which c gets flooded according to method *ProximityFlood* (the resistance value of c). Below we describe our new I/O-efficient algorithm that computes the resistance values on $\mathcal{G}$, which we call *ProximityResistance*.

As with *ProximityFlood*, the new algorithm consists of two main steps. In the first step, we compute for each cell $c \in \mathcal{G} \setminus R(\mathcal{G})$ the source cell source(c) and the obstruction obst(c). In the second step, we calculate the flood resistance values of all cells in $\mathcal{G} \setminus R(\mathcal{G})$.

Computing the source cells and obstruction values. For the first step, the main task is to compute the source cell for each non-river cell c; given this cell, it is straightforward to compute the obstruction obst(c). Calculating the source cells in $\mathcal{G}$ is equivalent to computing a Voronoi diagram on the xy-domain of $\mathcal{G}$; the sites of the Voronoi diagram are the center-points of the river cells in $\mathcal{G}$ and for any cell $c \in \mathcal{G} \setminus R(\mathcal{G})$ it holds that source(c) = c' if the center of c falls in the Voronoi region of $p(c')$. Computing the Voronoi diagram of the river cells can be done in $O(\text{sort}(N))$ I/Os [2,11]. Then we sweep simultaneously, from top to bottom, the diagram and grid $\mathcal{G}$. During the sweep, we maintain the diagram edges that intersect the sweep line, sorted according to the x-coordinate of their intersection point with this line. For every row of $\mathcal{G}$ that we encounter, we scan the edges that intersect the sweep line to determine the Voronoi region (and therefore the corresponding source cell) where each cell in the row belongs to. Notice that the number of edges in the sweep line is at most two times the number of cells in a row. This is because there cannot be more than two river cells per column whose Voronoi regions intersect the same horizontal line. Therefore, scanning the raster and updating the sweep line can be done efficiently in $O(\text{sort}(N))$ I/Os in total. From this we conclude that computing the source cells on the raster can be performed in $O(\text{sort}(N))$ I/Os. But we can do this more efficiently; in the full version of the paper we present an algorithm that computes the source cells using $O(\text{scan}(N))$ I/Os [4].

Computing the flood resistance values. In the second step of method *ProximityResistance* we compute for each cell $c \in \mathcal{G}$ its flood resistance f(c). Recall that for every cell c this resistance value is equal to the minimum rise value h_{rise} such that obst(c) $\leqslant h_{rise}$ and c is connected to the river by a path of cells with obstruction $\leqslant h_{rise}$. Based on this definition, we can reduce the computation of the flood resistance values to the problem of computing the *raise elevations* on a terrain, that was described by Danner *et al.* as part of their partial flooding algorithm [7]. This problem is defined as follows; let $\mathcal{G}$ be a raster and let $\zeta_1, \ldots, \zeta_k$ be a set of cells in $\mathcal{G}$ that we call *sinks*. For any path of cells π in $\mathcal{G}$ the height of π is defined as the height of the highest cell on this path. The *raise elevation* of a cell $c \in \mathcal{G}$ is the minimum height among all paths that connect c to ζ_i for any $1 \leqslant i \leqslant k$. Arge *et al.* provide an algorithm that computes the raise elevations for all the cells on the terrain in $O(\text{sort}(N))$ I/Os [7].

We can reduce the problem of computing the flood resistance values of the cells in $\mathcal{G}$ to an instance of the raise elevation problem as follows; we create a raster $\mathcal{G}'$ that has the same number of rows and columns as $\mathcal{G}$. For any river cell $\mathcal{G}(i,j) \in R(\mathcal{G})$ we let the corresponding cell $\mathcal{G}'(i,j)$ to be a sink. For any non-river cell $\mathcal{G}(i,j)$ we let cell $\mathcal{G}'(i,j)$ have elevation equal to the obstruction value of $\mathcal{G}(i,j)$. It is now easy to see that the raise value of any cell $\mathcal{G}'(i,j)$ is equal to the flood resistance value that we want to compute for $\mathcal{G}(i,j)$. By applying the I/O-efficient algorithm of Arge et $al.$ on $\mathcal{G}'$ we can compute the described flood resistance values in $O(\text{sort}(N))$ I/Os, which is the total I/O-efficiency of this algorithm.

2.2   Our New Method

In *ProximityResistance*, a cell c can only get flooded from the closest river cell source(c) in the xy-plane. Intuitively, this is very unnatural since the flow of water on the terrain is obviously influenced by the terrain topography. Therefore, we introduce a novel method which instead chooses source(c) based on a model that represents how water flows on the terrain. We refer to this new method as *UpstreamResistance*. Next, we describe how source(c) is chosen in *UpstreamResistance*, and then we show how to compute this I/O-efficiently.

For a raster $\mathcal{G}$ let $\mathcal{F}(\mathcal{G}) = (V, E)$ be the graph such that for each cell $c \in \mathcal{G}$ there exists exactly one vertex $v(c)$ in V, and there exists a directed edge in E from $v(c)$ to $v(c')$ if cells $c, c' \in \mathcal{G}$ are adjacent and $h(c) > h(c')$. We call this graph the *flow graph* of $\mathcal{G}$. For now let us assume that no adjacent cells in $\mathcal{G}$ have the same elevation value. Hence, there exists exactly one directed edge in $\mathcal{F}(\mathcal{G})$ for each pair of adjacent cells in $\mathcal{G}$, and $\mathcal{F}(\mathcal{G})$ is a DAG. The concept of the flow graph was introduced in previous works to model how water flows between cells on a DEM [7]. It is naturally assumed that water on a cell can flow only to neighbour cells with lower height; that is modeled with a directed edge in the flow graph.

For any cell $c \in G$ water from c may flow following different routes on the raster until reaching one or more cells on the boundary of river $R(\mathcal{G})$. In method *UpstreamResistance* we choose source(c) to be one of these cells on the river boundary, that is, the river cells where the water from c reaches. More formally, let c be a cell in $\mathcal{G}$. Consider a path in $\mathcal{F}(\mathcal{G})$ that starts from vertex $v(c)$ and ends at a vertex $v(c')$ where c' is a river cell, such that the path does not contain a vertex corresponding to any other river cell. We call such a path a *downstream path* of c. Let $DC(c)$ denote the set of all river cells that belong to some downstream path of c. In method *UpstreamResistance*, source(c) is the cell in $DC(c)$ with the highest elevation value. The flood resistance of c is then defined as the height difference $h(c) - h(\text{source}(c))$.

When it comes to implementing *UpstreamResistance* I/O-efficiently, the two key tasks for computing the flood resistances are constructing the flow graph $\mathcal{F}(\mathcal{G})$, and computing the source cell for every cell in $\mathcal{G}$. If no flat areas exist on $\mathcal{G}$, we can construct $\mathcal{F}(\mathcal{G})$ straightforwardly in $O(\text{scan}(N))$ I/Os. On the other hand, when $\mathcal{G}$ contains flat areas the construction of the flow graph becomes a more involved process. Yet, even in this case we can construct the flow graph efficiently, this time in $O(\text{sort}(N))$ I/Os. A detailed description of this process is provided in the full version of the paper [4]. As for computing the source cells, observe that for any cell c it holds that $\text{source}(c) = \text{source}(c')$ for some c' such that there exists an edge in $\mathcal{F}(\mathcal{G})$ from $v(c)$ to $v(c')$. Therefore, we can compute $\text{source}(c)$ by first computing the source cells for those neighbours of c that appear downstream in $\mathcal{F}(\mathcal{G})$, and then use these to infer $\text{source}(c)$. Arge $et\ al.$ describe an I/O-efficient algorithm that computes the number of upstream cells for every cell on a raster in $O(\text{sort}(N))$ I/Os [5]. Their algorithm can be easily modified for computing the source cells in $\mathcal{G}$. Therefore, we can perform this computation in $O(\text{sort}(N))$ I/Os, which defines the I/O-efficiency of the entire algorithm.

3   Implementations and Experiments

We implemented both algorithms described in Sect. 2, and we evaluated how fast they perform in practice, as well as how accurately they model real flood events. We implemented both algorithms in C++, using the open source library TPIE that provides I/O-efficient algorithms for sorting and scanning data [14]. We used the GNU g++ compiler (version 4.8.2), and the experiments were ran on a Linux Ubuntu operating system (release 14.04).

When implementing $ProximityResistance$ we made two modifications compared to the description in Sect. 2. First, when computing the source cells on $\mathcal{G}$ using a sweepline approach, we used the $O(\text{scan}(N))$ approach (described in the full version of this paper) and we made the practically realistic assumption that a constant number of rows in $\mathcal{G}$ can fit in main memory. Thus, instead of performing an external scan of each row and maintaining an I/O-efficient stack during the sweep, we simply store the two last rows that we swept in memory and perform all computations internally. Second, when computing the raise elevations we did not use the $O(\text{sort}(N))$ batched union-find algorithm by Agarwal $et\ al.$ [1] (that is quite involved), but instead a much simpler $O(\text{sort}(N)\log(N/M))$ algorithm also proposed by Agarwal $et\ al.$ Both Danner $et\ al.$ [7] and Agarwal $et\ al.$ showed that this simple union-find algorithm performs very well in practice. When implementing $UpstreamResistance$ we accurately followed the description in Sect. 2. The only difference was that we again used the practical union-find algorithm of Agarwal $et\ al.$ (that requires $O(\text{sort}(N)\log(N/M))$ I/Os), this time for computing the connected components of flat areas in $\mathcal{G}$, and for removing flat areas that correspond to spurious pits.

Measuring I/O-Efficiency in Practice. To measure the practical efficiency of each method, we ran our implementations on a massive raster dataset that represents the terrain surface of the entire country of Denmark. Publicly available through the website of the Danish Ministry of Environment [10], this raster consists of roughly 66.4 billion cells, arranged in 287,500 rows and 231,250 columns. Each cell represents a region of $1.6 \times 1.6\,\text{m}$ on the terrain and is assigned an elevation value which is a 4-byte floating point number. The total size of the uncompressed dataset is 268 GB. We refer to this dataset as `denmark`.

Raster **denmark** does not include any river data, and therefore we had to extract the river cells before conducting the experiments. To do so, we first preprocessed the raster by removing all shallow pits. Then, we selected the river cells based on the size of their upstream area. For this reason, we computed the flow graph of **denmark** as described in Sect. 2 except that for each cell c we included at most one outgoing edge. This outgoing edge points to the vertex $v(c')$ such that c' is a neighbour of c and the vector from $p(c)$ to $p(c')$ has the steepest downward slope. Then, we computed for each cell c the size of its upstream area; this is the area that is covered by all cells c' such that there exists a path from $v(c')$ to $v(c)$ in the flow graph. We extracted the river cells by selecting all cells whose upstream area was larger than $12.5\,\mathrm{km}^2$. We picked this threshold since the resulting river network resembles better the actual shape of the rivers in Denmark, according to available orthophotos.

We ran both of our algorithms on the **denmark** raster and the computed river cells, using a workstation that has a Xeon CPU (W3565), a four-core processor with 3.2 GHz per core. The workstation had 48 GB of main memory, and a raid (redundant array of independent disks) that consists of nineteen disks, with 3 Terabytes capacity in total. To showcase the I/O-efficiency of our algorithms, we reduced the size of working memory on this computer (maximum memory size available at any point during execution) to 22 GB. The total time taken by the implementation of *ProximityResistance* was roughly 24.2 h; only 2.4 h were used for computing the source cell for each non-river cell, and the remaining 21.8 h were spent on computing the resistance values. For the implementation of *UpstreamResistance*, the total execution time was approximately 31.1 h. The first stage of this method, where the flow graph of the input raster is computed, took 12.5 h. The remaining 18.6 h were spent for delineating the flat areas on the terrain, and computing the resistance values. On the same machine and for only a fraction of the same dataset, we attempted to run standard existing, non I/O-efficient, implementations of the original Berg Sonne's method. Yet, even after several days of execution these implementations could not produce an output. This is because these older implementations induce a very large number of I/O operations. Consequently, there is a huge amount of time spent for data transfers between the memory and the disk, while the CPU remains idle. On the contrary, the execution time for both of our I/O-efficient implementations is elegantly distributed between CPU processing and I/O operations. For both of our algorithms, measurements showed that roughly 60 % of the execution time is devoted to CPU activity, and the remaining to disk usage. From the above, it is clear that the implementations of both methods have a very good performance even for a dataset which is much larger than the available main memory. Each method took less than 1.5 days to process this dataset, using memory size which corresponds to roughly 8 % of the dataset's total size.

Evaluating the Quality of Flood Modeling. In the second set of experiments we used an actual flood event to evaluate the quality of the output produced by the two methods. This event is the catastrophic flood of the Indus river that took place in Pakistan in 2010 [9].

For the experiments we used a raster terrain extracted from the SRTM grid, a DEM that represents the earth surface from 60° North to 56° South [13]. The extracted raster covers a square region of approximately $2,160 \times 2,160$ km and includes the entire Indus river basin–see Fig. 1. The raster consists of $24,000 \times 24,000$ cells, and the dimension of each square cell is approximately 90 m. We refer to this dataset as `indus`.

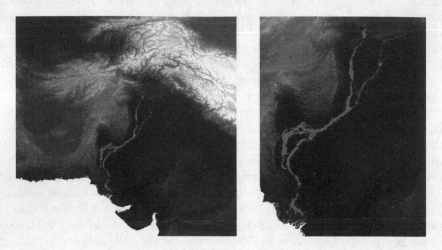

Fig. 1. Left: an illustration of the `indus` DEM together with the `flood` vector dataset. The cells of the DEM appear in grayscale colours, shaded according to their elevation values; cells of higher elevation are indicated by lighter shades. The polygons of the `flood` dataset appear in red colour. Right: a closer view of the flooded regions. (Color figure online)

Since the `indus` DEM does not contain any river data, we extracted the river cells based on the upstream area of each cell, in the same way as we did for the `denmark` dataset. In this case we used a threshold of 300 km^2 of upstream area since it produces a visual result that matches the shape of the local river network, as it appears in orthophotos acquired before the flood.

To evaluate the ability of our algorithms to accurately model floods, we used a vector dataset that shows the actual flooded regions around the river during the Indus river flood. This dataset was released by the Dartmouth Flood Observatory, and contains data acquired with MODIS (Moderate-resolution Imaging Spectroradiometer) technology [8]. We refer to this dataset as `flood`. The `flood` dataset was constructed based on several satellite photos of the Indus region, acquired during the period from the 1st to the 5th of August of 2010. It represents with polygons all the regions that were flooded in at least one day during this period. The bounding box of `flood` covers a rectangular region that spans approximately 1,118 and 911 km on the longitudinal and the longitudinal axes respectively. It contains 4,294 polygons, and the total area covered by these polygons is approximately 30,483 km^2. Refer to Fig. 1.

We ran our implementations of the two flood-modeling algorithms on the indus DEM and the extracted river cells, and we evaluated the output of each algorithm using a method that resembles the *Area-Under-the-Curve* (also known as AUC) measure, which is one of the most popular measures for model testing [6]. In particular, we overlayed flood with indus and extracted the cells in indus whose centers lie in the interior of a polygon in flood. We refer to these cells as the *flooded* cells of indus. In total, we identified slightly more than four million flooded cells. Next we selected at random a large set of pairs of cells. Each pair was selected so that it consists of one flooded cell and one non-flooded cell. We denote this set of pairs by $\mathcal{P}$. For each of our methods, we determined for each pair $pr \in \mathcal{P}$ if the flooded cell in pr scores a higher resistance value than the non-flooded cell, and calculated the percentage of the pairs in $\mathcal{P}$ for which this condition holds. We call this percentage the *output quality* of the method. The value of the output quality is an estimation of the AUC measure; the output quality value is equal to the AUC if $\mathcal{P}$ consists of all possible pairs of flooded/non-flooded cells in the region of interest. For our study, we chose 10^5 pairs, considering that this is a sufficient number for estimating the value of the AUC. For method *ProximityResistance* the output quality is 87 %, while for *UpstreamResistance* the output quality is 92 %. This shows clearly that both of the methods produce flood resistances that are highly consistent with the actual event.

To measure how the two methods perform on a more local scale, we calculated their output quality within several smaller regions. More specifically, within the xy-region covered by flood we extracted three sets of square windows, each set consisting of windows of certain size. In the first set each window is a square with dimension 20 km, in the second set each window has dimension 40 km, and the third set consists of windows of 80 km dimension. The windows of each set were picked in the following way. Within the region covered by flood we extracted at random 500 windows of the same size. Then we used a greedy algorithm to select a subset of these windows, so that there is no pair of windows in the subset that overlap with each other, and so that each window contains at least 500 flooded and at least 500 non-flooded cells. Thus, we ended up with a subset of 119 windows for the first set, and forty-five and twenty-two windows for the second and third set respectively. From each window, we selected 10^5 cell pairs, again so that each pair contains one flooded and one non-flooded cell. We then calculated the output quality of our methods for each window. Figure 2 shows the results for the windows of 20 km dimension, where the mean output quality was 61 % for *ProximityResistance* and 71 % for *UpstreamResistance*. For windows of 40 km dimension, *ProximityResistance* attained mean output quality 69 % and *UpstreamResistance* mean output quality 81 %. For the third set of windows, the values were 76 % and 85 %, respectively.

Therefore, for each window size *UpstreamResistance* has higher mean output quality than *ProximityResistance*. For both methods the output quality increases as the window size becomes larger. Yet, we observed that for all examined window sizes, there exist windows where at least one of the methods has an output quality value of less than 50 %.

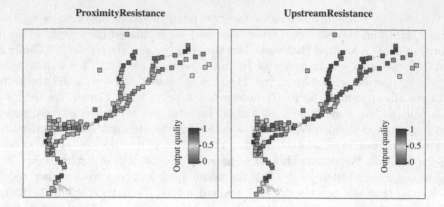

Fig. 2. The locations for the selected windows of 20 km dimension. Windows are represented by colored boxes, and each box is colored according to the method's output quality for this window. The relative size of the boxes in the figure is larger than the size of the original windows, to make each box more visible. The xy-regions of the original windows do not overlap with each other. Left: boxes colored based on the output quality values for `ProximityResistance`. Right: boxes colored according to the output quality values of `UpstreamResistance`. (Color figure online)

To examine the above further, we investigated if there is a correlation between the output quality values and the two following factors: heterogeneity of the terrain (variability of elevation values) and the number of flooded cells inside each window. To measure the heterogeneity of the terrain within each window w, we computed the logarithm of the standard deviation for the elevations of the cells in w. We call this value the *topographic heterogeneity* of w. In order to examine visually the relation between the output quality and the topographic heterogeneity among the different windows, we created a scatter plot for each method. Each scatter plot contains a 2-dimensional point $p(w)$ for every window w; the horizontal coordinate of $p(w)$ is equal to the topographic heterogeneity of w, and the vertical coordinate of this point is equal to the output quality of the method for w. Figure 3 shows the scatter plots that we produced for windows of 20 km dimension. It becomes evident that both of the methods score higher output quality values for windows of intermediate topographic heterogeneity. Most of the low output quality values appear on windows of small heterogeneity. Regions that consist mainly of flat areas belong to this category. In a similar way, we created a plot for each method where the horizontal coordinates of the presented points are equal to the number of flooded cells in the windows that we examine. However, the latter plots do not indicate any relation between this number and the output quality–see Fig. 3. The visualisations that we produced for the windows of larger size showed similar patterns.

Comparing the Output of the Methods. To gain more insight about methods *ProximityResistance* and *UpstreamResistance*, we visually examined the output that the two methods produced for the **denmark** dataset. For various rise values ρ

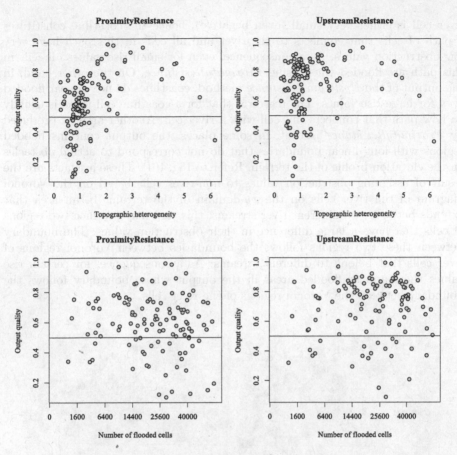

Fig. 3. Top: scatter plots that show the relation between the output quality of each method and topographic heterogeneity. **Bottom:** plots that show the relation between the output quality of each method and the number of flooded cells in each window.

we extracted the regions in the output of each method which consisted of all cells with flood resistance $\leqslant \rho$. Our first observation was that for the same rise value the flooded area that is computed by *ProximityResistance* is larger than in the output produced by *UpstreamResistance*. Refer to Fig. 4(a) and (b). This is an outcome of how the two methods estimate river floods around coastlines; in the output of *ProximityResistance*, almost the entire coastline of the terrain appears flooded even for very small rise values. Refer to Fig. 4(c). Recall that with *ProximityResistance* a cell c gets flooded for a rise value ρ if a) this cell has a height difference $\leqslant \rho$ from the closest river cell on the xy-domain (the obstruction value), and b) if there is a path from c to any river cell such that the obstruction values of all cells in the path is $\leqslant \rho$. The terrain cells close to the coastline have low height values, since they lie almost on sea level. Therefore, for each such cell the height difference from the closest river cell is either very small (even negative), hence the coastline constitutes a path of cells that connects to the river and all cells in this path have very low obstruction values. As a consequence, even for small rise values all cells in this path are flooded when using *ProximityResistance*. On the other hand, in the output of the *UpstreamResistance* method, coastlines do not appear flooded even for large rise values. The reason is that for a coastline cell there is usually no flow path that connects this cell with a river cell. Another artifact produced by *ProximityResistance* is that, in some places, the output contains flooded regions with long linear boundaries that do not correspond to actual obstacles on the elevation profile of the terrain. Refer to Fig. 4(d). These artifacts are the result of assigning obstruction values to non-river cells based on the Voronoi diagram of the river cells on the xy-domain of the terrain. In an area that extends between two different river streams, this step may produce two regions of cells that have a large difference in their obstruction values. The boundary between these two regions follows the boundaries between Voronoi regions of river cells that belong to different streams. As a consequence, for certain rise values there appear flooded areas in the output whose boundary follows the boundary between the Voronoi regions of the river cells.

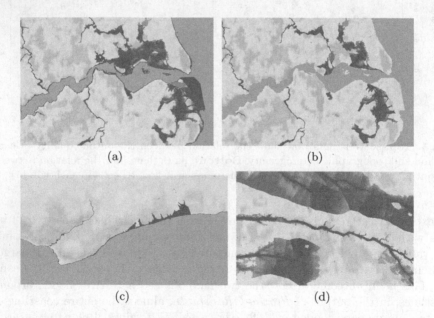

(a)   (b)

(c)   (d)

Fig. 4. An illustration of the outputs of *ProximityResistance* and *UpstreamResistance* for the denmark dataset. Flooded regions are indicated by dark blue color. (a) The output of the *ProximityResistance* around Hadsund town (northeast Jutland) for a rise value of half a meter. (b) The output of *UpstreamResistance* for the same region and rise value. (c) The output of *ProximityResistance* close to Vejle city with a rise value of just one milimeter. The entire coast appears flooded, with wide flooded areas at certain places. (d) The output of the *ProximityResistance* on a region with several river streams, for a rise value of 2.8 m. (Color figure online)

References

1. Agarwal, P.K., Arge, L., Yi, K.: I/O-efficient batched union-find and its applications to terrain analysis. In: Proceedings of the 22nd Annual Symposium on Computational Geometry, pp. 167–176 (2006)
2. Agarwal, P.K., Arge, L., Yi, K.: I/O-efficient construction of constrained delaunay triangulations. In: Proceedings of the 13th Annual European Symposium on Algorithms, pp. 355–366 (2005)
3. Aggarwal, A., Vitter, J.S.: The input/output complexity of sorting and related problems. Commun. ACM **31**(9), 1116–1127 (1988)
4. Alexander, C., Arge, L., Bøcher, P.K., Revsbæk, M., Sandel, B., Svenning, J.-C., Tsirogiannis, C., Yang, J.: Computing River Floods Using Massive Terrain Data. http://madalgo.au.dk/fileadmin/madalgo/OA_PDF_s/C401.pdf/
5. Arge, L., Toma, L., Vitter, J.S.: I/O-efficient algorithms for problems on grid-based terrains. ACM J. Exp. Algorithmics **6**(1) (2001)
6. Brefeld, U., Scheffer, T.: AUC maximizing support vector learning. In: Proceedings of the ICML Workshop on ROC Analysis in Machine Learning (2005)
7. Danner, A., Mølhave, T., Yi, K., Agarwal, P.K., Arge, L., Mitasova, H.: TerraStream: from elevation data to watershed hierarchies. In: Proceedings of the 15th ACM International Symposium on Advances in Geographic Information Science (ACM GIS), pp. 212–219 (2007)
8. The Dartmouth Flood Observatory Webpage. http://floodobservatory.colorado.edu/
9. Encyclopaedia Britannica Webpage, Pakistan Floods of 2010. http://www.britannica.com/EBchecked/topic/1731329/Pakistan-Floods-of-2010
10. Elevation Model of Denmark, Geodata Agency of the Danish Ministry of Environment. http://gst.dk/emner/frie-data/hvilke-data-er-omfattet/hvilke-data-er-frie/dhm-danmarks-hoejdemodel/
11. Goodrich, M.T., Tsay, J.J., Vengroff, D.E., Vitter, J.S.: External-memory computational geometry. In: Proceedings of the 34th IEEE Annual Symposium on Foundations of Computer Science (FOCS), pp. 714–723 (1993)
12. Kronvang, B., Søndergaard, M., Hoffmann, C.C., Thodsen, H., Bering Ovesen, N., Stjernholm, M., Nielsen, C.B., Kjærgaard, C., Schønfeldt, B., Levesen, B.: Etablering af P-Ådale. Technical report 840, National Environmental Research Institute of Denmark (2011)
13. Rodriguez, E., Morris, C.S., Belz, J.E.: A global assessment of the SRTM performance. Photogram. Eng. Remote Sens. **72**(3), 249–260 (2006)
14. TPIE, the Templated Portable I/O Environment. http://www.madalgo.au.dk/tpie/

Partitioning Polygons via Graph Augmentation

Jan-Henrik Haunert[1]([✉]) and Wouter Meulemans[2]

[1] Institute of Computer Science, University of Osnabrück, Osnabrück, Germany
janhhaunert@uni-osnabrueck.de
[2] giCentre, City University London, London, UK
wouter.meulemans@city.ac.uk

Abstract. We study graph augmentation under the dilation criterion. In our case, we consider a plane geometric graph $G = (V, E)$ and a set C of edges. We aim to add to G a minimal number of nonintersecting edges from C to bound the ratio between the graph-based distance and the Euclidean distance for all pairs of vertices described by C. Motivated by the problem of decomposing a polygon into natural subregions, we present an optimal linear-time algorithm for the case that P is a simple polygon and C models an internal triangulation of P. The algorithm admits some straightforward extensions. Most importantly, in pseudopolynomial time, it can approximate a solution of minimum total length or, if C is weighted, compute a solution of minimum total weight. We show that minimizing the total length or the total weight is weakly NP-hard.

Finally, we show how our algorithm can be used for two well-known problems in GIS: generating variable-scale maps and area aggregation.

1 Introduction

Polygons representing geographic objects can contain millions of vertices and thus can be difficult to handle. Often, they consist of multiple regions that are connected only via narrow bottlenecks, such as isthmuses in the case of land or straits in the case of water areas. To ease the handling of such polygons and to identify natural subregions, such as the Iberian Peninsula as a part of Europe, one often seeks a partition of a polygon into multiple smaller polygons of a certain type (e.g., into convex polygons). A triangulation of a polygon is the most common type of a polygon partition, yet often one is interested in larger (non-triangular) subregions. We present new algorithms for partitioning a polygon based on an internal triangulation of it: every output region is the union of a set of triangles of that triangulation. We consider our algorithm a useful tool for shape manipulation and demonstrate its effectiveness on two use cases: the generation of variable-scale maps and the aggregation of areas.

Our basic idea is to consider the polygon partitioning problem as a special *graph augmentation problem*. The vertices and edges of the input polygon P define a geometric graph G, which we augment with a selection of edges from a set C of candidate edges (that is, diagonals of P) to split P into multiple pieces. After the augmentation, the graph shall be well connected. More precisely,

© Springer International Publishing Switzerland 2016
J.A. Miller et al. (Eds.): GIScience 2016, LNCS 9927, pp. 18–33, 2016.
DOI: 10.1007/978-3-319-45738-3_2

for each candidate edge $\{u, v\} \in C$ we require that the *dilation* for u and v in the augmented graph is bounded by a user-set parameter. For any two vertices u, v of a geometric graph G, the dilation (sometimes also called stretch factor or detour factor) is defined as the ratio between the shortest u-v path via G and the Euclidean distance between u and v. By selecting a minimum number of edges from C we obtain a nice decomposition of the input polygon. As an alternative optimization objective we consider minimizing the total weight of the selected edges, assuming that for each edge in C a weight is given as part of the input.

Contributions. We introduce terminology and a general problem definition with three primary variants (unweighted, length-weighted and general weights) in Sect. 2. We review related work in Sect. 3. In Sect. 4 we consider the problem variants for the case that the graph to be augmented is a simple polygon without holes and the edges that can be added are an internal triangulation. We provide an optimal linear-time algorithm for the unweighted case, and present some extensions. We prove that both the general-weights case and the length-weighted case are weakly NP-hard, present a pseudopolynomial-time algorithm for the general-weights case, and show that it can provide a $(1+\varepsilon)$-approximation algorithm for the length-weighted case. We discuss our two use cases in Sect. 5.

2   Preliminaries

Graphs. Let $G = (V, E)$ denote a graph defined by its vertices V and edges $E \subseteq \{\{u, v\} \mid u, v \in V\}$. We call G a *geometric* graph if every vertex is assigned a position in $\mathbb{R}^2$ and each edge is represented by the line segment connecting its endpoints. A geometric graph is *plane* if vertices have unique positions and no two edges intersect, except at common endpoints.

Dilation. Let $G = (V, E)$ be a geometric graph and $u, v \in V$ be two vertices of G. We denote the Euclidean distance between u and v as $\|u - v\|$; we use $\|e\|$ to denote the length of edge e. The length of the shortest path in G between u and v is denoted by $d_G(u, v)$. We define the (vertex) *dilation* between u and v as $\Delta_G(u, v) = d_G(u, v)/\|u - v\|$; the dilation of the entire graph is $\Delta_G = \max_{u, v \in V, u \neq v} \Delta_G(u, v)$. If G is disconnected, its dilation is infinite.

Problem Statement. In this paper, we consider graph augmentation problems, where the augmentation is constrained to a prescribed set of vertex pairs. We call such vertex pairs *candidate edges*. Hence, a *problem instance* comprises

- a plane geometric graph $G = (V, E)$,
- a set $C \subseteq \{\{u, v\} \mid u, v \in V\}\backslash E$ of candidate edges, and
- a real number $\tau \geq 1$.

Consider $S \subseteq C$ to be a subset of the candidate edges. We denote by $G_S = (V, E \cup S)$ the graph obtained by augmenting G with the candidate edges in S. We call a candidate edge $\{u, v\} \in C$ *satisfied* with respect to S if $\Delta_{G_S}(u, v) \leq \tau$. A simple path in G_S whose length is sufficiently small to prove that $\Delta_{G_S}(u, v) \leq \tau$ is called a *witness* of $\{u, v\}$. Set S is a solution to the problem if all edges in C are satisfied (with respect to S). Note that we ask to satisfy only the pairs specified by the candidate edges; we do not guarantee that the dilation between all vertices is bounded by τ. This is a trade-off that we make to guarantee that solutions exist. In particular, $S = C$ is a solution for any problem instance.

However, we want to find a "good" solution. A primary criterion, in the context of polygon partitioning, is that the edges in S do not intersect each other or existing edges of G. Furthermore, we consider optimizing three different objective functions, resulting in the following problems:

- MinSize: minimize $|S|$.
- MinLength: minimize $\sum_{e \in S} \|e\|$.
- MinWeight: minimize $\sum_{e \in S} w(e)$, given weights $w : C \to \mathbb{R}^+$.

In the above, we provide an upper bound on the allowed dilation and minimize the cost (size, length or weight) of the solution. The dual variants instead bound the allowed cost and ask to minimize the dilation. We focus on the stated variants; our algorithms can solve the dual variant by a binary search on τ. This is possible since the problem is monotonic: any solution for τ is also a solution for $\tau' > \tau$, and thus increasing the dilation can only reduce the minimal cost.

3   Related Work

Partitioning. Partitions of polygons into triangles, monotone polygons, or convex polygons are common in the context of GIS [19] and have intensively been studied in computational geometry. For example, for the case that no additional vertices (i.e., *Steiner points*) are allowed, Keil and Snoeying [13] have shown that a simple polygon with n vertices and r reflex vertices can be partitioned into a minimum number of convex polygons in $O(n + r^2 \min\{r^2, n\})$ time. In the case that Steiner points are allowed, the problem can be solved in $O(n + r^3)$ time [5]. For polygons with holes, the problem is NP-hard in both cases [16].

Often motivated by problems in computer vision and pattern recognition, researchers have developed methods for partitioning polygons into "natural and intuitive" [17], "simpler" [8], or "approximately convex" [15] pieces, which need not be convex. However, these methods do not provide any guarantee of optimality with respect to the number of output pieces or a different measure.

Dilation. Algorithmic work involving dilation is motivated mostly by applications in infrastructure design (e.g. road or electricity networks). Much research has been done without planar considerations, e.g. [2]. Considering our use cases, we focus here on results with such planar considerations; see [4] for a survey.

Giannopoulos *et al.* [9] prove that, given a point set Q, computing a graph $G = (Q, E)$ with $\Delta_G \leq 7$ is NP-hard, if $|E|$ is bounded to $O(|Q|)$. They also prove that adding $O(|E|)$ edges to a geometric graph to bound the dilation to 7 is NP-hard. Both claims hold with and without requiring planarity. This supports the investigation of our variant, where we do not consider satisfying all pairs, but only those provided in a (constrained) candidate set.

Farshi *et al.* [7] show that it is possible to compute, for a given geometric graph, the edge that results in the largest dilation reduction in $O(n^4)$ time. This was later improved by Wulff-Nilsen [20] to $O(n^3 \log n)$ time. Note that repeatedly applying this greedy choice does not yield an optimal result. Aronov *et al.* [1] present algorithms for the following problem: given a point inside a polygon, compute a segment from the point to the boundary of the polygon such that the dilation from the given point to any point on the boundary is minimized.

If we measure dilation via the geodesic distance and only between vertices of which one is contained in a given small set, an FPTAS exists to compute a minimal-dilation triangulation of a simple polygon [14]. Klein *et al.* [14] attribute to folklore that a constrained Delaunay triangulation of a simple polygon has dilation at most $\pi(1 + \sqrt{6})/2 < 5.09$. This readily implies that our algorithms— run with τ and using as C the constrained Delaunay triangulation—compute a small set of edges such that *all* vertex pairs have dilation less than 5.09τ (in the geodesic model). A similar result was proven by Bose and Keil [3], stating that a constrained Delaunay triangulation (not necessarily of a polygon) has dilation at most $4\pi\sqrt{3}/9 \approx 2.42$, though only between pairwise visible points.

4 Triangulated Polygons

Here we study the dilation problem restricted to instances where G is a simple polygon P and C is an inner triangulation of P. We denote the resulting problems by MINSIZEPOLY, MINLENGTHPOLY and MINWEIGHTPOLY.

We present a linear-time optimal algorithm for MINSIZEPOLY in Sect. 4.1. In Sect. 4.2 we show how to deal with any nonintersecting set of internal diagonals as candidate edges; and in Sect. 4.3 we present a heuristic for dealing with holes. Finally, in Sect. 4.4 we prove that MINLENGTHPOLY and MINWEIGHTPOLY are weakly NP-hard; we present a pseudopolynomial-time algorithm for MIN-WEIGHTPOLY with integer weights and, via rounding, obtain an approximation algorithm for MINLENGTHPOLY.

4.1 Minimizing the Number of Selected Edges

To solve MINSIZEPOLY, we apply a recursive algorithm. Its recursion is structured using a rooted binary tree $\mathcal{T}$ on the edges of P and C. By maintaining three possible subsolutions for each node in $\mathcal{T}$, we show that we compute an optimal subsolution for each node based only on its children in $\mathcal{T}$.

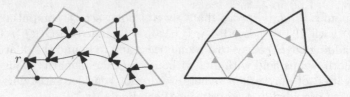

Fig. 1. (left) A binary tree T with root r. (right) A feasible role assignment for $\tau = 3$; the solid black diagonal is the only selected candidate edge in C, but allows a shorter path for another candidate edge.

Building a Tree. We define a directed binary tree T with nodes corresponding to the edges $P \cup C$ as follows. First, we pick an arbitrary edge of polygon P as root r. Then, we add the two edges incident to the same unprocessed triangle as children to r and recurse on each child. The result is a tree on the edges and candidate edges, rooted at r; see Fig. 1. If the embedding is given—the cyclic order of candidate edges at each vertex—we can compute T in $O(n)$ time, where n is the number of polygon edges. Otherwise, $O(n \log n)$ time suffices.

Components of T. Every edge $e \in P \cup C$ (a node in the tree) partitions T into two components[4]. The component that contains r is referred to as T_e^{ROOT}, the other as T_e^{LEAF}. Both of these components exclude e itself. For root r we define $T_r^{\text{ROOT}} = \emptyset$ and $T_r^{\text{LEAF}} = T \setminus \{r\}$. For uniformity of presentation, we also define a component T_e^{SELF} containing only edge e.

In a solution $S \subseteq C$ for MinSizePoly, each candidate edge $e = \{u, v\} \in C$ must have a witness: a simple u-v path of length at most $\tau \|e\|$. A witness of e lies fully within one of the three components of T defined by e.

Role Assignment. With our algorithm we compute a *role assignment* $\alpha \colon C \to \{\text{SELF}, \text{LEAF}, \text{ROOT}\}$ for all candidate edges. The role assignment indicates which component must contain a witness; we call α *feasible* if $T_e^{\alpha(e)}$ indeed contains a witness for all $e \in C$. A role assignment α directly prescribes the set S^α of edges that are part of the solution: $S^\alpha = \{e \mid e \in C \wedge \alpha(e) = \text{SELF}\}$. Hence, we refer to $|S^\alpha|$ as the size of α, using $|\alpha|$ as a shorthand. For uniformity, we define $\alpha(e) = \text{SELF}$ for all edges $e \in P$, but these are not part of S^α.

Figure 1 shows an instance with a role assignment. Every edge $e \in C$ is displayed according to its role: SELF-edges are black; ROOT- and LEAF-edges are gray with a small triangle indicating the direction of their shortest path.

As an edge can play three different roles, there are up to $3^3 = 27$ configurations of a role assignment for a triangle; see Fig. 2. We reduce this to 20 configurations as follows. Consider two edges e_1 and e_2. We call e_1 and e_2 *conflicting* in α if either: e_1 is the parent of e_2 in T, $\alpha(e_1) = \text{LEAF}$ and $\alpha(e_2) = \text{ROOT}$; or

---

[4] In this paper "edge" always indicates an element of $P \cup C$—a node in T—and never an edge between nodes (parent-child relation) in T.

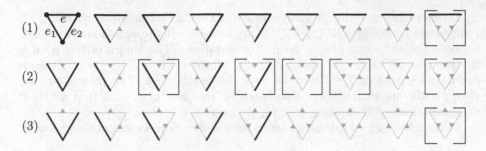

Fig. 2. The 27 configurations of roles for a triangle of an edge e and its children e_1 and e_2 in $\mathcal{T}$. The bracketed roles are not needed for an optimal solution.

e_1 and e_2 are siblings in $\mathcal{T}$ and $\alpha(e_1) = \alpha(e_2) = \text{ROOT}$. The following lemma implies that we may indeed discard the bracketed configurations in Fig. 2.

Lemma 1. *There exists a feasible role assignment with minimal size that does not contain any conflict.*

Proof. Consider a solution S with minimal size. Let α be the role assignment obtained by assigning SELF to $e \in S$ and ROOT or LEAF to the remaining edges, depending on which component of $\mathcal{T}$ contains the shortest path between the endpoints of e. To derive a contradiction, assume α contains a conflict between e_1 and e_2. This implies that $e_2 \in \mathcal{T}_{e_1}^{\alpha(e_1)}$ and vice versa. By construction, the shortest path π_1 for e_1 is contained in $\mathcal{T}_{e_1}^{\alpha(e_1)}$. Hence, π_1 must pass through the endpoints of e_2. However, this implies that the shortest path for e_2 is a subpath of π_1, and thus not in $\mathcal{T}_{e_2}^{\alpha(e_2)}$ as this component contains e_1. This is a contradiction, thus α cannot contain a conflict. □

Partial Assignments. Our algorithm computes role assignments for subtrees of $\mathcal{T}$. A *partial* role assignment α_e is an assignment on $\{e\} \cup \mathcal{T}_e^{\text{LEAF}}$. Its partial solution S^{α_e} is defined as $\{e' \mid e' \in C \cap (\{e\} \cup \mathcal{T}_e^{\text{LEAF}}) \wedge \alpha_e(e') = \text{SELF}\}$; again we use $|\alpha_e|$ as a shorthand for the size of S^{α_e}. A partial assignment for the root r corresponds to a (full) role assignment. Assignment α_e is feasible if one of the following holds for all $e' \in \{e\} \cup \mathcal{T}_e^{\text{LEAF}}$:

1. $\alpha_e(e') = \text{SELF}$; or
2. $\alpha_e(e') = \text{LEAF}$ and $(S^{\alpha_e} \cup P) \cap \mathcal{T}_{e'}^{\text{LEAF}}$ contains a witness for e'; or
3. $\alpha_e(e') = \text{ROOT}$ and either:
   (a) $(S^{\alpha_e} \cup P) \cap \mathcal{T}_{e'}^{\text{ROOT}} \cap (\{e\} \cup \mathcal{T}_e^{\text{LEAF}})$ contains a witness for e';
   (b) the combined length of the two shortest paths in $S^{\alpha_e} \cup P$ from the end-points of e' to the endpoints of e is at most $\tau \cdot \|e'\| - \|e\|$.

The rationale for case 3 is that either the edge is already satisfied (3a) or it is to be satisfied by what has yet to come (3b). However, the latter must ensure that there is still some length "to be spent" in order to complete the solution.

Lemma 1 and the triangle inequality imply that, for a feasible α_e with $\alpha_e(e) \in \{\text{SELF}, \text{LEAF}\}$, all edges in $\{e\} \cup \mathcal{T}_e^{\text{LEAF}}$ are satisfied. It presents a shortest path between the endpoints of e to future computations. The length of this path is the *front-length* of α_e, denoted by $L(\alpha_e)$. Moreover, if $\alpha_e(e) = \text{ROOT}$, then a contiguous subset of $\mathcal{T}_e^{\text{LEAF}}$ may all have this assignment. The *front-allowance* $R(\alpha_e)$ is the maximal allowed length on the root side of e, such that all these assignments are still satisfied. If $\alpha_e(e) \neq \text{ROOT}$, it is infinite.

In the following, all role assignments are feasible, unless mentioned otherwise.

Algorithm. The algorithm relies on a postorder recursive traversal of $\mathcal{T}$ to compute the partial assignment α_e for each edge e. Calling this with r hence results in the full role assignment α. However, to do the recursion correctly, we cannot simply compute a single partial assignment, but compute three instead:

Definition 1. *The following three partial role assignments are defined:*

- α_e^{SELF}: *the smallest partial role assignment with $\alpha_e(e) = \text{SELF}$.*
- α_e^{LEAF}: *the partial role assignment with minimal front-length among the smallest partial role assignments with $\alpha_e(e) = \text{LEAF}$.*
- α_e^{ROOT}: *the partial role assignment with maximal front-allowance, among the smallest partial role assignments with $\alpha_e(e) = \text{ROOT}$ and $R(\alpha_e) \geq \|e\|$.*

We compute these assignments based on the partial assignments of the child nodes. The base case, a leaf of $\mathcal{T}$, corresponds precisely to an edge of P. For these, we consider only α_e^{SELF} to be defined, with size 0 and front-length $L(\alpha_e) = \|e\|$. For root r, again corresponding to an edge of P, we are interested only in computing α_r^{SELF}, the size of which (not counting r) is the size of the solution. Any other node of $\mathcal{T}$ is a candidate edge e, with precisely two children in $\mathcal{T}$: e_1 and e_2. To compute the partial assignments in this case, we simply try the 20 cases of Fig. 2 and find those that satisfy Definition 1. By storing the size of the partial assignments, the size of a new partial assignment is simply the sum of the sizes of the children's partial assignments, increased by 1 if e is assigned SELF. However, not all cases may lead to feasible assignments. We therefore check the feasibility as follows, where row numbers refer to the labels in Fig. 2.

Cases in the first row correspond to computing α_e^{SELF}. For cases involving $\alpha_{e_1}^{\text{ROOT}}$ (and analogously for e_2), the front-allowance is met if $L + \|e\| \leq R(\alpha_{e_1}^{\text{ROOT}})$ holds, where L is the front-length provided by sibling.

Cases in the second row correspond to computing α_e^{LEAF}. Cases with a ROOT assignment for a child can be ignored by Lemma 1. We must ensure that the combined front-length of e_1 and e_2 is at most $\tau \cdot \|e\|$.

Cases in the third row correspond to computing α_e^{ROOT}. We check and compute front-allowances. Since e is not part of the solution, a front-allowance of a child is "propagated". For a child with a ROOT assignment, its propagated front-allowance is its front-allowance minus the front-length of its sibling. The minimum of this propagated front-allowance (if any) and $\tau \cdot \|e\|$ is the new front-allowance for e in this case and we check whether it is longer than $\|e\|$.

Note that α_e^{SELF} always exists, but α_e^{LEAF} and α_e^{ROOT} need not exist. Only cases for which both partial assignments for the children exist are computed.

Correctness. To prove the algorithm correct, we shall prove that the computed partial assignments, α_e^{SELF}, α_e^{LEAF} and α_e^{ROOT}, indeed are the smallest feasible partial assignments according to Definition 1. The lemma below is at the heart of this proof. Essentially, it states that we can always get a partial assignment with infinite front-allowance and minimal front-length by increasing the size of an assignment by at most one.

Lemma 2. *For any edge e in $\mathcal{T}$, we know that $|\alpha_e^{\text{SELF}}| \le 1+\min\{|\alpha_e^{\text{LEAF}}|, |\alpha_e^{\text{ROOT}}|\}$, where the size of a partial assignment is considered infinite if it does not exist.*

Proof. Consider α_e^{LEAF} or α_e^{ROOT}. If we change the assignment of e to SELF, we obtain again a feasible partial assignment. The lemma readily follows. □

Lemma 3. *The computed partial assignments correspond to Definition 1.*

Proof. We prove this lemma via structural induction. In the base case, e is a leaf of $\mathcal{T}$. Hence, it is an edge of P and the only partial role assignment is α_e^{SELF} with size zero (since e is not in C). Trivially, this has minimal size.

In the inductive case, e is not a leaf of $\mathcal{T}$. It has two children, e_1 and e_2. Let β_e be an optimal partial assignment, according to Definition 1. It implies partial assignments β_{e_1} and β_{e_2} for its two subtrees. Let $\alpha_{e_1} = \alpha_{e_1}^{\beta_e(e_1)}$ be a shorthand for the partial assignment computed by our algorithm, for the given case; α_{e_2} is defined analogously. We use $*$ to consistently indicate either e_1 or e_2.

If $|\beta_*| < |\alpha_*|$, we arrive at a contradiction with the induction hypothesis, which implies that α_* has minimal size.

To argue about the case that $|\beta_*| \ge |\alpha_*|$ holds for both children, we first make the following observations. If $|\beta_*| = |\alpha_*|$, then we can replace β_* with α_* without making the solution worse: by the induction hypothesis, α_* cannot have a greater front-length or a lower front-allowance. If $|\beta_*| > |\alpha_*|$, we cannot make this replacement as α_* may have a greater front-allowance or lower front-length. However, by Lemma 2, we now know that $|\alpha_*^{\text{SELF}}| \le |\beta_*|$ and this assignment has overall minimal front-length and infinite front-allowance. Hence, replacing β_* with α_*^{SELF} does not make the solution worse.

When we carry out both replacements as described above, we obtain a partial assignment that is not worse than β_e and thus adheres to Definition 1. Due to exhaustive case analysis, our algorithm computes this partial assignment. □

The computed partial assignment α_r^{SELF} corresponds to a full role assignment; it is minimal by Lemma 3. This readily implies the following theorem.

Theorem 1. *The algorithm computes an optimal solution to MinSizePoly.*

Complexity. After building $\mathcal{T}$, the straightforward implementation of this algorithm runs in optimal $O(n)$ time, for a polygon with n edges. Keeping track of which cases give the best result in the computation of each partial assignment, allows the recovery of the optimal solution in $O(n)$ time as well.

4.2   Fewer Diagonals

Suppose we require only that C is a nonintersecting set of diagonals inside P. Our algorithm can be modified to also deal with such a case. The most significant change is that T is no longer binary: nodes may have higher degree. Lemmas 1 and 2 straightforwardly generalize to this case. Hence, we may conclude that an optimal partial assignment can be obtained by using LEAF assignments of those children of e that have the smallest front-length. Thus, we sort the children according to front-length of their LEAF assignment. Testing every child with a ROOT assignment, we can do a binary search to find the best selection of other children to use a LEAF assignment, the rest using SELF. Hence, processing a single edge e takes $O(d_e \log d_e)$ time, where d_e is the degree in T. The total execution time is $O(n \log d)$ where d is the maximal degree in T.

4.3   Polygons with Holes

Let us consider a simple polygon P *with* holes; C is an inner triangulation of P. To bound the dilation, we need at least some edges to connect the outer boundary of P and each hole. We thus proceed as follows, similar to [8]. First, we compute a minimal-length set $T \subseteq C$ that connects these boundaries, i.e., a minimal spanning tree on the boundary components of P. We use these edges to carve open P into a polygon P_T without holes (see Fig. 3). We then run our algorithm on P_T; let S_T denote its solution. The solution S to P is given by $S_T \cup T$. This heuristic does not provide an approximation guarantee, since the distance along the boundary of P_T can be higher than the distance in the graph $P \cup T$; this may result in adding edges to the solution unnecessarily.

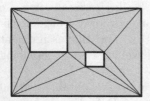 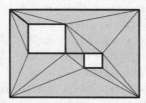 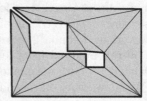

Fig. 3. (left) A polygon with two holes. (middle) Two edges are used as T, to connect the boundaries. (right) T is used to define a single polygon without holes.

4.4   Minimizing the Total Weight or Length of the Selected Edges

We now analyze the computational complexity of MinLengthPoly and MinWeightPoly and, thereafter, present algorithms for their solution.

Theorem 2. MinLengthPoly *is weakly NP-hard.*

Proof. Our proof is by reduction from the weakly NP-complete problem PARTITION, defined as follows: let $\mathcal{A} = \{a_1, \ldots, a_n\}$ be a set of positive integers and let $A = \sum_{a_i \in \mathcal{A}} a_i$; is there a set $I \subseteq \mathcal{A}$ such that $\sum_{a_i \in I} a_i = A/2$? For a PARTITION instance, we construct a MINLENGTHPOLY instance $\mathcal{M}$ with $\tau = 3$ and the polygon P and triangulation C as shown in Fig. 4, using one last point at distance $7A$ to the right of v_{n+1}. We prove that $\mathcal{M}$ admits a solution S of total length at most $3A/2$ if and only if $\mathcal{A}$ is a yes-instance of PARTITION.

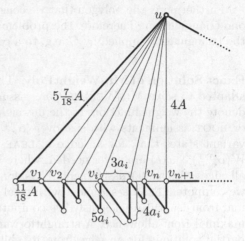

Fig. 4. MINLENGTH instance constructed for instance $\{a_1, a_2, \ldots, a_n\}$ of PARTITION.

Let $\mathcal{A} = \{a_1, \ldots, a_n\}$ be a yes-instance of PARTITION and let $I \subseteq \mathcal{A}$ be such that $\sum_{a_i \in I} a_i = A/2$. We show that $S = \{\{v_i, v_{i+1}\} \mid i \in I\}$ is a solution to MINLENGTH instance $\mathcal{M}$ with total length at most $3A/2$. Every edge $\{v_i, v_{i+1}\} \in C$ with $i \in \{1, \ldots, n\}$ (i.e., every *horizontal* edge) is trivially satisfied as P already contains a path of length $3a_i$. The vertical edge $\{u, v_{n+1}\}$ is exactly satisfied: walking in counter-clockwise direction along P yields a u-v_{n+1} path of length $15A$ and every horizontal edge $\{v_i, v_{i+1}\} \in S$ reduces the length of this path by $5a_i + 4a_i - 3a_i = 6a_i$; therefore, the shortest u-v_{n+1} path has total length $15A - 6A/2 = 12A = \tau 4A = \tau\|\{u, v_{n+1}\}\|$. Every other edge $\{u, v_i\}$ incident to u is satisfied, because it is longer than $\{u, v_{n+1}\}$, while at the same time the shortest u-v_i path is shorter than the shortest u-v_{n+1} path. By construction, the selected edges have total length $3A/2$.

Now, let $S \subseteq C$ be a solution to $\mathcal{M}$ of total length at most $3A/2$. Because every non-horizontal edge has a length of at least $4A$, S contains only horizontal edges. The edge $\{u, v_{n+1}\}$ can be satisfied only if the total length of horizontal edges in S is at least $3A/2$: hence, the total length of S is exactly $3A/2$. Therefore, the numbers in $\mathcal{A}$ corresponding to the edges in S sum up to $A/2$.

The coordinates of the vertices of the input polygon are rationals—or integer if we scale by a factor of 18—and polynomial in the sum A of $\mathcal{A}$. Therefore, the reduction can be computed in pseudopolynomial time. □

Theorem 3. MINWEIGHTPOLY *is weakly NP-hard.*

Proof. We use the same reduction as in the proof of Theorem 2, except that we define the weights as a part of the MINWEIGHTPOLY instance: we set the weight of each horizontal edge to its length and of each other edge to $4A$. All weights are polynomial in A. With this the argument works as before. □

Since MINLENGTHPOLY and MINWEIGHTPOLY are weakly NP-hard, the more general problems MINLENGTH and MINWEIGHT are weakly NP-hard too.

Furthermore, the polygon that we constructed for our reduction admits only one triangulation. Therefore, the problems do not become easier, if we restrict the triangulation implied by C, e.g. to a constrained Delaunay triangulation [6].

Exact Solution of MinWeightPoly. The algorithm for MINSIZEPOLY can be adapted to solve MINWEIGHTPOLY, assuming integer weights. Let $w: C \to \mathbb{N}$ denote the weight function. In the unweighted case, Lemma 2 implies that LEAF or ROOT assignments with size over $|\alpha_e^{\text{SELF}}| - 1$ are never needed. Its weighted variant states that, for an edge e, LEAF or ROOT assignments with size over $W(\alpha_e^{\text{SELF}}) - w(e)$ are never needed, where $W(\cdot)$ denotes the sum of weights over all edges with a SELF assignment. Thus, for each diagonal e and $i \in \{1, \ldots, w(e)\}$, we compute a LEAF assignment with total weight exactly $w(\alpha_e^{\text{SELF}}) - i$ and minimal front-length. Analogously, we compute up to $w(e)$ ROOT assignments, with maximal front-allowance. A straightforward implementation for computing the partial solutions for an edge from its children's solutions thus takes $O(w(e)^2)$ time. Therefore, this algorithm takes $O(\sum_{e \in P \cup C} w(e)^2) \subseteq O(w_{\text{max}} \cdot w_{\text{sum}}) \subseteq O(nw_{\text{max}}^2)$ time, where $w_{\text{max}} = \max_{e \in C} w(e)$ and $w_{\text{sum}} = \sum_{e \in C} w(e)$.

Approximating MinLengthPoly. If edge lengths are integer or fixed-point numbers, the weighted algorithm can compute the solution in pseudopolynomial time. Otherwise, rounding yields an approximate solution, as detailed below.

Let λ denote a small constant and assume $1 + \lambda \le \min_{e \in C} \|e\|$. We define two weight functions: $w(e) = 2\lambda \cdot round(\|e\|/(2\lambda))$ and $w'(e) = round(\|e\|/(2\lambda))$. We run the weighted algorithm using w' as its integer weight function. However, w and w' are identical up to scaling and thus produce the same optimal results. The rounding in w implies $\|e\| - \lambda < w(e) \le \|e\| + \lambda$ and $w(e) > 1$ by assumption.

Let S denote the result of the algorithm; it has weight $w(S) = \sum_{e \in S} w(e)$ and length $l(S) = \sum_{e \in S} \|e\|$. We find that $w(S) > l(S) - \lambda|S| > l(S) - \lambda w(S)$, implying $l(S) < (1 + \lambda)w(S)$. Let S^* denote an optimal solution to MINLENGTHPOLY; we find $w(S^*) \le l(S^*) + \lambda|S^*| \le l(S^*) + \lambda w(S^*)$ and thus $l(S^*) \ge (1 - \lambda)w(S^*)$. The approximation ratio obtained by our algorithm is $l(S)/l(S^*) < (1 + \lambda)w(S)/((1 - \lambda)w(S^*))$. Since S is optimal in terms of weight, this simplifies to $(1 + \lambda)/(1 - \lambda)$.

The running time of this approach is $O(nW'^2)$ where $W' = \max_{e \in C} w'(e)$. As $w'(e) \le \|e\|/(2\lambda) + \frac{1}{2}$, we find that this is $O(nL^2/\lambda^2)$ where $L = \max_{e \in C} \|e\|$. We thus get a (pseudo)PTAS to approximate MINLENGTHPOLY, that computes a $(1 + \varepsilon)$-approximation in $O(nL^2(\frac{2+\varepsilon}{\varepsilon})^2) = O(nL^2/\varepsilon^2)$ time.

5 Use Cases

Two vertices lying on opposite sides of a narrow part of a polygon typically have a very large dilation: a connection across the strait of Gibraltar, for example, is much shorter than a path along the coast, all around the Mediterranean Sea. Hence, our dilation-based method may find natural subregions of a polygon.

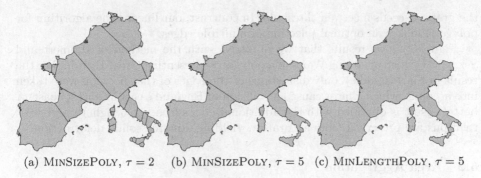

(a) MINSIZEPOLY, $\tau = 2$ (b) MINSIZEPOLY, $\tau = 5$ (c) MINLENGTHPOLY, $\tau = 5$

Fig. 5. Results of our algorithms for a part of Europe. Especially the MINLENGTHPOLY solution (c) nicely reflects the Iberian and Italian peninsulas.

This general hope is confirmed by the results that we obtained with implementations of our algorithms; see Fig. 5. Here we apply our method to two specific problems: computing distorted maps (Sect. 5.1) and aggregating areas (Sect. 5.2).

5.1   Computing Distorted Maps

Several methods exist to distort a map, for example, to resolve spatial conflicts or to emphasize certain information. Such methods often rely on constraints that are defined based on a geometric graph representing the map [10,11]. An edge in this graph may represent a line segment of a map object, but usually additional edges are needed to model the constraints for the output map. We consider our graph augmentation method as a useful tool for finding such relevant edges.

The method of Harrie and Sarjakoski [10] for the resolution of conflicts relies on a constrained Delaunay triangulation of the map objects. A constraint for the length of a triangle edge $e = \{u, v\}$ is introduced if e is shorter than a threshold ε and the map does not contain a u-v path of less than a number k of line segments. Similarly, our method selects edges of a triangulation based on a geometric distance and a graph-theoretical distance between two vertices of the map. However, while the method of Harrie and Sarjakoski measures the graph-theoretical distance in the input map, our method considers the graph-theoretical distance *after* augmenting the map with the selected edges. We consider our approach promising as it avoids redundant constraints.

The method of Haunert and Sering [11] enlarges a user-selected focus region in a map while minimizing the distortion, which is measured at the edges of a graph representing the map objects, for example, a network of roads or country borders. Additional edges are necessary if the relative position should be maintained for some pairs of vertices: e.g., vertices on opposite sides of a strait. To make a good selection of edges, Van Dijk et al. [18] have developed a greedy heuristic that iteratively augments the map with an edge of maximal dilation (among all edges of a constrained Delaunay triangulation) while the dilation of the graph exceeds a certain threshold. In contrast, our linear-time algorithm for polygons makes an optimal selection of multiple edges.

Figure 6 shows results that we obtained with the method of Haunert and Sering [11] when enlarging Wales in a polygon representing Great Britain. For the result in Fig. 6(middle), only distortions of the edges of that polygon were taken into account, which almost caused a collision of England's east and west coast. A better result is obtained with the additional edges (see Fig. 6(right)): east-west relations are preserved more accurately, yielding a more "solid" deformation.

5.2   Area Aggregation

Information on land cover is often given as a planar subdivision that consists of regions of different classes (urban, rural, forest, etc.). To generalize such data, one often aggregates the areas into larger regions such that many-to-one relationships arise. Usually, every output area must have at least a certain minimal size. Subject to this requirement, Haunert and Wolff [12] suggested minimizing a cost function that combines two objectives: the overall weighted class change should be small and the resulting areas should be geometrically compact. They showed that the problem is NP-hard and developed an exact method based on integer linear programming and a heuristic method based on simulated annealing.

Figure 7(a) shows a sample from the German digital landscape model ATKIS DLM 50, corresponding to a topographic map of scale 1:50 000. We processed this sample with the simulated-annealing-based aggregation method of Haunert and Wolff [12]; see Fig. 7(c). Each output polygon has at least $400\,000\,\mathrm{m}^2$, which is a requirement for the scale 1:250 000. Observe that several settlement areas (red) are lost. To obtain a better solution, we apply our algorithm for MINSIZEPOLY

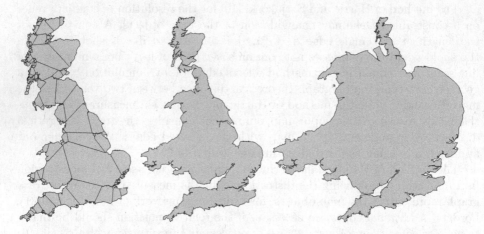

Fig. 6. (left) A polygon representing Great Britain with the edges of a MINSIZEPOLY solution with $\tau = 2$. Variable-scale maps computed (middle) without and (right) with consideration of the selected diagonals. The method of Haunert and Sering [11] was used with a scale factor of 2 for Wales; the results were scaled to the same height.

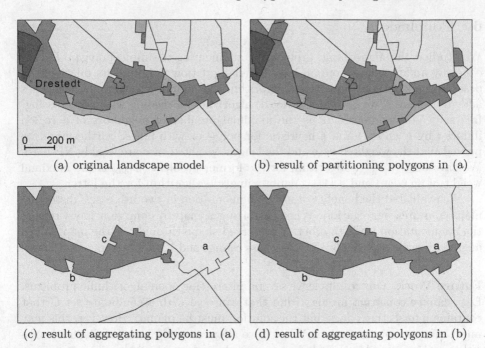

(a) original landscape model   (b) result of partitioning polygons in (a)

(c) result of aggregating polygons in (a)   (d) result of aggregating polygons in (b)

Fig. 7. Results of the simulated-annealing-based aggregation method of Haunert and Wolff [12] when applied to an example from the landscape model ATKIS DLM 50 (c) without and (d) with application of our MinSizePoly algorithm with $\tau = 4$ for pre-processing. Three corresponding parts in the solutions are labeled with a, b, and c. (Color figure online)

with $\tau = 4$ and use its result (Fig. 7(b)) as input for the aggregation method. The solution that we obtain (Fig. 7(d)) is clearly better with respect to the total class change: the relatively large settlement labeled with a is retained. Moreover, more compact shapes have been produced, for example, by filling small concavities in the polygons; see the labels b and c. Based on the objective function defined by Haunert and Wolff we can quantify this improvement: for a sample of $n_1 = 325$ polygons from ATKIS DLM 50 the aggregation method yielded a solution of 7.1 % less total cost when using the polygon partitioning algorithm, which resulted in $n_2 = 881$ polygons. The cost for class change was reduced by 3.2 % and the cost for non-compactness by 12.2 %. The higher quality comes at the cost of an increased number of input polygons for the aggregation method. Hence, fast heuristics for aggregation are needed and it is reasonable to minimize the number of output polygons when using our polygon partitioning method. In our experiments, we ran simulated annealing with the same very large number ($8\,810\,000 = n_2 \cdot 10^4$) of iterations to produce near-optimal solutions; this took slightly more than half an hour on a desktop PC.

6   Conclusion

We studied the algorithmic problem of augmenting a simple polygon P of n edges by adding edges from an internal triangulation to bound its dilation. We described an optimal linear-time algorithm to minimize the number of edges added. Moreover, we gave an $O(n \log d)$ algorithm for dealing with any crossing-free set C of candidates (d is the maximal number of neighbors of a region induced by P and C) and a heuristic for polygons with holes. Furthermore, we proved that the weighted case and the length-weighted case are weakly NP-hard. We gave an $O(nw_{max}^2)$ algorithm for the former problem (w_{max} is the maximal weight of an edge) and a $(1 + \varepsilon)$-approximation algorithm for the latter.

We evaluated the benefits of using augmentation in two use cases: distorting maps and area aggregation. When distorting a map to enlarge a focus region, the augmentation leads to a better preserved shape throughout the map. When aggregating areas, it yields 3.2 % less class change and 12.2 % better compactness.

Future Work. Our results leave several interesting open algorithmic problems. E.g., can we construct an algorithm that can deal with a candidate set C that contains intersecting edges, but the solution must be planar? However, this may imply that no solution exists. What if we allow not only internal diagonals of a polygon, but any edge that does not cross the polygon boundary?

We plan to run extensive experiments to further explore graph augmentation for our use cases, to provide guidelines for parameter and weight selection and model the trade-offs between computation time and quality more explicitly.

Acknowledgments. The authors would like to thank Johannes Oehrlein for helpful discussions on the topic of this paper. W. Meulemans is supported by Marie Skłodowska-Curie Action MSCA-H2020-IF-2014 656741.

References

1. Aronov, B., Buchin, K., Buchin, M., Jansen, B., de Jong, T., van Kreveld, M., Löffler, M., Luo, J., Silveira, R.I., Speckmann, B.: Connect the dot: computing feed-links for network extension. J. Spat. Inf. Sci. **3**, 3–31 (2011)
2. Aronov, B., de Berg, M., Cheong, O., Gudmundsson, J., Haverkort, H., Smid, M., Vigneron, A.: Sparse geometric graphs with small dilation. Comput. Geom. **40**, 207–219 (2008)
3. Bose, P., Keil, J.: On the stretch factor of the constrained Delaunay triangulation. In: Proceedings 3rd International Symposium on Voronoi Diagrams in Science and Engineering, pp. 25–31 (2006)
4. Bose, P., Smid, M.: On plane geometric spanners: a survey and open problems. Comput. Geom. **47**(7), 818–830 (2013)
5. Chazelle, B., Dobkin, D.: Decomposing a polygon into its convex parts. In: Proceedings of the 11th Annual ACM Symposium on Theory of Computing, pp. 38–48 (1979)
6. Chew, L.P.: Constrained Delaunay triangulations. Algorithmica **4**(1–4), 97–108 (1989)

7. Farshi, M., Giannopoulos, P., Gudmundsson, J.: Improving the stretch factor of a geometric network by edge augmentation. SIAM J. Comput. **38**(1), 226–240 (2008)
8. Feng, H.Y.F., Pavlidis, T.: Decomposition of polygons into simpler components: feature generation for syntactic pattern recognition. IEEE Trans. Comput. **24**(6), 636–650 (1975)
9. Giannopoulos, P., Klein, R., Knauer, C., Kutz, M., Marx, D.: Computing geometric minimum-dilation graphs is NP-hard. Int. J. Comput. Geom. Appl. **20**(2), 147–173 (2010)
10. Harrie, L., Sarjakoski, T.: Simultaneous graphic generalization of vector data sets. GeoInformatica **6**(3), 233–261 (2002)
11. Haunert, J.-H., Sering, L.: Drawing road networks with focus regions. IEEE Trans. Vis. Comput. Graph. **17**(12), 2555–2562 (2011)
12. Haunert, J.-H., Wolff, A.: Area aggregation in map generalisation by mixed-integer programming. Int. J. Geogr. Inf. Sci. **24**(12), 1871–1897 (2010)
13. Keil, J.M., Snoeyink, J.: On the time bound for convex decomposition of simple polygons. Int. J. Comput. Geom. Appl. **12**(3), 181–192 (2002)
14. Klein, R., Levcopoulos, C., Lingas, A.: A PTAS for minimum vertex dilation triangulation of a simple polygon with a constant number of sources of dilation. Comput. Geom. **34**, 28–34 (2006)
15. Lien, J.-M., Amato, N.M.: Approximate convex decomposition of polygons. Comput. Geom. Theory Appl. **35**(1), 100–123 (2006)
16. Lingas, A.: The power of non-rectilinear holes. In: Nielsen, M., Schmidt, E.M. (eds.) Automata, Languages and Programming. LNCS, vol. 140, pp. 369–383. Springer, Heidelberg (1982)
17. Siddiqi, K., Kimia, B.B.: Parts of visual form: computational aspects. IEEE Trans. Pattern Anal. Mach. Intell. **17**(3), 239–251 (1995)
18. van Dijk, T.C., van Goethem, A., Haunert, J.-H., Meulemans, W., Speckmann, B.: Accentuating focus maps via partial schematization. In: Proceedings of the 21st ACM SIGSPATIAL International Conference on Advances in Geographic Information Systems, pp. 418–421 (2013)
19. Voisard, A., Scholl, M.O., Rigaux, P., Databases, S.: With Application to GIS. Morgan Kaufmann, Burlington (2002)
20. Wulff-Nilsen, C.: Computing the dilation of edge-augmented graphs in metric spaces. Comput. Geom. **43**(2), 68–72 (2010)

Hierarchical Prism Trees for Scalable Time Geographic Analysis

Carson J.Q. Farmer[1]([⊠]) and Carsten Keßler[2]

[1] Geography Department, University of Colorado at Boulder, Boulder, CO, USA
carson.farmer@colorado.edu
[2] Department of Development and Planning, Aalborg University Copenhagen,
Copenhagen, Denmark
kessler@plan.aau.dk

Abstract. As location-aware applications and location-based services continue to increase in popularity, data sources describing a range of *dynamic processes* occurring in near real-time over multiple spatial and temporal scales are becoming the norm. At the same time, existing frameworks useful for understanding these dynamic spatio-temporal data, such as time geography, are unable to scale to the high volume, velocity, and variety of these emerging data sources. In this paper, we introduce a computational framework that turns time geography into a scalable analysis tool that can handle large and rapidly changing datasets. The Hierarchical Prism Tree (HPT) is a dynamic data structure for fast queries on spatio-temporal objects based on time geographic principles and theories, which takes advantage of recent advances in moving object databases and computer graphics. We demonstrate the utility of our proposed HPT using two common time geography tasks (finding similar trajectories and mapping potential space-time interactions), taking advantage of open data on space-time vehicle emissions from the EnviroCar platform.

Keywords: Time geography · Dynamic indexing · Spatio-temporal queries · Scalability

1 Introduction

Decision making in the corporate, private, and public spheres is increasingly based on spatio-temporal information. These information sources include real-time traffic counts, location-based social-media interactions, environmental sensor networks, as well as space-time trajectories of humans, animals, and vehicles. At the same time, modern advances in information and communication technology have converged with popular culture (*e.g.*, geo-tagging, location-based services, crowd-sourcing, *etc.*) to create an environment that is overflowing with new forms of spatial data [1,2]. Many of these emerging data sources contain details about movements and flows of individuals, objects, or information over geographic space, and are part of a growing list of dynamic spatio-temporal data sources.

© Springer International Publishing Switzerland 2016
J.A. Miller et al. (Eds.): GIScience 2016, LNCS 9927, pp. 34–47, 2016.
DOI: 10.1007/978-3-319-45738-3_3

Existing frameworks for understanding dynamic processes *are* available, including the rich conceptual and theoretical frameworks of *time geography* [3,4]. Hägerstrand's time geography was originally developed to understand how human migration activities are constrained at the individual level, and provides an ideal framework within which to explore modern spatio-temporal data sources. Indeed, there has been renewed interest in time geography concepts for geospatial research [5,6], including for location-based services [7,8], accessibility [9,10], trip planning [11,12], and health [13]. Despite this increasing interest, issues of scalability and applicability to emerging data sources are limiting time geography's use in data-intensive research.

While time-geography is useful for thinking about many types of spatio-temporal movements, much of the existing literature focuses on a limited number of individuals or features, and does not generally scale to larger problems. In this paper, we present a computational framework for time geographic analysis that aims to preserve the underpinnings of time geography (and in particular, Miller's [3] time geographic measurement theory), while at the same time increasing the scalability and applicability of the framework to meet the needs of a data-intensive research agenda.

In the following section (Sect. 2), a brief background on time geography is presented, followed by a presentation of dynamic (spatio-temporal) data structures and bounding volume hierarchies (Sect. 3) as a potential means of scaling time geographic concepts. Building on these ideas, a framework (Sect. 4) for the development of time geographic data structures which takes advantage of recent advances in moving object databases and computer graphics research is introduced. Following this (Sect. 5), two examples of this framework applied to common time geographic analysis tasks are presented, using space-time data on vehicle emissions from the EnviroCar platform [14]. We conclude (Sect. 6) with a discussion of the proposed framework and directions for future research.

2   Background

The basic concepts of time geography are the *space-time path*, describing changes in an object's location with time, and the *space-time prism*, describing an object's travel *potential*. This potential is constrained by the speed at which the object can travel (v_{max}), as well as locations at which the object must be present (*e.g.*, home and work when the object in question is a person). In general, a space-time *path* (see Fig. 1a) consists of a sequence of control points and a corresponding sequence of path segments connecting these points. In this definition, control points are observed or measured locations in space and time, and segments connect temporally adjacent control points. A space-time *prism* (see Fig. 1b) may exist between any pair of temporally adjacent control points, creating a time interval during which unrecorded (or future) travel may occur. An object may thus occupy locations in space other than the straight-line segment between two adjacent control points. The outline of the prism represents the limits of the locations that can be visited, as defined by the known space-time control points, and the object's maximum velocity, v_{max}, which defines the prism's diameter.

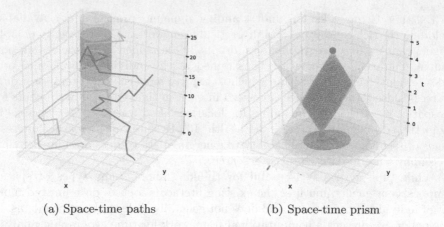

(a) Space-time paths   (b) Space-time prism

Fig. 1. Features of Hägerstrand's time-geography. Concepts include (a) space-time paths, bundling and stations (cylinder), as well as (b) space-time prisms, and accompanying start and end points, future and past cones (semi-transparent), and potential path areas (projected onto base).

The time-geographic concepts above have been formalized by Miller [3], who introduces a rigorous measurement theory based on three key assumptions: (1) the metric space satisfies the notions of identity, non-negativity, and triangular inequality about distance, (2) data are recorded at specific points in time, and (3) analysts have perfect information about the system (although relaxations of this assumption have been explored to some degree [3,15]). Building on these relatively simple assumptions, Miller has developed mathematical (and geometrical) definitions for space-time paths, prisms, stations, bundles (convergence of two or more paths for some shared activity over some given length of time), and intersections (two or more features sharing the same location(s) in space and time). Miller also provides strict conditions within which space-time paths are bundled and where intersections may occur between paths and prisms.

Research areas that typically employ time geography as an analysis tool deal with different aspects of mobility (*e.g.*, location-based services, accessibility, trip planning, health). The proliferation of mobile devices, sensor networks, and new developments such as the Internet of Things create an abundance of new data sources for these domains, which have traditionally dealt with small, easily tractable, and carefully selected samples. In the following sections, we will introduce a computational framework that turns time geography into a scalable analysis tool that can handle large and rapidly changing datasets, allowing the aforementioned domains to leverage these new data sources. We will argue that dynamic spatial indexes are not sufficient in this context, and that dynamically updated bounding volume hierarchies present a viable solution.

3 Dynamic Spatial Indexes

A wide range of data structures have been proposed for efficient queries on spatial and spatio-temporal data [16], including indexing strategies geared towards location-based services [17], real-time data [18], or more general spatio-temporal data [19,20]. However, for objects that may move in space and time, these indexes have to be *continually* updated, which can limit their utility in many cases. In order to address this issue, a number of dynamic indexing algorithms [21,22], including dynamic *spatial* indexes [23–25] have been developed, many of which are designed specifically for keeping track of moving objects [26–28]. These efforts have lead to a number of useful data structures and indexing schemes for static *and* dynamic spatial data, with a particular focus on 2D geometries (although some innovative exceptions have been proposed [29]). Because time geography embeds objects in 3D *space-time*, it is prudent (and useful) to query and perform analysis on objects in this space directly. For example, while conceptually similar to 2D space plus 1D time, a 3D index allows us to query and explore the *joint* space-time in a more efficient way (rather than querying space and then time or vice versa) and allows us to work directly with 3D volumes, rather than 2D time slices. For this, one can turn to the computer graphics literature, where data models for static and continuously moving 3D objects are required to speed up the rendering process [30,31].

3.1 Bounding Volume Hierarchies

Many 3D spatial indexing (or space partitioning) algorithms, such as R-trees, octrees, and kd-trees, slice 3D space with a flat 3D plane [32] to create sub-volumes. This is efficient to search, but presents a problem when objects overlap the split boundary. In dynamic applications of octrees or kd-trees [21,33], objects may be placed into all sub-volumes they touch. This requires extra overhead when working with moving objects, and extra tests when traversing the space to handle duplicate occurrences. As such, while kd-trees have excellent performance for *static* geometries [34], when it comes to dynamic settings with multiple moving objects, a different approach is required[1].

Instead of selecting a split-plane to divide volumes, a bounding volume hierarchy (BVH) tree of arbitrary enclosing volumes (*e.g.*, bounding boxes, capsules, cylinders, spheres, *etc.*) can be used [30,36] (Fig. 3). Here, the sub-volumes of a node don't have a particular split plane dividing them, and instead, the objects are divided to minimize some feature of the sub-volumes (generally the surface-area or volume, estimated by a heuristic). This approach has been shown to display superior construction performance over kd-trees [34], and because objects need not be split across sub-volumes, it also allows for *dynamic* object updates, insertions, and deletions [37], which facilitate dynamic BVH implementations. Furthermore, because the tree contains arbitrary enclosing volumes (*i.e.*, there is no clear split plane), sub-volumes are allowed to overlap. Indeed, the ability

---

[1] Although some parallel versions of kd-trees [35] do show promise.

for sub-volumes to overlap is one of the main reasons that BVHs can handle efficient dynamic updates. When objects only move a short distance, the only adjustment required is a simple adjustment of the bounds of their enclosing volume(s). Even if the volumes overlap other volumes, the BVH will still function correctly (although at slightly reduced efficiency). Furthermore, the arbitrary enclosing volumes provide a significant level of flexibility, even facilitating nested (or multi-scale) BVHs (*i.e.*, a BVH of BVHs is possible).

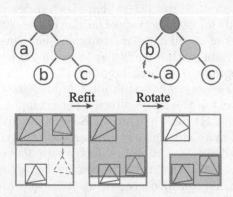

Fig. 2. Tree rotations are local restructuring operations that modify subtrees of a binary tree by swapping direct child and grandchild nodes [38]. In this case, as triangle (c) moves, the bounding volume expands, but rather than splitting the modified node into separate nodes containing triangles (b) and (c), tree rotations allow the BVH to identify and perform helpful merges and splits, such as merging (a) and (c) into a new leaf node.

When there *is* significant overlap, the BVH tree generally needs to be restructured [39]. To perform this re-structuring efficiently, [38] have developed a method based on localized updates to the BVH structure via tree-'rotations' which has proven extremely useful [37,40]. Combining these tree-rotations with the ability to have overlapping volumes, handling moving objects in a BVH works in two ways: (1) if the movement is minimal, the BVH can be quickly and conservatively expanded to handle the new location (at the cost of efficiency), or (2) if movement is significant (*i.e.*, overlaps are large), tree-rotations to optimize the BVH structure can be performed (Fig. 2).

4   Hierarchical Prism Trees

As mentioned previously, one of the key features of time geography is the space-time prism, a 3D geometric construct that defines *potential* space-time accessibility. Using a BVH tree, prism intersection tests can now be performed on the actual prism volume, rather than at discrete time slices, which is common practice in GIS-based time geography *e.g.* [41,42]. Algorithms for intersection detection of cones and bounding boxes are readily available, many of which are well-tested and efficient[2]. For most time geography analysis, simple bounding box intersection tests provide a quick test of intersection, with more computational tests (cone/cone and cone/cylinder) reserved for intersecting prisms (though in most cases, only a 2D projection of prism intersections is required, not the actual intersection of the prisms).

The above BVH techniques can be implemented in a time geographic framework, where the 3D space represents location on the x and y axes, and time on the t (or z) axis (see Fig. 3). The concepts of cones and prisms from time geography mean that *approximate* location queries can be handled using relatively simple collision tests and *predictive* location queries [43] can take advantage of the uncertain nature of cones/prisms. While these types of queries require *a priori* information about an object's behavior (v_{max}), when the prism shape and size are unknown (or likely to be variable), existing methods are available that can be used to estimate features of the space-time prism/cone [42,44,45]. Furthermore, because the BVH only requires an estimate of the bounding *volume* (*i.e.*, not the geometry of the object itself) to facilitate efficient updates and queries, more fine-grained analyses and queries are able to *lazily* [46] evaluate an object's position and shape, leading to further efficiency gains. Now, space-time intersections, searches, and analyses can be efficiently implemented for a large number of continually moving space-time objects, with minimal computational overhead, within a hierarchical tree of space-time prisms, or a Hierarchical Prism Tree (HPT).

When an even larger number of objects are being tracked, a nested approach to structuring space-time paths and prisms may be required. For instance, rather than tracking each individual space-time prism in a HPT, it may be preferable to track the overall space-time path instead; using the HPT to handle updates of the overall trajectory. When finer-grained details are needed (*i.e.*, to compute joint potential path areas), a nested HPT of space-time prisms can be lazily generated and queried.

5   Examples

In this section, we present two common time geographic analysis tasks which take good advantage of the proposed HPT framework. The first (Sect. 5.2), based on finding similar space-time movement patterns, is somewhat simplistic given the nature of the binary tree solution proposed here. The second (Sect. 5.3), based on computing joint potential path areas [44,45] for multiple space-time paths, is more complex, and requires multiple levels of queries and calculations. For these examples, we take advantage of vehicle trajectories from the EnviroCar[3] project's RESTful API. The (preliminary) Python code implementing the examples discussed in this section is available at https://github.com/carsonfarmer/hypt.

---

[2] See for example, http://www.realtimerendering.com/intersections.html.
[3] https://www.envirocar.org.

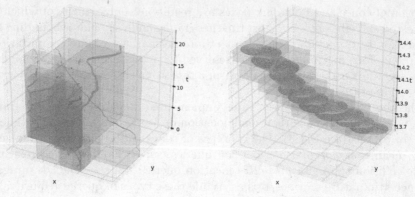

(a) HPT with space-time paths   (b) HPT showing space-time prisms

Fig. 3. Dynamic HPT techniques map ideally onto a time geography framework. A HPT of space-time paths (a) can be dynamically built and efficiently queried via time-slice, nearest-neighbor, or bounding box queries, and results can be filtered (b) using more complex intersection tests at the level of space-time prisms (in this case, the HPT in (b) is a subset of the largest space-time path traveling west-to-east in (a)). Tree leafs and nodes are denoted by semi-transparent and empty boxes respectively, with nested trees (a) and space-time prisms (b) as solid objects.

5.1   EnviroCar

EnviroCar is a community-based data collection platform for gathering vehicle-borne sensor data and producing environmental information [14]. EnviroCar uses standard Bluetooth OBD-II adapters[4], which are connected to a vehicle via the standard OBD connection that allows it to read parameters such as speed or revolutions per minute. From there, a smartphone records the data at regular time intervals, augmented with GPS information from the EnviroCar smartphone app. The EnviroCar app automatically calculates further information such as fuel consumption and CO_2 emissions, which can then be uploaded to the Enviro-Car platform server for subsequent analysis and sharing with the wider research and citizen-science communities.

EnviroCar trajectories provide an ideal test-bed for exploring some of the concepts presented in this paper. For each control point in a series of EnviroCar trajectories, we have several measures that can be used to determine the shape and size of its corresponding space-time prism. For instance, the recorded speed at each point in the trajectory can be used to determine the value of v_{max} (maximum velocity), which is of relevance when computing dynamic potential path areas or other metrics that are dependent on the space-time prism. Additional variables such as CO_2, can be stored along with the control point and associated prism to answer queries such as "how much CO_2 was produced by vehicles in

---

[4] http://www.obdii.com/background.html.

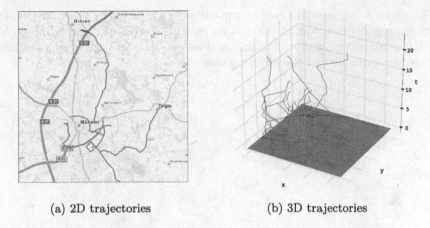

(a) 2D trajectories   (b) 3D trajectories

Fig. 4. A random selection of 43 EnviroCar space-time paths, encorporating ~ 4050 space-time prisms (see Fig. 3b). Note that times have been scaled from 0 to $\sim 20\,\text{min}$ for demonstration purposes. Basemap data, imagery, and map information provided by MapQuest, OpenStreetMap and contributors, ODbL. Trajectories data provided by EnviroCar [14], ODbL.

this area over this time period?" or "which locations (joint potential path areas) have the highest number of CO_2 measurements in this region?" (Fig. 4).

5.2 Similarity Analysis

Similarity analysis across space-time paths is a common task in time geography research. The ability to identify similar space-time paths can aide researchers in locating space-time stations and bundling, improve visualization though path clustering (grouping similar paths), and path aggregation (forming composite paths) [41], as well as identifying similar geospatial 'lifelines' for discovering the environmental factors responsible for hot-spots and clusters of certain diseases [47]. Additionally, a common task in animal movement analysis is to identify areas of (potential) spatio-temporal overlap (or separation) between different animal species [48] or individual animals of the same species [45] (see Sect. 5.3). These types of analysis are generally aided by first identifying *similar* space-time paths.

A number of similarity measures exist in the literature (see [41,47] for examples), and while it is not the goal of this paper to present a new comprehensive method for similarity analysis, frequently, the task of space-time path similarity search (or clustering) is a first step in an analytical workflow, designed to reduce complexity and aid pattern recognition. As such, the HPT framework presented here provides a useful heuristic for grouping similar space-time paths – with little to no additional effort on the part of the analyst. This is because the goal of the HPT algorithm is to minimize the size of the sub-volumes, and by doing so, they are also implicitly minimizing the 'distance' between space-time paths. Additionally, due to the incremental nature of the HPT update algorithm used here (see Fig. 2 for a discussion of tree restructuring via rotations), the addition of a new trajectory (or new control point in an existing trajectory) simply integrates with the existing trajectory 'clusters', and subsequent updates can potentially improve the optimality of the grouping over time.

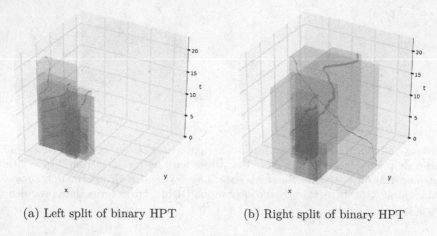

(a) Left split of binary HPT   (b) Right split of binary HPT

Fig. 5. Top-level split of a binary HPT (see Fig. 3) into left (a) and right (b) components. Within each split, the right and left components of the second-level split are denoted by different shading.

Figure 5 provides an example of the implicit 'grouping' of similar space-time paths using the previous EnviroCar trajectories example from Fig. 3: a top-level split of the tree into left (Fig. 5a) and right (Fig. 5b) components. This simple two-stage split separates trajectories into similar path-types, with paths circling Münster's downtown core in Fig. 5a and cross-/inter-town paths in Fig. 5b. Further similarity breakdowns can be observed, including two separate, temporally-offset, spatially-similar groupings in Fig. 5a with one showing travel between the University's geosciences building in the north-west and the Loddenheide area in the south (see Fig. 5b for reference).

5.3   Joint Potential Path Areas

In time geography analysis, it is often of interest to identity areas where interactions in space-time could occur. For example, researchers working with animal telemetry data may be interested in mapping regions where inter- or intra-species interactions may have occurred in an effort to better understand animal movement behaviors (avoidance, attraction, *etc.*) [44]. Similarly, it may be useful to highlight potential contact points for infectious disease transmission, or to identify regions of high or low densities of space-time interactions [49]. For the current example, we are interested in addressing the second question presented in Sect. 5.1, where we are trying to identify locations in the study region that have the highest number of CO_2 estimates. By determining these regions of overlap in space and time, we can potentially identify regions where we can have more confidence in our estimated CO_2 values.

To identify regions where multiple estimates have been made around the same space-time, we need to identify potential 'contact' points between vehicular trajectories, and then map their corresponding joint potential path areas (jPPA) [50]. A potential path area describes the elliptical region in space that a moving object or person could potentially reach given fixed start and end points. It can be conceptualized as the projection of the spacetime prism between two control points onto the geographical plane [3,50]. As such, a jPPA is simply the 2D projection of the intersection of two space-time prisms. Previously, this type of analysis involved two steps: (1) determining potential space-time contacts by temporally syncing trajectories and performing distance-based queries at various time slices (space-time prisms can be used at the cost of additional computation), and (2) computing the intersection of identified prism-pairs at various time slices to compute the jPPA.

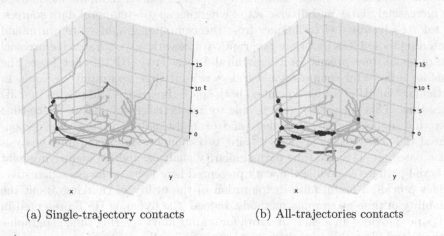

(a) Single-trajectory contacts   (b) All-trajectories contacts

Fig. 6. Interaction patterns of a subset of the EnviroCar trajectories (see Fig. 5a), with potential contact points (PCP) for a single trajectory with all other trajectories in the subset (a), and the PCP between *all* trajectories in the subset and their corresponding PPAs. Note that we are showing overlapping PPAs (with darker regions representative of the jPPAs) that have been increased in size (×5) to aid in visualization.

A Naive version of the first step requires $\mathcal{O}(n^2)$ queries across a pair of trajectories, making it nearly impossible to scale to more than a handful of trajectories or control-points. Some efficiencies can be gained by using spatial indexing systems in a GIS-framework, however, this is often not done in practice. Because the HPT presented here is a binary tree (with a query time of $\mathcal{O}(\log n)$), we are able to reduce the time complexity of this process to $\mathcal{O}(n \log n)$ (additional speed gains are possible via more efficient 'dual-tree' approaches [51,52]). Figure 6a shows an example result for this type of query for a single trajectory to all other trajectories in a subset of the EnviroCar trajectories used previously. In this case, space-time contact is based on *potential* contact using the space-time prisms along the trajectories. Building on this, Fig. 6b shows potential contacts between *all* pairs of trajectories in the subset, along with their corresponding PPAs (projected onto the x/y plane). With the contact points identified, it is relatively straightforward to compute the relevant PPAs of the interacting space-time prisms by projecting their intersecting portions (portions that share the same space-time volume) onto the 2D geographic space. The jPPAs are then simply the geometric intersection of these PPAs (not shown), which can be computed using standard computational geometry techniques.

6   Conclusions

The primary goal of this paper is to introduce methodological and technical improvements based on time-geographic theories and methods. To this end, we have presented an extensible framework for scaling time geographic methods to the increasingly large and diverse set of emerging spatio-temporal data sources. By taking advantage of techniques from the computer graphics literature, and combining these ideas in a time geography framework, we outline a hierarchical tree of space-time prisms, or Hierarchical Prism Tree (HPT), that forms the basis for a powerful computational framework for time geography research. In particular, our HPT is able to embed both space-time paths and prisms in a 3D space-time. This space-time tree is able to handle large volumes of space-time data that are potentially dynamic (and/or real-time) in nature. We demonstrated the utility of our approach using two common time geography analysis tasks, based on (1) space-time path similarity analysis, and (2) identifying joint potential path areas. While the work presented here is by no means exhaustive, it does provide a useful initial exploration of the utility of thinking about the scalability of time geographic methods. Indeed, the dynamic HPT presented in this paper provides an ideal framework for scaling and exploring time-geographic methods and ideas in an intuitive and computationally efficient manner.

The development of the HPT presented in this paper offers many avenues for further development. Currently, we are exploring ways to scale various space-time intersection queries in order to facilitate the data-driven generation of space-time prisms for data integration, as discussed in [42]. Additionally, the dynamic nature of the HPT is designed to facilitate tracking and analysis of real-time spatio-temporal data sources, such as those generated by the recently launched ICARUS initiative[5] or the long-established Argos system[6]. In order to make time geography methods accessible to the research communities working with such platforms, we are currently developing a suite of tools for working with

---

[5] ICARUS analyzes the migratory behavior of animals such as birds and bats: http://icarusinitiative.org.

[6] Argos is a global, satellite-based platform widely used in animal tracking: http://www.argos-system.org/.

space-time data using the Python programing language. Python is continuing to gain favor among data scientists and academic researchers, and implementation of various time geography methods within our HPT framework should facilitate increased adoption of time geography concepts and methods throughout the social and environmental sciences. A computational framework that is able to scale time geographic analysis from working with small, localized samples, to large, globally-distributed (possibly real-time) data sources has the potential to increase the utility of time geography concepts and methods to new domains and research questions significantly.

References

1. Batty, M.: Smart cities, big data. Environ. Plan. **39**(2), 191–193 (2012)
2. Yang, C., Raskin, R., Goodchild, M., Gahegan, M.: Geospatial cyberinfrastructure: past, present and future. Comput. Environ. Urban Syst. **34**(4), 264–277 (2010). Geospatial Cyberinfrastructure
3. Miller, H.J.: A measurement theory for time geography. Geogr. Anal. **37**(1), 17–45 (2005)
4. Hägerstrand, T.: What about people in regional science? Papers Reg. Sci. Assoc. **24**, 7–21 (1970)
5. Miller, H.J.: What about people in geographic information science? In: Fisher, P., Unwin, D. (eds.) Representing GIS, pp. 215–242. Wiley, Hoboken (2005)
6. Shaw, S.L.: Guest editorial introduction: time geography - its past, present and future. J. Transp. Geogr. **23**, 1–4 (2012). Special Issue on Time Geography
7. Crease, P., Reichenbacher, T.: Linking time geography and activity theory to support the activities of mobile information seekers. Trans. GIS **17**(4), 507–525 (2013)
8. Raubal, M., Miller, H.J., Bridwell, S.: User-centred time geography for location-based services. Geogr. Ann.: Ser. B Hum. Geogr. **86**(4), 245–265 (2004)
9. Kwan, M.P.: Gender and individual access to urban opportunities: a study using space-time measures. Prof. Geogr. **51**(2), 210–227 (1999)
10. Miller, H.J.: Modelling accessibility using space-time prism concepts within geographical information systems. Int. J. Geogr. Inf. Syst. **5**(3), 287–301 (1991)
11. Raubal, M., Winter, S., Teßmann, S., Gaisbauer, C.: Time geography for ad-hoc shared-ride trip planning in mobile geosensor networks. ISPRS J. Photogramm. Remote Sens. **62**(5), 366–381 (2007)
12. Winter, S., Raubal, M.: Time geography for ad-hoc shared-ride trip planning. In: 7th International Conference on Mobile Data Management 2006, MDM 2006 (2006)
13. Rainham, D., McDowell, I., Krewski, D., Sawada, M.: Conceptualizing the health-scape: contributions of time geography, location technologies and spatial ecology to place and health research. Soc. Sci. Med. **70**(5), 668–676 (2010)
14. Bröring, A., Remke, A., Stasch, C., Autermann, C., Rieke, M., Möllers, J.: Enviro-Car: a citizen science platform for analyzing and mapping crowd-sourced car sensor data. Trans. GIS **19**(3), 362–376 (2015)
15. Winter, S., Yin, Z.C.: The elements of probabilistic time geography. GeoInformatica **15**(3), 417–434 (2011)
16. Samet, H.: Applications of Spatial Data Structures. Addison-Wesley, Boston (1990)
17. Myllymaki, J., Kaufman, J.: High-performance spatial indexing for location-based services. In: Proceedings of 12th International Conference on World Wide Web, WWW 2003, pp. 112–117. ACM, New York (2003)

18. Gustafsson, T., Hansson, J.: Dynamic on-demand updating of data in real-time database systems. In: Proceedings of 2004 ACM Symposium on Applied Computing, SAC 2004, pp. 846–853. ACM, New York (2004)
19. Papadias, D., Tao, Y., Kanis, P., Zhang, J.: Indexing spatio-temporal data warehouses. In: Proceedings of 18th International Conference on Data Engineering 2002, pp. 166–175 (2002)
20. Theodoridis, Y., Sellis, T., Papadopoulos, A., Manolopoulos, Y.: Specifications for efficient indexing in spatiotemporal databases. In: Proceedings of 10th International Conference on Scientific and Statistical Database Management 1998, pp. 123–132, Jul 1998
21. Wang, W., Yang, J., Muntz, R.: Pk-tree: a spatial index structure for high dimensional point data. In: Tanaka, K., Ghandeharizadeh, S., Kambayashi, Y. (eds.) Information Organization and Databases: Foundations of Data Organization. SISECS, vol. 579. Springer, Berlin (2000)
22. Tayeb, J., Ulusoy, Ö., Wolfson, O.: A quadtree-based dynamic attribute indexing method. Comput. J. 41(3), 185–200 (1998)
23. Navarro, G., Reyes, N.: Dynamic spatial approximation trees for massive data. In: 2nd International Workshop on Similarity Search and Applications, SISAP, pp. 81–88, August 2009
24. Navarro, G., Reyes, N.: Dynamic spatial approximation trees. J. Exp. Algorithmics 12, 1.5:1–1.5:68 (2008)
25. Bo, Z., Fu-ling, B.: Dynamic quadtree spatial index algorithm for mobile GIS. Comput. Eng. 33(15), 86 (2007)
26. Xia, Y., Prabhakar, S.: Q+rtree: efficient indexing for moving object databases. In: Proceedings of 8th International Conference on Database Systems for Advanced Applications 2003 (DASFAA 2003), pp. 175–182, March 2003
27. Myllymaki, J., Kaufman, J.H.: DynaMark: a benchmark for dynamic spatial indexing. In: Chen, M.-S., Chrysanthis, P.K., Sloman, M., Zaslavsky, A. (eds.) MDM 2003. LNCS, vol. 2574, pp. 92–105. Springer, Heidelberg (2003)
28. Myllymaki, J., Kaufman, J.: Locus: a testbed for dynamic spatial indexing. IEEE Data Eng. Bull. Spec. Issue Index. Mov. Objects 25, 48–55 (2002)
29. Zhu, Q., Gong, J., Zhang, Y.: An efficient 3D r-tree spatial index method for virtual geographic environments. J. Photogramm. Remote Sens. 62(3), 217–224 (2007)
30. Ize, T., Wald, I., Parker, S.G.: Asynchronous BVH construction for ray tracing dynamic scenes on parallel multi-core architectures. In: Proceedings of 7th Eurographics Conference on Parallel Graphics and Visualization, EGPGV 2007, pp. 101–108. Eurographics Association, Aire-la-Ville (2007)
31. Glassner, A.S.: An Introduction to Ray Tracing. Academic Press Ltd., London (1989)
32. Stich, M., Friedrich, H., Dietrich, A.: Spatial splits in bounding volume hierarchies. In: Proceedings of Conference on High Performance Graphics 2009, HPG 2009, pp. 7–13. ACM, New York (2009)
33. Maneewongvatana, S., Mount, D.M.: Analysis of approximate nearest neighbor searching with clustered point sets. CoRR cs.CG/9901013 (1999)
34. Vinkler, M., Havran, V., Bittner, J.: Bounding volume hierarchies versus kd-trees on contemporary many-core architectures. In: Proceedings of 30th Spring Conference on Computer Graphics. SCCG 2014, pp. 29–36. ACM, New York (2014)
35. Shevtsov, M., Soupikov, A., Kapustin, A.: Highly parallel fast kd-tree construction for interactive ray tracing of dynamic scenes. Comput. Graph. Forum 26(3), 395–404 (2007)

36. He, L., Ortiz, R., Enquobahrie, A., Manocha, D.: Interactive continuous collision detection for topology changing models using dynamic clustering. In: Proceedings of 19th Symposium on Interactive 3D Graphics and Games, i3D 2015, pp. 47–54. ACM, New York (2015)
37. Stein, C., Limper, M., Kuijper, A.: Spatial data structures for accelerated 3D visibility computation to enable large model visualization on the web. In: Proceedings of 19th International ACM Conference on 3D Web Technologies, Web3D 2014, pp. 53–61. ACM, New York (2014)
38. Kopta, D., Ize, T., Spjut, J., Brunvand, E., Davis, A., Kensler, A.: Fast, effective BVH updates for animated scenes. In: Proceedings of ACM SIGGRAPH Symposium on Interactive 3D Graphics and Games, I3D 2012, pp. 197–204. ACM, New York (2012)
39. Yoon, S.E., Curtis, S., Manocha, D.: Ray tracing dynamic scenes using selective restructuring. In: Proceedings of 18th Eurographics Conference on Rendering Techniques, EGSR 2007, pp. 73–84. Eurographics Association, Aire-la-Ville (2007)
40. Karras, T., Aila, T.: Fast parallel construction of high-quality bounding volume hierarchies. In: Proceedings of 5th High-Performance Graphics Conference, HPG 2013, pp. 89–99. ACM, New York (2013)
41. Miller, H., Raubal, M., Jaegal, Y.: Measuring space-time prism similarity through temporal profile curves. In: 19th AGILE Conference on Geographic Information Science - Geospatial Data in a Changing World, p. 19 (2016)
42. Keßler, C., Farmer, C.J.Q.: Querying and integrating spatial-temporal information on the web of data via time geography. Web Semant.: Sci. Serv. Agents World Wide Web 35(1), 25–34 (2015)
43. Schwesinger, U., Siegwart, R., Furgale, P.: Fast collision detection through bounding volume hierarchies in workspace-time space for sampling-based motion planners. In: 2015 IEEE International Conference on Robotics and Automation (ICRA), pp. 63–68, May 2015
44. Long, J., Nelson, T.: Home range and habitat analysis using dynamic time geography. J. Wildl. Manag. 79(3), 481–490 (2015)
45. Long, J.A., Nelson, T.A.: Measuring dynamic interaction in movement data. Trans. GIS 17(1), 62–77 (2013)
46. Larsson, T., Akenine-Möller, T.: A dynamic bounding volume hierarchy for generalized collision detection. Comput. Graph. 30(3), 450–459 (2006)
47. Sinha, G., Mark, D.M.: Measuring similarity between geospatial lifelines in studies of environmental health. J. Geogr. Syst. 7(1), 115–136 (2005)
48. Gao, P., Kupfer, J.A., Zhu, X., Guo, D.: Quantifying animal trajectories using spatial aggregation and sequence analysis: a case study of differentiating trajectories of multiple species. Geogr. Anal. 48, 275–291 (2016)
49. Demšar, U., Virrantaus, K.: Space-time density of trajectories: exploring spatiotemporal patterns in movement data. Int. J. Geogr. Inf. Sci. 24(10), 1527–1542 (2010)
50. Long, J.A., Webb, S.L., Nelson, T.A., Gee, K.L.: Mapping areas of spatial-temporal overlap from wildlife tracking data. Mov. Ecol. 3(1), 1–14 (2015)
51. Ram, P., Lee, D., March, W., Gray, A.G.: Linear-time algorithms for pairwise statistical problems. In: Advances in Neural Information Processing Systems (NIPS), December 2009, vol. 22. MIT Press (2010)
52. Gray, A.G., Moore, A.W.: N-body problems in statistical learning. In: Leen, T.K., Dietterich, T.G., Tresp, V. (eds.) Advances in Neural Information Processing Systems (NIPS), December 2000, vol. 13. MIT Press (2001)

Network Analysis

Mining Network Hotspots with Holes: A Summary of Results

Emre Eftelioglu[1]([⊠]), Yan Li[1], Xun Tang[1], Shashi Shekhar[1], James M. Kang[2], and Christopher Farah[2]

[1] Department of Computer Science, University of Minnesota, Minneapolis, USA
{eftel003,lixx4266,tangx456,shekhar}@umn.edu
[2] National Geospatial-Intelligence Agency, Springfield, USA
{James.M.Kang,Christopher.C.Farah}@nga.mil

Abstract. Given a spatial network and a collection of activities (e.g. crime locations), the problem of Mining Network Hotspots with Holes (MNHH) finds network hotspots with doughnut shaped spatial footprint, where the concentration of activities is unusually high (e.g. statistically significant). MNHH is important for societal applications such as criminology, where it may focus the efforts of officials to identify a crime source. MNHH is challenging because of the large number of candidates and the high computational cost of statistical significance test. Previous work focused either on geometry based hotspots (e.g. circular, ring-shaped) on Euclidean space or connected subgraphs (e.g. shortest path), limiting the ability to detect statistically significant hotspots with holes on a spatial network. This paper proposes a novel Network Hotspot with Hole Generator (NHHG) algorithm to detect network hotspots with holes. The proposed algorithm features refinements that improve the performance of a naïve approach. Case studies on real crime datasets confirm the superiority of NHHG over previous approaches. Experimental results on real data show that the proposed approach yields substantial computational savings without reducing result quality.

Keywords: Hotspot detection · Crime hotspots · Spatial scan statistics

1 Introduction

Given a spatial network and a collection of activities (i.e. crime locations), the problem of Mining Network Hotspots with Holes (MNHH) finds hotspots with doughnut shaped spatial footprint on a spatial network (i.e. road network), where the concentration of activities is unusually high (i.e. statistically significant).

The problem of Mining Network Hotspots with Holes (MNHH) has important societal applications in criminology, where identifying crime hotspots may improve police response [1]. In environmental criminology, domain experts create geographic profiles of criminals using the locations of crimes and try to find where a serial criminal frequently commutes, thereby focusing the efforts of police forces in the field [2]. Our notion of Network Hotspots with Holes originates from

© Springer International Publishing Switzerland 2016
J.A. Miller et al. (Eds.): GIScience 2016, LNCS 9927, pp. 51–67, 2016.
DOI: 10.1007/978-3-319-45738-3_4

two key concepts in criminology, namely inner buffer zone (e.g. comfort zone) and distance decay [2]. Inner buffer zone is an area around a criminal's frequently visited locations, where crimes are less likely due to the risks caused by reduced anonymity. Distance decay relates to a least effort principle, where crimes occur relatively close to criminal's frequently visited locations, since traveling long distances requires time and money. The opposing effects of inner buffer zone and distance decay create an activity zone with a doughnut shaped spatial footprint around a path that a criminal usually travels. Figure 1 illustrates these concepts where the green squares represent activities (i.e. crime), the blue line shows a path between home and work (blue squares), the black road segments represent the inner buffer zone where the activities are less likely and the red road segments create the activity zone (i.e. outer buffer) that we define as network hotspot with hole (*NHH*) in this paper.

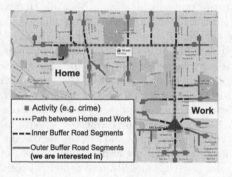

Fig. 1. A path between home and work, an inner buffer zone where the activities are sparse and the activity zone (i.e. outer buffer) that we are interested in (best in color). (Color figure online)

Informally, the problem of Mining Network Hotspots with Holes (MNHH) can be defined as follows: given a spatial network (e.g. road network), an activity set associated with road segments (e.g. street robberies), a log likelihood ratio threshold (θ), a *p-value* threshold (α_p), a maximum outer buffer distance ($\widehat{t_{max}}$) and a unit distance (ω), find network hotspots with holes where the concentration of activities is significantly higher than outside ($p\text{-}value \leq \alpha_p$).

Challenges: MNHH is challenging due to the potentially large number of candidate network hotspots with holes ($O(N^4)$) in a given dataset of millions of road network nodes (N). For large road networks (e.g. 10^8 road segments in the U.S.), this causes exorbitant computation times as well as a prohibitively large enumeration space. Moreover, the interest measure, "log likelihood ratio (*Log LR*)", does not have a monotonicity property, meaning that there is no order between the *Log LR* of a network hotspot with hole (*NHH*) and another *NHH* it may contain. Thus, interest measure cannot be used for computational speed-up. In addition, the statistical significance test multiplies the cost.

Related Work and Their Limitations: Statistically significant hotspot detection approaches can be classified into two categories depending on the study area: Euclidean space based and Network based approaches. Euclidean approaches include spatial scan statistics and are widely used for the detection and evaluation of circular [3,4], elliptical [5,6], rectangular [7] and ring-shaped hotspots [8]. These techniques are useful for understanding the distribution of disease [9], or detecting a disease outbreak or even identifying the location of a criminal (e.g. through ring-shaped hotspot detection). However, criminal activities and other human activities diffuse along road networks [10] and therefore Euclidean distances do not reflect actual travel distances causing biased results. For example in Fig. 2(a), the traveling distance from $E5$ node to $C7$ node will not be the same as Euclidean distance due to the lake in between. In addition, people's activities are mostly dependent on their routine commutes (i.e. home-work-recreation) instead of a single place. In Fig. 3(a), SaTScan [3] outputs a circular hotspot with a large space without activities with a low log likelihood ratio. Similarly, in Fig. 3(b), ring-shaped hotspot detection (RHD) outputs a hotspot with low log likelihood ratio due to using Euclidean distance as well as assuming a single center (i.e. crime base of a criminal). Thus, geometry-based techniques may not be appropriate for modeling hotspots on road networks. A more detailed comparison of the recent related work can be found in [9,11].

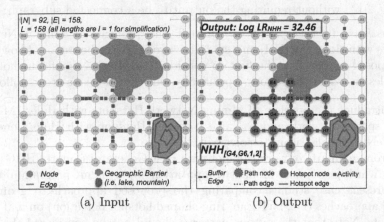

(a) Input   (b) Output

Fig. 2. An example input and output of our proposed approach for Mining Network Hotspots with Holes. Edges represent streets and nodes represent road intersections. (Color figure online)

A second category of hotspot detection is network-based. These methods leverage the underlying spatial network, which improves the detection of activities that diffuse along the spatial network [12–14]. However, these often focus on detecting paths or road segments which have unusually high activities and require a hotspot to be a connected subgraph (e.g. shortest path), causing them to miss network hotspots with holes [15–17]. Figure 3(c) shows the output of a

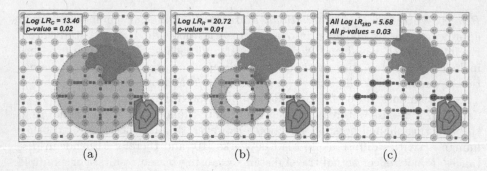

Fig. 3. Output of the related work for the input in Fig. 2(a). SaTScan (a), ring-shaped hotspot detection (b) and significant route discovery (c)

significant route discovery approach which enumerates shortest paths between nodes and returns those that have a significantly high number of activities [15]. The output fails to identify the significant region of interest and includes 4 hotspots with low log likelihood ratios and high *p-values*, indicating lack of significance.

In contrast to previous methods, our Mining Network Hotspots with Holes (MNHH) method can find statistically significant network hotspots with holes (e.g. Fig. 2(b)) without requiring the output to be a connected subgraph.

Contributions: In this paper, we present the problem of Mining Network Hotspots with Holes (MNHH) on a spatial network. To the best of our knowledge, the proposed approach is the first to consider statistically significant hotspots with holes on a spatial network. Specifically, our contributions are as follows:

- We introduce the problem of Mining Network Hotspots with Holes (MNHH) on a spatial network and a Naïve Network Hotspot with Hole Generator (NaïveNHHG) algorithm to solve MNHH.
- We propose Smart Network Hotspot with Hole Generator (SmartNHHG) algorithm which prevents redundant computations by dynamic programming.
- We present case studies comparing the proposed approach to geometry-based approaches (i.e., SaTScan, ring-shaped hotspot detection) on real crime datasets. Note that the output patterns should not be considered the same (e.g. circles and rings vs. network sub-graphs).
- Experimental results on real data show that SmartNHHG yields substantial computational savings over NaïveNHHG without sacrificing result quality.

Scope: This work focuses on finding hotspots with holes on road networks where each activity (i.e. crime event) is associated with a road segment (i.e. edge). This does not imply that the original activities must necessarily have occurred at edges. Each activity set is pre-processed to associate activities to the closest edge on the road segment. In addition, other properties of road networks (e.g. speed limit, traffic density) are not considered. In this work, the number of activities on the road network is fixed and does not change over time. Finally, this paper does not provide guidance on parameter (e.g. t_{min}, t_{max}) value selection. However, users may evaluate the spatial distribution of events using centrographic statistics [18] and select parameters accordingly.

Outline: This paper is organized as follows: Sect. 2 presents the basic concepts and problem statement for MNHH. Section 3 presents the Naïve and Smart Network Hotspot with Hole Generator (SmartNHHG) algorithms. Section 4 presents case studies which qualitatively evaluate the output of SmartNHHG on real crime datasets. Experimental evaluation is in Sect. 5. Section 6 presents a discussion. Section 7 concludes the paper and previews future work.

2 Basic Concepts and Problem Statement

2.1 Basic Concepts

Definition 1. *A **spatial network** $G = (N, E)$ is a set of nodes (N) and edges (E) where each node $n_v \in N$ is associated with coordinates (x, y) representing its location in an Euclidean space. E is a subset of the cross product $N \times N$ and an edge $e_i \in E$, which joins nodes n_u and n_v, is associated with a length $l_{u,v} \geq 0$.*

In Fig. 2(a) grey circles represent nodes (e.g. intersections), grey lines represent edges (e.g. streets) and there are two geographic barriers (e.g. lake, mountain). The length of the network is the sum of all edge lengths $L_{total} = \sum l_{e \in G}$.

Definition 2. *An **activity set** A is a collection of activities. An activity $a \in A$ is an object of interest associated with only one edge $e \in E$.*

For example in Fig. 2(a), the edge between n_{G2} and n_{G3} has 3 activities.

Definition 3. *A **shortest path** $p_{u,v}$ is a sequence of nodes $[n_1, n_2, ..., n_i]$ such that $[e_1, e_2, ..., e_i] \in E$ and $n_i \in N$ are distinct and the sum of edge lengths is minimized. The length of a shortest path is $L_p = \sum l_{e \in p}$.*

For example, $p_{A0,B2} = [A0, A1, A2, B2]$ and $L_p = 3$ in Fig. 2(a).

Definition 4. *__Distance__ between a node n_i and a path $p_{u,v}$ is $d(n_i, p_{u,v}) = \min(L_p(n_i, n_j \in p_{u,v}))$.*

For example, $d(D3, p_{A0,B2}) = 3$ in Fig. 2(a).

Definition 5. *A **Network Buffer** $(NB_{u,v,t})$ is a closed set of nodes $N_{NB} \subset N$ and edges $E_{NB} \subset E$ such that $d(n_i, p_{u,v}) \leq t$, $\forall n_i \in N_{NB}$ and $t = k\omega$ for some $k \in \mathbb{R}^+$ and a unit distance ω.*

For example , in Fig. 2(b), $NB_{G4,G6,2}$ is the set of all blue/red nodes and all blue/red/black edges.

Definition 6. *A **Network Hotspot with Hole** ($NHH_{u,v,t_{min},t_{max}}$) is the closure [19] (a set and its limit points that is denoted by Cl) of the set difference of outer buffer $NB_{u,v,t_{max}}$ and inner buffer $NB_{u,v,t_{min}}$, where t_{min} is the inner and t_{max} is the outer buffer distance and the distance interval is closed i.e., inclusive of t_{min} and t_{max}.*

Thus, $NHH_{u,v,t_{min},t_{max}} = Cl(NB_{u,v,t_{max}} \setminus NB_{u,v,t_{min}})$. The sum of the length of the edges in NHH is denoted by $L_{NHH} = \sum l_{e \in NHH}$.

In Fig. 2(b), black edges represent the $NB_{G4,G6,1}$ and red nodes and edges represent the $NHH_{G4,G6,1,2}$ around the path $p_{G4,G6}$ with $t_{min} = 1$ and $t_{max} = 2$.

Definition 7. *Log Likelihood Ratio (Log LR_{NHH}) is the test statistic for a candidate NHH. Since a NHH is on a road network, it uses L_{NHH}, instead of the hotspot area as used in [20]. The equation can be shown as:*

$$Log\ LR_{NHH} = Log\left(\left(\frac{c}{B}\right)^c \times \left(\frac{|A|-c}{|A|-B}\right)^{|A|-c} \times I()\right) \quad (1)$$

$B = \frac{|A| \times L_{NHH}}{L_{total}}$ *and* $I() = \begin{cases} 1, & \text{if } c > B \\ 0, & \text{otherwise,} \end{cases}$

B is the expected and c is the observed number of points for a NHH, $|A|$ is the cardinality of A and $I()$ is an indicator function. $I() = 1$ when a candidate NHH has more points than expected ($c > B$); otherwise $I() = 0$ [3].

For example, the sum of the lengths of the edges of $NHH_{G4,G6,1,2}$ in Fig. 2(b) is $L_{NHH} = 19$ and the total length of the spatial network is $L_{total} = 158$. Thus $B = \frac{50 \times 19}{158} = 6.012$. In this *NHH*, there are $c = 30$ points. Thus, $I = 1$ since $30 > 6.012$.

Using Eq. 1, $Log\ LR_{NHH} = Log\left(\left(\frac{30}{6.012}\right)^{30} \times \left(\frac{50-30}{50-6.012}\right)^{50-30} \times 1\right) = 32.46$

Definition 8. *A **Hypothesis Test** determines whether a NHH occurred by chance or not. The null hypothesis H_0 states that the points are randomly distributed on a spatial network and the alternative hypothesis H_1 states that the candidate NHH has a significantly higher number of activities than outside. In order to determine the hypothesis test result, the significance level (p-value) of a NHH is computed by finding the order of the actual Log LR_{NHH} in the test statistic distribution (obtained by Monte Carlo simulations) and dividing that position by $m + 1$. If the p-value of a NHH is lower than the desired threshold (α_p), the H_1 cannot be rejected, and we say that the candidate NHH is a significant NHH. Note that these concepts are inherited from SaTScan [3].*

2.2 Problem Statement

Formally, Mining Network Hotspots with Holes (MNHH) problem is as follows:
Given:

1. A spatial network $G = (N, E)$ with activity count function $a(u, v) \geq 0$ and length function $l(u, v) > 0$ for each edge $e_i \in E$,

2. A log likelihood ratio threshold (θ) and a *p-value* threshold (α_p),
3. A number of Monte Carlo simulation trials (m),
4. A maximum outer buffer distance ($\widehat{t_{max}}$) and a unit distance (ω),

Find: Network hotspots with holes with $Log\ LR_{NHH} \geq \theta$ and *p-value* $\leq \alpha_p$.
Objective: Computational efficiency and correctness of the output.
Constraint: Nodes $n_j \in NHH$ may not be connected to each other.
Example: The graph in Fig. 2(a) can be viewed as a road network, composed of streets (edges) and intersections (nodes). The aim is to find network hotspots with holes (*NHH*) that meet the given log likelihood ratio and significance levels (*p-value* threshold θ). In environmental criminology, finding such a hotspot may have two benefits: (1) it may focus the search for a criminal to the path at the center of *NHH*. (2) it may help determine the locations to deploy new police patrols to prevent crime. In Fig. 2(b), $NHH_{G4,G6,1,2}$ is returned since $Log\ LR_{NHH} = 32.46$ and *p-value* $= 0.01$. Although the output includes more *NHHs*, since the other *NHHs* were overlapping, we show only the *NHH* with highest $Log\ LR_{NHH}$ to reduce the visual clutter.

3 Proposed Approach

In this section, we first describe a naïve version of our network hotspot with hole generator algorithm (NaïveNHHG). Then we present our SmartNHHG algorithm with refinements that include two novel dynamic programming approaches and a Monte Carlo simulation speed-up. The proposed algorithms present steps for candidate enumeration, candidate evaluation using $Log\ LR$ and statistical significance test. It should be noted that in some communities these steps are practiced separately. However, in this work, we present algorithms that describe these processes together for the sake of self-containment.

3.1 Naïve NHH Generator Algorithm

Algorithm 1 presents the pseudocode for the NaïveNHHG approach. The algorithm begins by creating all pair shortest paths, P_{apsp}, in the spatial network (step 1). Next, each shortest path is used as a center to enumerate *NHHs* with different inner and outer buffer distances (t_{min}, t_{max}) (step 2). Finally, the statistical significance of each *NHH* is evaluated by m Monte Carlo simulations and the significant *NHHs* are returned (as the output).

NaïveNHHG Example: Table 1 shows a sample execution trace of NaïveNHHG. The spatial network has 92 nodes, 158 edges, and 50 activities (green squares on the edges). All edge lengths are set to 1 for illustration purposes. Inputs are set to log likelihood ratio threshold $\theta = 30$, *p-value* threshold $\alpha_p = 0.01$, maximum outer buffer distance $\widehat{t_{max}} = 5$ and unit distance $\omega = 1$.

In step 1 of Table 1, all pairs of shortest paths are computed as shown in the first column (4 out of 2.5×10^4 are shown). In step 2, the *NHHs* are enumerated by using the set difference of NB with t_{min} and t_{max}. Then, $Log\ LR_{NHH}$

Algorithm 1. NaïveNHHG Algorithm

Input:
   1) A spatial network $G = (N, E)$ with activity count function $a(u, v) \geq 0$ and length function $l(u, v) > 0$ for each edge $e_i \in E$,
   2) A log likelihood ratio threshold (θ) and a *p-value* threshold (α_p),
   3) A maximum outer buffer distance $(\widehat{t_{max}})$ and a unit distance (ω)

Output:
   Network hotspots with holes (*NHH*) with *p-value* $\leq \alpha_p$

Algorithm:
1: **Step 1:** Generate all pair shortest paths P_{apsp}
2: **For each** shortest path $p_{u,v} \in P_{apsp}$
3:     **Step 2:** Enumerate candidate *NHH* with t_{min} and t_{max}
4: **Step 3:** Significant *NHH* ← candidate *NHH* with *p-value* $\leq \alpha_p$ using m Monte Carlo simulations
5: **Return** Significant *NHH* with $Log\ LR_{NHH} \geq \theta$

---

are computed for each *NHH* and *NHH*s with $Log\ LR_{NHH} \geq 30$ are stored as candidates. In step 3, the significance of candidate *NHH*s are determined and significant *NHH*s are returned (as the output) as shown in Fig. 2(b). Although many *NHH*s were evaluated as significant, only the *NHH* on the top row of Table 1 is returned since $\theta = 30$. If a user is interested in all significant *NHH*, θ threshold can be set 0. Also, one may notice that many *NHH*s were similar in the output. This issue is discussed in Sect. 6.

Table 1. An example execution trace of NaïveNHHG.

Step 1		Step 2					Step 3
Start - End	Path	t_{min}	t_{max}	CNHH	l	Log LR$_{NHH}$	p-value$_{NHH}$
<G4,G6>	[G4, G5, G6]	1	2	30	19	32.46	0.01
<G4,H6>	[G4, G5, G6, H6]	1	2	25	21	20.39	0.01
<G4,F6>	[G4, G5, G6, F6]	1	2	27	19	24.59	0.01
<G4,G7>	[G4, G5, G6, G7]	1	2	25	20	21.17	0.01
...	...	...	...	...	...	...	...

Enumerating Candidate NHH: Algorithm 2 shows the steps of candidate *NHH* enumeration on NaïveNHHG. For each shortest path $(p_{u,v})$ in the set of all pair shortest paths (P_{apsp}) (line 1), candidate *NHH*s are enumerated as follows: First, inner and outer NB are defined by t_{min} and t_{max} (line 2–3). Note that these values are changed by unit distance ω on every iteration and even for a single $p_{u,v} \in P_{apsp}$, *NHH*s with different t_{min} and t_{max} are enumerated. Next, for each node n_j of $p_{u,v}$, single source shortest paths from that node to all other nodes in the spatial network are enumerated (line 5). If the length of any of these shortest paths is less than the t_{min}, it is saved in $NB_{u,v,t_{min}}$. Similarly, if the length of any of these shortest paths is less than t_{max}, it is saved in $NB_{u,v,t_{max}}$ (line 6–9). Finally, $NHH_{u,v,t_{min},t_{max}} = Cl(NB_{u,v,t_{max}} \setminus NB_{u,v,t_{min}})$ (line 10) and its $Log\ LR_{NHH}$ is computed using $L_{NHH} = \sum l_{e \in NHH}$ and its activity count. If $Log\ LR_{NHH} \geq \theta$ threshold, then the NHH is saved as a candidate (line 11). This process is repeated for all paths in P_{apsp} and t_{min} and t_{max} until $\widehat{t_{max}}$.

In NaïveNHHG, candidate NHHs are enumerated by varying t_{min} and t_{max} for all pairs of shortest paths. However, enumeration becomes exorbitant even for road networks of 10^2 nodes. To improve the scalability of NaïveNHHG, we analyzed NaïveNHHG and determined the redundant computations. Next, we propose refinements to reduce redundant computations but increase scalability.

Algorithm 2. Enumerating Candidate NHH - NaïveNHHG

1: **for each** $p_{u,v} \in P_{apsp}$ **do**
2:     **for each** $t_{max} = 2\omega$ to $\widehat{t_{max}}$ **do**
3:         **for each** $t_{min} = \omega$ to t_{max} **do**
4:             **for each** $n_j \in p_{u,v}$ **do**
5:                 **for each** Single Source Shortest Path $p_{n_j,n} \in P_{sssp}$ **do**
6:                     **if** $t_{min} > L_{p_{n_j},n}$ **then**
7:                         $NB_{u,v,t_{min}} \leftarrow$ Edges and Nodes from $p_{n_j,n}$
8:                     **if** $t_{max} \geq L_{p_{n_j},n}$ **then**
9:                         $NB_{u,v,t_{max}} \leftarrow$ Edges and Nodes from $p_{n_j,n}$
10:                 $NHH_{u,v,t_{min},t_{max}} \leftarrow Cl(NB_{u,v,t_{max}} \setminus NB_{u,v,t_{min}})$
11:                 Candidate $NHH \leftarrow NHH$ with $Log\ LR_{NHH} \geq \theta$

3.2 Smart NHH Generator Algorithm

This section explains our smart approach for solving the MNHH problem. Our algorithm features three key ideas for achieving computational savings while maintaining result quality: Distance based dynamic programming, edge stitching and Monte Carlo simulation speed-up.

Distance Based Dynamic Programming (DP) Approach: Algorithm 3 shows the steps of distance based DP approach, which avoids redundant calculation of NHH with different t_{min} and t_{max} by enumerating NHH with $t_{max} - t_{min} = \omega$ (line 2–9) and then using the set union of these to create NHHs with different inner and outer NB (line 10–12). A simplified example can be seen in Fig. 4. In this example, in order to enumerate $NHH_{F4,F5,1,3}$ (on the right), the set union of $NHH_{F4,F5,1,2}$ (on the left) and $NHH_{F4,F5,2,3}$ (in the middle) is used. Thus, instead of running a new enumeration process for $NHH_{F4,F5,1,3}$, the algorithm simply uses the previously computed $NHH_{F4,F5,1,2}$ and $NHH_{F4,F5,2,3}$.

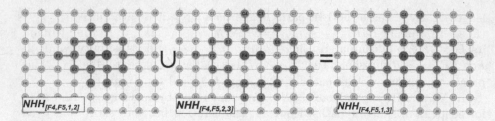

Fig. 4. Distance based dynamic programming approach. To determine $NHH_{F4,F5,1,3}$, the set union of $NHH_{F4,F5,1,2}$ and $NHH_{F4,F5,2,3}$ is used (best in color).

Algorithm 3. Enumerating Candidate NHH - Distance Based DP Approach

1: **for each** $p_{u,v} \in P_{apsp}$ **do**
2:    **for each** $t_{max} = 2\omega$ to $\widehat{t_{max}}$ and $t_{min} = t_{max} - \omega$ **do**
3:       **for each** $n_j \in p_{u,v}$ **do**
4:          **for each** Single Source Shortest Path $p_{n_j,n} \in P_{sssp}$ **do**
5:             **if** $t_{min} > L_{p_{n_j},n}$ **then**
6:                $NB_{u,v,t_{min}} \leftarrow$ Edges and Nodes from $p_{n_j,n}$
7:             **if** $t_{max} \geq L_{p_{n_j},n}$ **then**
8:                $NB_{u,v,t_{max}} \leftarrow$ Edges and Nodes from $p_{n_j,n}$
9:          $NHH_{u,v,t_{min},t_{max}} \leftarrow Cl(NB_{u,v,t_{max}} \setminus NB_{u,v,t_{min}})$
10:    **for each** $t_{max} = 2\omega$ to $\widehat{t_{max}}$ **do**
11:       **for each** $t_{min} = \omega$ to t_{max} **do**
12:          $NHH_{tmin,tmax} = NHH_{tmin+\omega,tmax} \bigcup NHH_{tmin,tmax-\omega}$

Edge Stitching Approach: Edge Stitching exploits a basic property of paths, i.e. every path consists of edges. Thus, NHHs around single edges can be enumerated, then these can be stitched to create NHHs around longer paths (avoid Line 1 of Algorithm 2). In Algorithm 4, first, all NHHs around single edges are enumerated (Line 1–9). Next, these NHHs are stitched to create NHHs for longer paths (Line 10–14) as illustrated in Fig. 5. In this example, in order to create $NHH_{G4,G6,1,2}$, $NHH_{G4,G5,1,2}$ and $NHH_{G5,G6,1,2}$ are stitched together. Once we create the set union of these NHHs, edges and nodes that belong to $NB_{G4,G6,1}$ are removed to determine $NHH_{G4,G6,1,2}$ (in Fig. 2(b)).

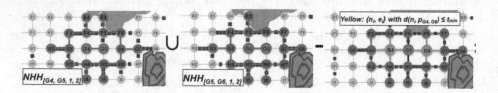

Fig. 5. Edge stitching approach. To determine $NHH_{G4,G6,1,2}$ (Fig. 2(b)); $NHH_{G4,G5,1,2}$ and $NHH_{G5,G6,1,2}$ are stitched together and then the nodes and edges of $NB_{G4,G6,1}$ are removed (best in color).

Algorithm 4. Enumerating Candidate NHH - Edge Stitching Approach

1: **for each** $e_i \in E$ **do**
2:     **for each** $t_{max} = 2\omega$ to $\widehat{t_{max}}$ **do**
3:         **for each** $t_{min} = \omega$ to t_{max} **do**
4:             **for each** Single Source Shortest Path from n_u and n_v $p_{n_j,n} \in P_{sssp}$ **do**
5:                 **if** $t_{min} > L_{p_{n_j},n}$ **then**
6:                     $NB_{u,v,t_{min}} \leftarrow$ Edges and Nodes from $p_{n_j,n}$
7:                 **if** $t_{max} \geq L_{p_{n_j},n}$ **then**
8:                     $NB_{u,v,t_{max}} \leftarrow$ Edges and Nodes from $p_{n_j,n}$
9:                 $NHH_{u,v,t_{min},t_{max}} \leftarrow Cl(NB_{u,v,t_{max}} \setminus NB_{u,v,t_{min}})$
10: **for each** $p_{u,y} \in P_{apsp}$ **do**
11:     **for each** $t_{max} = 2\omega$ to $\widehat{t_{max}}$ **do**
12:         **for each** $t_{min} = \omega$ to t_{max} **do**
13:             $NHH_{u,y,t_{min},t_{max}} = Cl(NHH_{u,v,t_{min},t_{max}} \bigcup NHH_{v,y,t_{min},t_{max}} \setminus$
    $NB_{u,y,t_{min}})$
14:             Candidate $NHH \leftarrow NHH_{u,y,t_{min},t_{max}}$ with $Log\ LR_{NHH} \geq \theta$

Monte Carlo Simulation Speed-Up: The following three refinements are used to speed-up Monte Carlo simulations. First refinement is to create all pair shortest paths once, and use them for each simulation trial. NaïveNHHG runs for m times for Monte Carlo simulations. However, the spatial network does not change between iterations. Thus, we prevent redundant shortest path calculations in Monte Carlo Simulations. Second, if any NHH^{random} has $Log\ LR_{NHH}^{random} \geq Log\ LR_{NHH}^{actual}$ then that iteration terminates since there is no reason to keep looking at all NHHs in that random dataset if a NHH^{random} beats the maximum of $Log\ LR_{NHH}$ from the actual dataset. Third, Monte Carlo simulation is terminated if the $p\text{-}value \geq \alpha_p$ because the α_p threshold won't be met at the end.

It should be noted that these and similar refinements are often used in related work to speed-up the Monte Carlo simulation process [3,15,21]. Therefore, details of the execution trace of those speed-up approaches are omitted from this paper.

4   Case Study

We conducted two case studies to evaluate SmartNHHG qualitatively comparing its output with the output of SaTScan [3] and a ring-shaped hotspot detection [8] method using two real crime datasets (Figs. 6(a) and 7(a)). For both of the case studies, we matched activities to edges as counts. The road network was obtained from the US Census Bureau Tiger/Line Shapefile [22]. The map visualizations were prepared using QGIS and Open Layers Plugin (www.qgis.org).

The first crime dataset in Fig. 6(a) consists of 64 theft committed between 2013 and 2014 in South Side Neighborhood of Chicago, Illinois [23]. We set $\omega = 0.04\,\text{km}$, $\widehat{t_{max}} = 0.6\,\text{km}$, $\theta = 20$ and $\alpha_p = 0.01$.

The second crime dataset in Fig. 7(a) consists of 128 burglary crimes committed between 2013 and 2014 in Caballo Hills Neighborhood of Oakland, California [24]. We set $\omega = 0.5$ km, $\widehat{t_{max}} = 2.6$ km, $\theta = 20$ and $\alpha_p = 0.03$.

For the first crime dataset in Fig. 6(a), SaTScan produced a small circular hotspot as shown in Fig. 6(b). This is due to the fact that those activities occurred close to each other. For the second crime dataset in Fig. 7(a), SaTScan's output was a large circular hotspot. Since SaTScan uses Euclidean distances to enumerate circles, none of the outputs for the case studies reflected the effect of the road network.

In Figs. 6(c) and 7(c), ring-shaped hotspots returned by RHD indicate single center rings (i.e. a single crime source location) due to its enumeration method and space (i.e. Euclidean). Although the output of RHD in Fig. 6(c) aligns with the activities in the study area due to the street morphology in Chicago, it produced very different results for the burglary crimes in Oakland, California.

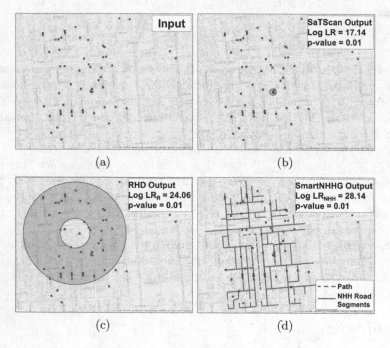

(a)   (b)

(c)   (d)

Fig. 6. Case study 1: Theft crimes in Chicago, Illinois. (a) shows the input, (b) shows the output of SaTScan, (c) shows the output of ring-shaped hotspot detection and (d) shows the output of SmartNHHG. *Log LR* values are not comparable due to the Euclidean and network spaces. (best in color).

As noted earlier, criminals are known to commit crimes around the routes they often commute as described in environmental criminology [2]. Therefore, when we take a look at the output of SmartNHHG in both case studies

(Figs. 6(d), 7(d) and (e)), we see that the output aligns with such crime patterns. For example, blue paths at the centers of the *NHH*s in Fig. 7(d) and (e) are the only routes to reach those houses that burglary crimes occurred (i.e. activities) which may make sense in the context of environmental criminology. Finally, it should be noted that our tool should be considered as a decision support tool for the analysts and the results should be analyzed by them in the context of additional domain information to prevent potentially misleading results.

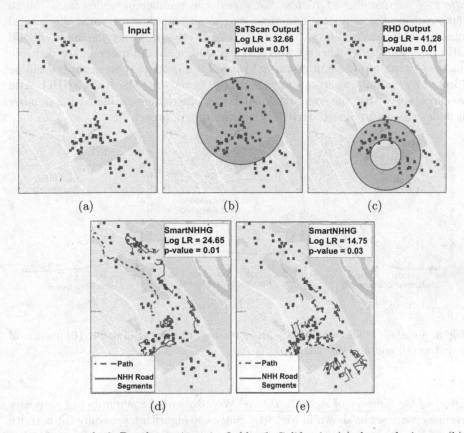

Fig. 7. Case study 2: Burglary crimes in Oakland, California. (a) shows the input, (b) shows the output of SaTScan, (c) shows the output of ring-shaped hotspot detection and (d) and (e) shows the output of SmartNHHG. *Log LR* values are not comparable due to the Euclidean and network spaces. (best in color).

5   Experimental Evaluation

We also conducted an experimental evaluation to observe the effect of the algorithmic refinements compared with the Naïve approach. The experiments were performed on real-world data obtained from the City of Chicago portal [23].

The dataset contained 676 theft crimes that were committed in Chicago, Illinois, between 2013 and 2014. The road network was obtained from the US Census Bureau Tiger/Line Shapefile [22]. For each edge on the road network, activities were matched and their counts on edges were aggregated. In the experiments, the number of Monte Carlo simulation trials was set to $m = 0$, since we did not perform any experiments on m due to the fact that our Monte Carlo simulation speed-up approaches were trivial and previously used in [15,21].

Effect of the Number of Nodes: We varied the number of nodes from 750 to 1500, causing the asymptotic increase on the all pair shortest paths, since total number of all pairs will be $\binom{750}{2}$ and $\binom{1500}{2}$ respectively. We set the log likelihood ratio threshold to $\theta = 20$. We also selected the maximum outer buffer distance ($\widehat{t_{max}} = 5$ km) and the unit distance to $\omega = 1$ km (note that these inputs will be selected by domain experts). SmartNHHG is faster than the NaïveNHHG. Also we can observe that the computational savings increase with increasing number of nodes thanks to SmartNHHG's edge stitching approach.

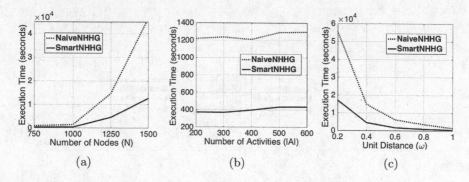

Fig. 8. Scalability of SmartNHHG with increasing (a) number of nodes, (b) number of activities (c), and unit distance (ω).

Effect of the Number of Activity Points: We also varied the number of activities in the activity set as shown in Fig. 8(b). Since the algorithm uses only the activity count on each edge, this experiment did not affect the execution times of either algorithm. However, SmartNHHG performs around three times faster than the NaïveNHHG. In the future, we plan to leverage activity counts on edges to improve the scalability of SmartNHHG.

Effect of Unit Distance ω: In this experiment, inputs are the number of nodes $|N| = 1000$ and the maximum outer buffer distance $\widehat{t_{max}} = 5$ km. The unit distance ω was varied by 0.2, 0.4, 0.6, 0.8 and 1 km. In Fig. 8(c), SmartNHHG is faster and computational savings increase with smaller ω thanks to SmartNHHG's distance based dynamic programming approach.

In summary, experiments confirm that SmartNHHG performs faster than NaïveNHHG thanks to the proposed algorithmic refinements.

6   Discussion

Techniques Without Significance Test: This paper focuses on hotspot detection techniques that use statistical significance to remove chance patterns but there are also techniques that do not test for statistical significance. These techniques (i.e. DBSCAN [25], K-Means [26], KMR [27], Clumping [12]) are state-of-the-art to detect clusters (i.e. a set of objects partitioned into a set of meaningful sub-classes) in a point process. However, since they do not test for statistical significance, they are not suitable for applications where false positive results may cause harm. For example, a neighborhood falsely identified as a crime hotspot may become stigmatized, causing residents' property values to drop. In addition, adding significance test to these approaches is often non-trivial since they lack a metric (e.g. log likelihood ratio test) for ranking candidate clusters. Thus, techniques without statistical significance test were not considered in our work.

Post-processing of the Output: Our proposed approach returns all possible *NHH* given an activity set and a spatial network. However, during our experiments, we often observed that multiple overlapping hotspots were returned on the same subgraph of the spatial network. To reduce the visual clutter, we used two simple rules in our visualizations: *(1) For two p_k and p_l:* If $p_l \subset p_k$ and there are two significant NHH_k and NHH_l and $t_{min}^k = t_{min}^l$ and $t_{max}^k = t_{max}^l$, then only NHH_k will be returned. *(2) For a path p:* If there are two significant NHH_i, NHH_j and $t_{min}^i \leq t_{min}^j$ and $t_{max}^i \geq t_{max}^i$ then only NHH_i will be returned.

7   Conclusion

This work explored the problem of mining network hotspots with holes in relation to important application domains such as crime analysis. We proposed a Smart Network Hotspot with Hole Generator algorithm that discovers multiple network hotspots with holes (*NHH*) on a spatial network. The proposed approach uses distance based dynamic programming and edge stitching approaches as well as Monte Carlo simulation speed-ups to enhance its performance. We presented two case studies using crime activity sets comparing our proposed approach with a ring-shaped hotspot detection method. Experimental evaluation using real data indicates that the proposed algorithmic refinements yield substantial computational savings without sacrificing result quality.

In future, we plan to explore refinements including sub-edge level *NHH* enumeration, active node filtering and dynamic segmentation. We also plan to explore "emerging" *NHH*s from spatiotemporal activity sets (i.e. time tags for the activities). Additionally, factors (e.g. demographics, activity relationships [28], urbanization [29,30]) that generate *NHH* will be explored.

Acknowledgments. This material is based upon work supported by the National Science Foundation under Grants No. 1029711, IIS-1320580, 0940818 and IIS-1218168, the USDOD under Grants No. HM1582-08-1-0017. We would like to thank Kim Koffolt and University of Minnesota Spatial Computing Research Group for their comments.

References

1. Fitterer, J., Nelson, T., Nathoo, F.: Predictive crime mapping. Police Pract. Res. **16**(2), 121–135 (2015)
2. Brantingham, P., et al.: Environmental Criminology. SAGE, Beverly Hills (1981)
3. Kulldorff, M.: SaTScan user guide for version 9.0 (2011)
4. Eftelioglu, E., Tang, X., Shekhar, S.: Geographically robust hotspot detection: a summary of results. In: ICDM International Workshop on Spatial and Spatiotemporal Data Mining (SSTDM) (2015)
5. Kulldorff, M., et al.: An elliptic spatial scan statistic. Stat. Med. **25**(22), 3929–3943 (2006)
6. Tang, X., et al.: Elliptical hotspot detection: a summary of results. In: ACM SIGSPATIAL Workshops (2015)
7. Neill, D.B., et al.: A fast multi-resolution method for detection of significant spatial disease clusters. In: Advances in Neural Information Processing Systems (2003)
8. Eftelioglu, E., et al.: Ring-shaped hotspot detection: a summary of results. In: IEEE International Conference on Data Mining, pp. 815–820 (2014)
9. Grubesic, T.H., Wei, R., Murray, A.T.: Spatial clustering overview and comparison: accuracy, sensitivity, and computational expense. Ann. Assoc. Am. Geogr. **104**(6), 1134–1156 (2014)
10. Beavon, D.J., Brantingham, P.L., Brantingham, P.J.: The influence of street networks on the patterning of property offenses. Crime Prev. Stud. **2**, 115–148 (1994)
11. Law, J., Quick, M., Chan, P.: Bayesian spatio-temporal modeling for analysing local patterns of crime over time at the small-area level. J. Quant. Criminol. **30**(1), 57–78 (2014)
12. Okabe, A., Okunuki, K.-I., Shiode, S.: The SANET toolbox: new methods for network spatial analysis. Trans. GIS **10**(4), 535–550 (2006)
13. Okabe, A., Sugihara, K.: Spatial Analysis Along Networks: Statistical and Computational Methods. Wiley, New York (2012)
14. Shiode, S., Shiode, N.: Network-based space-time search-window technique for hotspot detection of street-level crime incidents. Int. J. Geogr. Inf. Sci. **27**(5), 866–882 (2013)
15. Dev, O., et al.: Significant route discovery: a summary of results. In: Duckham, M., Pebesma, E., Stewart, K., Frank, A.U. (eds.) GIScience 2014. LNCS, vol. 8728, pp. 284–300. Springer, Heidelberg (2014)
16. Shi, L., Janeja, V.P.: Anomalous window discovery for linear intersecting paths. IEEE Trans. Knowl. Data Eng. **23**(12), 1857–1871 (2011)
17. Costa, M.A., Assunção, R.M., Kulldorff, M.: Constrained spanning tree algorithms for irregularly-shaped spatial clustering. Comput. Stat. Data Anal. **56**(6), 1771–1783 (2012)
18. Levine, N.: Crime mapping and the crimestat program. Geogr. Anal. **38**(1), 41–56 (2006)
19. Kuratowski, K.: Topology, vol. 1. Elsevier, Amsterdam (2014)
20. Kulldorff, M.: A spatial scan statistic. Commun. Stat.-Theor. Methods **26**, 1481–1496 (1997)
21. MacKay, D.J.: Information Theory, Inference and Learning Algorithms. Cambridge University Press, Cambridge (2003)
22. Us census bureau tiger/line shapefiles. http://www.census.gov/geo/maps-data/data/tiger-line.html. Accessed 9 Dec 2015

23. City of Chicago data portal. https://data.cityofchicago.org/Public-Safety/Crimes-2001-to-present/ijzp-q8t2. Accessed 01 Dec 2014
24. City of Oakland data portal. https://data.oaklandnet.com. Accessed 1 May 2016
25. Ester, M., et al.: A density-based algorithm for discovering clusters in large spatial databases with noise, pp. 226–231. AAAI Press (1996)
26. Hartigan, J.A., Wong, M.A.: Algorithm AS 136: a K-means clustering algorithm. Appl. Stat. **28**, 100–108 (1979)
27. Oliver, D., et al.: A k-main routes approach to spatial network activity summarization. IEEE Trans. Knowl. Data Eng. **26**, 1464–1478 (2014)
28. Guo, D.: Local entropy map: a nonparametric approach to detecting spatially varying multivariate relationships. Int. J. Geogr. Inf. Sci. **24**(9), 1367–1389 (2010)
29. Wolfe, M.K., Mennis, J.: Does vegetation encourage or suppress urban crime? Evidence from Philadelphia, PA. Landscape and Urban Planning **108**, 112–122 (2012)
30. Hirschfield, A., Birkin, M., Brunsdon, C., Malleson, N., Newton, A.: How places influence crime: the impact of surrounding areas on neighbourhood burglary rates in a British city. Urban Stud. **51**(5), 1057–1072 (2014)

Distance-Constrained k Spatial Sub-Networks: A Summary of Results

KwangSoo Yang[✉]

Florida Atlantic University, Boca Raton, USA
yangk@fau.edu

Abstract. Given a graph and a set of spatial events, the goal of Distance-Constrained k Spatial Sub-Networks (DCSSN) problem is to find k sub-networks that meet a distance constraint and maximize the number of spatial events covered by the sub-networks. The DCSSN problem is important for many societal applications, such as police patrol assignment and emergency response assignment. The problem is NP-hard; it is computationally challenging because of the large size of the transportation network and the distance constraint. This paper proposes a novel approach for finding k sub-networks that maximize the coverage of spatial events under the distance constraint. Experiments and a case study using Chicago crime datasets demonstrate that the proposed algorithm outperforms baseline approaches and reduces the computational cost to create a DCSSN.

Keywords: Spatial network query · Resource allocation · Spatial network database

1 Introduction

In this work, we propose a new problem of spatial network covering, called Distance-Constrained k Spatial Sub-Networks (DCSSN). Given a graph and a set of spatial events (e.g., crime incidents, traffic collisions, etc.), Distance-Constrained k Spatial Sub-Networks (DCSSN) finds k sub-networks that meet a distance constraint and maximize total number of spatial events covered by the sub-networks. Figure 1(a) shows an example input of DCSSN consisting of a graph with 14 nodes $(A, B, \ldots, N)$, 20 edges, and 25 locations of spatial events. Consider $k = 3$ and the shortest path distance between nodes in a sub-network is at most 2. Figure 1(b) shows an example output of DCSSN where distinct line styles show three sub-networks. The DCSSN problem is NP-hard (a proof is provided in Sect. 1.4). Intuitively, the problem is computationally challenging because of the large size of the transportation network and the distance constraint.

1.1 Application Domain

The DCSSN problem is important for critical applications such as identifying the most vulnerable areas within distance constraints. As an example, let us consider

© Springer International Publishing Switzerland 2016
J.A. Miller et al. (Eds.): GIScience 2016, LNCS 9927, pp. 68–84, 2016.
DOI: 10.1007/978-3-319-45738-3_5

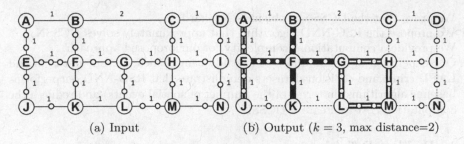

(a) Input   (b) Output ($k = 3$, max distance=2)

Fig. 1. Example of the input and output of DCSSN

police patrol district design. Since crimes are unpredictable in the city, it is important to provide a reliable police service that responds quickly to potential requests. DCSSN identifies concentrations of spatial events and allocates limited resources to districts that are in the highest risk areas. It also minimizes dispatch time to incidents by providing compact sub-networks. DCSSN can also be applied to discover groups of spatial locations that effectively interact and communicate with others because it ensures the travel time between two locations is within the constraint. Possible examples of such situations are provided in Table 1.

Table 1. Possible applications of DCSSN

Application	Benefit of DCSSN
Police patrol district	Manage police patrol to concentrate on high-crime areas
Fire prevention	Identify highly vulnerable areas and watch for early signs of fire
Disease surveillance and response	Detect and monitor topological high-risk areas to prevent the spread of infectious diseases
Road traffic control	Identify high traffic accident areas for special attention and enforcement

1.2 Our Contribution

In this paper, we propose a novel algorithm for finding k sub-networks based on the rooted sub-graph (RSG) and the nearest neighbor distribution (NND) function. Our approach follows three main steps: (1) construction of the rooted sub-graph and the nearest neighbor distribution function, (2) assignment of spatial events to sub-networks. (3) update of the nearest neighbor distribution function. Specifically, our contributions are as follows:

– We introduce a new spatial network covering problem, namely Distance-Constrained k Spatial Sub-Networks (DCSSN).

- We prove that the DCSSN problem is NP-hard.
- We propose the RSG-NND algorithm that approximately solves DCSSN.
- We provide a computational complexity for our proposed approach.
- Our experimental results and a case study using Chicago crime datasets and LAPD crime and collision datasets demonstrate that RSG-NND outperforms baseline algorithms (in terms of the number of spatial events) and reduces the computation cost to create a DCSSN.

1.3   Problem Definition

In our formulation of the DCSSN problem, a transportation network is represented and analyzed as a graph composed of nodes and edges. Each node represents a spatial location in geographic space (e.g., road intersections) and each edge between two nodes represents a road segment and has a travel distance. Each spatial event has a spatial location on edge. The $DCSSN(N, E, D, I, k, d_{max})$ problem is defined as follows:

Input: A transportation network G with
- a set of nodes N and a set of edges E,
- a set of non-negative integer lengths of edges $D : E \rightarrow \mathbb{Z}_{\geq 0}$
- a set of spatial event locations I,
- the number of sub-networks k, and
- the distance constraint d_{max}

Output: A set of Distance-Constrained k Spatial Sub-Networks SG_k

Objective:
- Maximize total number of spatial events covered by k sub-networks (SG_k).

Constraints:
- Distance Constraint: The shortest path distance between two nodes in $sg \in SG_k$ should be no greater than d_{max}.

1.4   Problem Hardness

The NP-hardness of DCSSN follows from a well-known result about the NP-hardness of the maximum clique problem.

Theorem 1. *The DCSSN problem is NP-hard.*

Proof. The NP-hardness of DCSSN can be proved by reduction from a well known NP-complete problem, the maximum clique problem (MCP). Given a graph G, MCP finds the largest clique. Let $A = (N, E)$ be an instance of MCP, where N is a set of nodes, E is a set of edges. Let $B(N, E, D, I, k, d_{max})$ be an instance of the DCSSN problem, where N is a set of nodes, E is a set of edges, D is a set of distances of E, I is a set of spatial events in E, k is the number of sub-networks, and d_{max} is the distance constraint. Let k be 1 and let d_{max} be 1. Then it is easy to show that the instance of MCP is a special case of DCSSN, where every edge has distance of 1 and contains exactly one spatial event $i \in I$. Since A is constructed from B in polynomial-bounded time, the proof is complete.

1.5   Related Work

To the best of knowledge, there is no prior work on distance constrained spatial sub-network covering. Circle covering problems have been studied to find complete and partial spatial covering on the geometric space [1,7,8,19,20,23,36]. However, geometrical approaches (e.g., Euclidean distance) are not ideal for spatial networks [4,37]. Metric k-center problems can find complete coverage of spatial events, which minimizes the longest edge between the center and spatial event locations [21,24,25]. However, these approaches do not honor distance constraints between two nodes in a sub-network and cover all spatial events, leading to a limitation of the detection of distance-constrained spatial sub-networks. Clustering methods have been widely used in related research about partial coverage problem [2,17,22,41]. However, these methods do not consider distance constraints to build sub-networks. There exists a significant body of research on spatial network analysis. The K-function has been applied to spatial networks to analyze the distribution of events and detect clusters [31,34,40,42]. Scan statistics has been used to detect anomalies on networks [28,35]. The concept of spatial auto-correlation has been used to analyze the correlation between two variables on spatial networks [5,10]. Network kernel density estimation has been studied to analyze probability distributions of events and provide visual patterns of relative density on spatial networks [16,30]. Network-based variable-distance clumping method has been proposed to discover multi-scale network-based clumps [38]. The network farthest-pair point clustering method uses the complete linkage method to discover clusters on networks [30,39]. Network Voronoi diagrams have been intensively studied to help spatial analysis on networks [30]. Farthest-point network Voronoi diagram partitions a network into sub-networks in each of which the farthest site is the same [30]. Capacity-constrained network Voronoi diagram has been proposed to create a set of contiguous service areas that meet service center capacities and minimize the sum of the distances from customers to allotted service centers [43]. There has been an interest in developing approaches for network path covering problems. Shortest path covering problems identify the minimal cost path that covers all of the nodes on the network [12,13,15,29]. In particular, the median tour problem and the maximal covering tour problem use bi-objective optimization models to minimize the total path length and maximize the accessibility of demand nodes [14]. FlowScan technique uses a density-based clustering algorithm to identify hot routes on spatial networks [27]. K-Main Routes technique discovers k shortest paths to summarize spatial activities on networks [32]. Many other type of spatial network covering problems are also studied [6,9,36]. However, these problems do not honor distance constraints. In this work, we propose a novel approach for creating Distance-Constrained k Spatial Sub-Networks (DCSSN) that honors distance constraints and maximizes the coverage of spatial events.

1.6   Outline

The rest of the paper is organized as follows: Sect. 2 describes our proposed approach. We provide correctness proofs of the proposed approach in Sect. 2.1.

Section 3 describes the experiment design and presents the experimental observations and results. Section 4 reports a case study using Chicago crime datasets and LAPD crime and collision datasets. Finally, Sect. 5 concludes the paper.

2   Proposed Approach for DCSSN

In this section, we introduce our Rooted Sub-Graph with the Nearest Neighbor Distribution (RSG-NND) approach to the DCSSN problem.

RSG-NND Algorithm: The RSG-NND algorithm starts with constructing both the rooted sub-graphs (RSG) and the nearest neighbor distribution (NND) functions and finds an attractor node located in the area with the highest event density. The key idea in RSG-NND is to construct a data-structure to index the nearest-neighbor events and assign the highest nearest-neighbor weighted edges to the attractor node to maximize the coverage of spatial events. This process can create sub-networks under distance constraints and iteratively find dense sub-networks in topological space.

The RSG-NND algorithm consists of the following three steps: (1) construction of the rooted sub-graphs and the nearest neighbor distribution functions, (2) assignment of spatial events to sub-networks that maximizes the coverage of spatial events and honors the distance constraint, and (3) update of the nearest neighbor distribution functions.

In the first step, RSG-NND creates the rooted sub-graph from every node ($r \in N$) and constructs the nearest neighbor distribution (NND) function. The idea of RSG is to create a sub-graph induced by all nodes within distance d of the root r.

Definition 1. *The rooted sub-graph RSG(r,d) is the sub-graph of G(N, E) spanned by all nodes of G at distance at most d from the root r.*

The nearest neighbor distribution (NND) function can be represented by

$$NND(r,d) = \sum\nolimits_{e \in RSG(r,d)} \#events(e), \qquad (1)$$

where r is the root of RSG, d is the distance constraint, and $\#events(e)$ is the number of spatial events in edge e.

Figure 2 shows an example of the NND function. Figure 2(a) illustrates the input with a transportation network (14 nodes, 20 edges, and 25 spatial events). Every edge is associated with a distance (e.g., travel time), as indicated by the number displayed alongside it. Consider NND for node F. Figures 2(b) to (e) show RSGs rooted at node F with different distance constraints. Figure 2(f) shows the data-structure to index the accumulated counts of nearest-neighbor events (i.e., $\#event(e)$) from node F. For example, 17 events are located within the distance of 2 from node F (see Fig. 2(c)). This model, which we refer to as

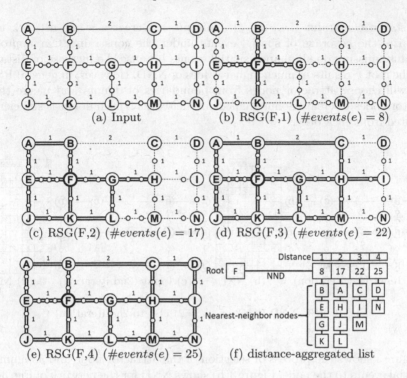

Fig. 2. Creation of Nearest Neighbor Distribution function of node F

distance-aggregated list, allows for storage of nearest neighbor nodes in every distance. This is useful for NND queries as it updates the NND in linear time (Lemma 1).

In the second step, RSG-NND searches the accumulated counts of events (i.e., values of NND) that lie within the closed interval between $d_{max}/2$ and d_{max} (i.e., $[d_{max}/2, d_{max}]$) from all nodes and chooses the node with the highest values of NND. There are three reasons to use the closed interval between $d_{max}/2$ and d_{max}. First, RSG with distance $d_{max}/2$ honors the distance constraint of DCSSN (Lemma 2). That means, RSG with distance $d_{max}/2$ can be a subset of the sub-network in DCSSN. If d_{max} is an odd number, we use the closed interval $[\lfloor d_{max}/2 \rfloor, d_{max}]$ because RSG with distance $\lceil d_{max}/2 \rceil$ may violate the distance constraint. Second, RSG with distance d_{max} may honor the distance constraint of DCSSN, but RSG with distance $d > d_{max}$ always violates the distance constraint. Finally, the sub-network in DCSSN is always a subset of the RSG with distance d_{max}. As a tie break, we choose the node that has higher values within the smaller distance. We refer to this node as an attractor node. Intuitively, the node with the highest values in this range has the largest expectation to become the attractor node which absorbs the next attractor node to form the DCSSN. After choosing the attractor node a, RSG-NND assigns the highest weighted edge (e.g., the number of spatial events) to the node a under the distance constraint (i.e., d_{max}). This process creates a sub-network that maximizes the coverage of spatial events under the constraint. Our approach may cause missing donut-shaped spatial events because the increase of distance from the root r results in much higher values of NND. However, in general RSG-NND will choose attractor nodes from boundaries of donuts and merge these attractor nodes to create donut-shaped sub-networks if each donut is sufficiently separated from each other.

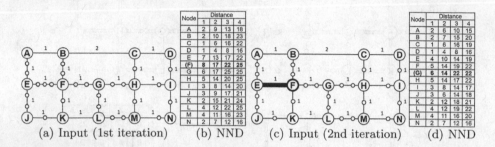

(a) Input (1st iteration)   (b) NND   (c) Input (2nd iteration)   (d) NND

Fig. 3. RSG-NND ($k = 3$, $d_{max} = 2$) (1st to 2nd iteration)

Figure 3 shows the process of selection of the attractor node and assignment of spatial events to the node. Figure 3(b) shows NND for the network of Fig. 3(a). Consider $d_{max} = 2$ and $k = 3$. In this example, node F has the highest weight in the closed interval between 1 and 2 (i.e., $[1, 2]$). After the selection of the attractor node (i.e., F), RSG-NND scans all edges under the distance constraint and allots the highest weighted edge (i.e., EF) to node F (see Fig. 3(c)).

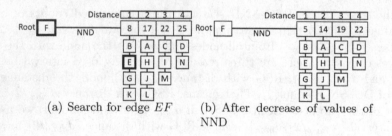

(a) Search for edge EF (b) After decrease of values of NND

Fig. 4. Update of distance-aggregated list

In the third step, RSG-NND removes all the spatial events on the allotted edge (i.e.,, EF) and updates the NND function. A naive approach for the update is to create the rooted sub-graphs and repeatedly count all spatial events. However, it takes $O(n^2 \cdot \log n)$ for all nodes. In order to minimize the computational cost, RSG-NND uses two key ideas. First it constructs the distance-aggregated list and updates NNDs in linear time (Lemma 1). Second it updates only the part of nodes affected by the removal of spatial events (i.e., the updates are applied only to nodes which are reachable from the allotted edge within the distance constraint). Figure 4 shows the update of NND for node F. After removing three spatial events on edge EF, RSG-NND decreases the value of 3 from d = 1 to 4 (Lemma 1). The result of update is shown in Fig. 3(d). This process continues until all the spatial events are removed and terminates in $O(m)$ iterations, where m is the number of edges.

Algorithm 1. Generalized rooted sub-graph with the nearest neighbor distribution function (RSG-NND) Algorithm (Pseudo-code)

Inputs:
  - A transportation network $(G(N, E))$ with a set of nodes N and edges E.
  - A set of spatial event locations I on E
  - Every edge has a distance $d(e)$
  - The number of sub-networks k
  - The distance constraint d_{max}

Outputs: Distance-Constrained k Spatial Sub-Networks ($DCSSN$)

Steps:
1: Construct **the rooted sub-graphs (RSGs) and the nearest neighbor distribution functions (NNDs).**
2: $DCSSN \leftarrow \emptyset$
3: **while** $|I| > 0$ **do**
4:    Find an attractor node a in the range between $d_{max}/2$ and d_{max}.
5:    Assign the highest weighted edge $e_h \in E$ to a to maximize the coverage of spatial events under the distance constraint (i.e., d_{max}).
6:    $DCSSN \leftarrow DCSSN \cup e_h$
7:    Remove spatial events $i \in I$ from G.
8:    Update NNDs according to removal of spatial events i.
9: **end while**
10: return $DCSSN$ (i.e., k sub-networks that maximize coverage of spatial events.)

Algorithm 1 presents the pseudo-code for a generalized version of RSG-NND. First, RSG-NND creates the rooted sub-graphs and the nearest neighbor distribution functions in the closed interval between $d_{max}/2$ and d_{max} (Line 1). Then, it chooses the highest NND node as an attractor node (Line 4). After that, it sorts edges in RSG based on the weights and assigns the highest weighted edge to the attractor node under the distance constraint (i.e., within distance d_{max}) (Line 5). If this sub-network can be added to an existing sub-network under the distance constraint, two individuals are combined (Line 6). If it cannot be combined with the existing sub-network, it becomes a new sub-network (Line 6). Then, it removes the spatial events in the network G (Line 7) and updates NNDs (Line 8). This process continues until all the spatial events are removed. Finally, the DCSSN is returned (Line 10).

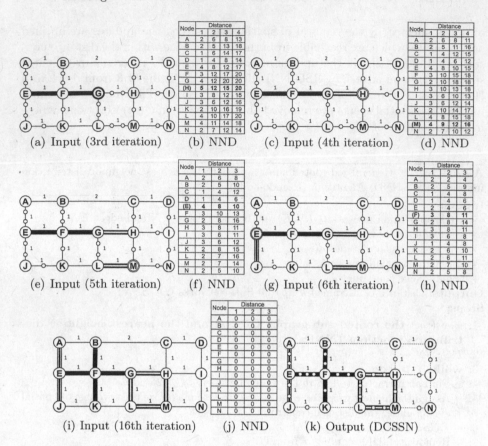

(a) Input (3rd iteration)   (b) NND   (c) Input (4th iteration)   (d) NND

(e) Input (5th iteration)   (f) NND   (g) Input (6th iteration)   (h) NND

(i) Input (16th iteration)   (j) NND   (k) Output (DCSSN)

Fig. 5. RSG-NND (k = 3, d_{max} = 2) (17th iteration produces a DCSSN)

Figures 3, 4 and 5 show the execution of the RSG-NND algorithm. RSG-NND starts with creating RSGs and constructs NNDs (Fig. 3(b)). In this example, node F becomes an attractor node (see Fig. 3(b)) and takes edge EF to construct the sub-network. Figure 3(c) shows the network after the allotment of edge EF to node F. Next, RSG-NND updates NNDs and finds the next attractor node (i.e., G) as well as edges with the highest weight (e.g., FG, GH, GL, and LM). As a tie break rule, the edge connecting to the incident with the highest values in NNDs (i.e., FG) will be allotted into node G. Since new allotment (i.e., FG) can be combined with the existing sub-network (i.e., EF) without breaching the distance constraint, both are combined into one sub-network. Figure 5(c) shows the result of the 4*th* iteration. In this example, three nodes have the highest NND value at the distance of 1 (i.e., E, L, and M). As a tie break rule, the node with the highest value at the distance of 2 (i.e., M) will be selected. After 17 iterations, RSG-NND allots all spatial events to sub-networks and returns the result of DCSSN (Fig. 5(k)).

2.1 Proof and Analysis

In this section, we prove that the RSG-NND is correct, i.e., the RSG-NND algorithm creates a DCSSN.

Lemma 1. *The update of NND takes linear time*

Proof. The node set for every distance in the distance-aggregated list can be constructed using hash-tables. The operation to test whether a node can be reachable from the root r within the distance d takes a constant time [18]. Assume that spatial events on edge $e \in E$ are removed. Let the number of removed events be i. Then RSG-NND scans the distance-aggregated list and finds the two nodes (e.g., n_1 and n_2) which are incident on edge e at a cost of $O(d_{max})$ (see Fig. 4). After that, RSG-NND chooses the farthest node (e.g., n_2) from the root and finds the distance (e.g., d_2) from the root r. It takes $O(d_{max})$. Finally, RSG-NND decreases the values of NND after the distance d_2 by i, which takes $O(d_{max})$. Therefore the update of NND takes linear time.

Lemma 2. $RSG(r, d_{max}/2)$ *meets the distance constraint.*

Proof. According to Definition 1, all nodes in $RSG(r, d_{max}/2)$ are reachable from the root r within distance $d_{max}/2$. Assume that there are two nodes n_1 and n_2 in $RSG(r, d_{max}/2)$. Then the shortest path distance between n_1 and n_2 is at most $d_{max}/2$ because there is a path consisting of $n_1 \rightarrow r$ and $r \rightarrow n_2$. Therefore, $RSG(r, d_{max}/2)$ meets the distance constraint.

2.2 Computational Complexity of the RSG-NND Algorithm

Let n be the number of nodes, let m be the number of edges, let k be the number of sub-networks, and let d_{max} be the distance constraint. RSG-NND uses Dijkstra's algorithm to create RSGs at a cost of $O(n \cdot (n \cdot \log n + m))$ [3]. Because the transportation network is a sparse graph (i.e., $m = O(n)$), the complexity becomes $O(n^2 \cdot \log n)$. The construction of distance-aggregated list takes $O(n \cdot d_{max})$. A hash-table is used to store the set of the nearest neighbor nodes within the distance constraint providing constant-time performance for retrieval operations [18]. The assignment of the most highest weighted edge to the attractor node takes $O(m \cdot \log m)$ because edges should be sorted based on weights. The update of distance-aggregated list takes $O(n \cdot d_{max})$ because RSG-NND examines every node and scans the distance-aggregated list to update the NND function. Assume that we consider the large-sized network (i.e., $d_{max} \ll \log m$). Since the number of iterations is bounded by $O(m)$, RSG-NND takes $O(n^2 \cdot \log n + m \cdot (m \cdot \log m))$. Then the computational complexity of RSG-NND is $O(n^2 \cdot \log n + m^2 \cdot \log m)$.

3 Experimental Evaluation

In this section, we present the experiment design and an analysis of the experiment results.

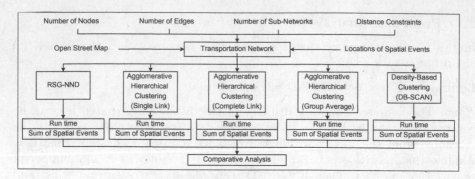

Fig. 6. Experiment layout

3.1   Experiment Layout

Figure 6 shows our experimental setup. For the transportation network, we used a Chicago, IL road map consisting of 19,075 nodes and 28,943 edges, taken from OpenStreetMap [33]. For the spatial events, we used real-world crime datasets from January to April. 2015 that consist of spatial locations on a Chicago, IL road map [11] and created Distance-Constrained k Sub-Networks (DCSSN). For simplicity, we mapped all incidents to the nearest edge (i.e., road segment).

Ideally we would test our proposed algorithm against comparable algorithms from related work. Unfortunately, we found no algorithms in the literature that honor distance constraints. The closest algorithm may be spatial partial clustering [2,17,22,41]. We used agglomerative hierarchical clustering (AHC) method and density-based clustering (DBC) method as our baseline partial clustering. Since both methods have no distance constraints, we modified the algorithms to group the spatial points within the distance constraints. In AHC methods, we only merged the highest weighted edges under the distance constraint. DBC method requires two parameters: the maximum radius of the neighborhood (Eps) and the minimum number of points required to form a dense region ($MinPts$) [17]. In DBC method, we used a recursive call of DB-SCAN to determine two parameters for creating better solution [17]. In the merge procedure, DB-SCAN was modified to honor the distance constraint. Both methods require the shortest path computation which takes $O(n \cdot (n \cdot \log n + m))$ [3]. Assume that $m = O(n)$. Since the process of AHC takes $O(n^2 \cdot \log n)$, the time complexity of AHC is bounded by $O(n^2 \cdot \log n)$ [41]. The process of DBC takes $O(n \cdot \log n)$; therefore the time complexity of DBC is bounded by $O(n^2 \cdot \log n)$ [22,41]. We tested five different approaches: (1) RSG-NND, (2) single-linkage clustering (AHC-Single), (3) complete-linkage clustering (AHC-Complete), (4) Group average linkage (AHC-Average), and (5) DB-SCAN (DBC). The algorithms were implemented in Java 1.8 with a 16 GB memory run-time environment. All experiments were performed on an AMD FX(tm)-8120 CPU machine running MS Windows 10 with 32 GB of RAM.

3.2   Experiment Result and Analysis

We experimentally evaluated five different approaches by comparing the impact on performance of (1) number of edges (i.e., $|E|$) and (2) distance constrains (i.e., d_{max}). Performance measurements were the number of spatial events covered by DCSSN and execution time.

Effect of the Number of Edges. The purpose of the first experiment was to evaluate the effect of number of edges on the performance of the algorithms. We fixed the number of sub-networks to 5 and the distance constraint (i.e., d_{max}) to 1 km, and incrementally increased the number of edges from $5,585$ to $28,943$ by enlarging the network size. The experiment was done using four different dataset from January to April, 2015 and execution times were averaged over 10 test runs. Figure 7(a) shows that the RSG-NND algorithm performs better than other approaches. This is because RSG-NND uses the distance-aggregated list to estimate the global density and groups the spatial events in terms of local density. As can be seen, the number of spatial events covered by sub-networks does not increase directly in proportion as the number of edges increases because the objective is to find the five most dense distance-constrained sub-networks. Figure 7(b) shows that the run-time increases as the number of edges increases. As can be seen, both DBC and RSG-NND are faster than agglomerative hierarchical clustering (AHC) methods. RSG-NND outperform other approaches in terms of solution quality because the spatial partial clustering methods (i.e., AHC, and DBC) are not designed for the objective of DSCSSN.

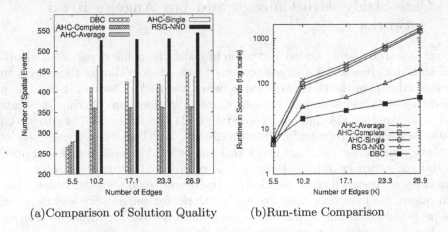

(a)Comparison of Solution Quality   (b)Run-time Comparison

Fig. 7. Effect of the number of edges ($d_{max} = 1$ km)

Effect of Distance Constraints. The second set of experiment evaluated the effect of distance constraints. We fixed the number of edges to $23,367$ and incrementally increased the distance constraint (i.e., d_{max}) from 0.5 km to 2.5 km.

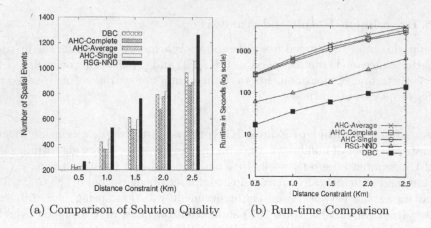

(a) Comparison of Solution Quality   (b) Run-time Comparison

Fig. 8. Effect of the distance constraints ($|E| = 23,367$)

The experiment was done using four different dataset from January to April. 2015 and execution times were averaged over 10 test runs. Figure 8(a) shows that RSG-NND outperforms other approaches in terms of solution quality. As the distance (size) constraint increases, so does the performance gap. Figure 8(b) shows that the run-time increases as the distance constraint increases. This is because all algorithms require the computation of shortest distance between nodes.

4   Case Study with Chicago and Los Angeles Road Networks

In our case study, we imagined a scenario in which the police department identified high-risk areas and encouraged officers to spend more time in these areas to prevent crime. For the transportation network, we used a Chicago, IL road map consisting of 23,165 nodes, 35,161 edges, and 4,463 locations of crime incidents and a Los Angeles, CA road map consisting of 23,576 nodes, 34,764 edges, and 5,400 locations of crime and collision incidents. We chose two different sets of distance constraints (i.e., 1 km and 2 km) and created five distance-constrained spatial sub-networks. Each sub-network is represented by different colors. For simplicity, we assigned crime locations to the nearest road segment. Due to time limitations, we have only used five sub-networks for preliminary evaluation of the proposed algorithm. In the future, we plan to test our algorithm on a larger number of sub-network to characterize the DCSSN.

4.1   Case Study Results and Analysis

In this section, the goal was to investigate the following questions: (1) Is the RSG-NND able to create dense sub-networks? and (2) How does the distance constraint affect DCSSN?

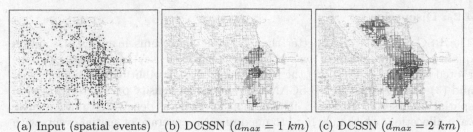

(a) Input (spatial events)   (b) DCSSN ($d_{max} = 1\ km$)   (c) DCSSN ($d_{max} = 2\ km$)

Fig. 9. Case study: Chicago, IL road map ($|N| = 23,165$, $|E| = 35,161$, $|I| = 4,463$, and $k = 5$) (Color shows sub-networks) (Color figure online)

Case Study 1: Chicago, IL Road Map. We used crime datasets in January 2015 that consist of $4,463$ locations of crime incidents on a Chicago, IL road map (Fig. 9(a)) [11]. We fixed the number of sub-networks to 5 and increased the distance constraint from 1 km to 2 km (Fig. 9). Figures 9(b) and (c) show that crime incidents are concentrated in eastern areas. DCSSN within 1 km (Fig. 9(b)) covers 563 locations and DCSSN with 2 km (Fig. 9(c)) covers $1,223$ locations. As the distance (size) constraint increases, sub-networks expand towards north and south direction.

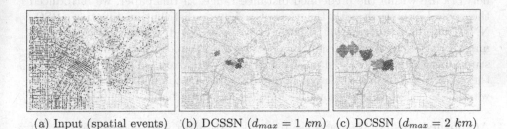

(a) Input (spatial events)   (b) DCSSN ($d_{max} = 1\ km$)   (c) DCSSN ($d_{max} = 2\ km$)

Fig. 10. Case study: Los Angeles, CA road map ($|N| = 23,576$, $|E| = 34,764$, $|I| = 5,400$, and $k = 5$) (Color shows sub-networks) (Color figure online)

Case Study 2: Los Angeles, CA Road Map. We used LAPD crime and collision datasets in January 2015 that consist of 5,400 locations of crime and collision incidents (Fig. 10(a)) [26]. We fixed the number of sub-networks to 5 and increased the distance constraint from 1 km to 2 km. Figure 10(b) and (c) show that crime incidents are concentrated in western areas. DCSSN within 1 km (Fig. 10(b)) covers 475 incident locations and DCSSN with 2 km (Fig. 10(c)) covers 1,045 incident locations. As the distance (size) constraint increases, sub-networks expand toward west and east direction.

4.2   Discussion

The RSG-NND performs better than baseline algorithms in terms of solution quality. This improvement was obtained using three key components: (1) the rooted sub-graph (RSG), (2) the nearest neighbor distribution (NND) function, and (3) update of NND. DB-SCAN uses the global density parameters to group spatial events [17]. However, it does not update the density estimation after each allocation, which may hard to maximize the coverage of spatial events with distance constraints. Single-linkage clustering outperforms other agglomerative hierarchical clustering methods. Even though complete-linkage clustering tends to minimize the increase in diameter of the clusters at each iteration, it also shows a limitation to maximize the coverage with distance constraints [2,22,41]. The experimental results show that RSG-NND outperforms than other approaches in terms of the coverage of spatial events and speeds up the computation of DCSSN.

5   Conclusion and Future Work

We presented the problem of creating distance-constrained k spatial sub-networks (DCSSN). Creating a DCSSN is challenging because of the large size of the transportation network and the constraint that any two nodes in the sub-network are within the predefined distance range. In this paper, we introduced the rooted sub-graph with the nearest neighbor distribution function (RSG-NND) approach for creating a DCSSN to meet the distance constraint while maximizing the coverage of spatial events. We presented experiments and a case study using Chicago crime datasets. In future work, we plan to further explore the DCSSN problem and design parallel algorithms to create a DCSSN. Also, we plan to study computational techniques for spatio-temporal sub-networks.

Acknowledgments. We would like to thank FAU under start-up funds for student support, equipment and travel.

References

1. Agarwal, P.K., Procopiuc, C.M.: Exact and approximation algorithms for clustering. Algorithmica **33**(2), 201–226 (2002)
2. Aggarwal, C.C., et al.: Data Clustering: Algorithms and Applications. CRC Press, Boca Raton (2013)
3. Ahuja, R., Magnanti, T., Orlin, J., Weihe, K.: Network Flows: Theory, Algorithms and Applications. Prentice-Hall, Englewood Cliffs (1993)
4. Barthélemy, M.: Spatial networks. Phys. Rep. **499**(1), 1–101 (2011)
5. Black, W.R., Thomas, I.: Accidents on Belgium's motorways: a network autocorrelation analysis. J. Transp. Geogr. **6**(1), 23–31 (1998)
6. Buchin, K., et al.: Detecting hotspots in geographic networks. In: Sester, M., Bernard, L., Paelke, V. (eds.) Advances in GIScience. LNGC, pp. 217–231. Springer, Berlin (2009)

7. Carmi, P., Katz, M.J., Lev-Tov, N.: Covering points by unit disks of fixed location. In: Tokuyama, T. (ed.) ISAAC 2007. LNCS, vol. 4835, pp. 644–655. Springer, Heidelberg (2007)
8. Chazelle, B.M., et al.: On a circle placement problem. Computing **36**(1–2), 1–16 (1986)
9. Chen, Z., et al.: Discovering popular routes from trajectories. In: 2011 IEEE 27th International Conference on Data Engineering, pp. 900–911. IEEE (2011)
10. Chun, Y.: Modeling network autocorrelation within migration flows by eigenvector spatial filtering. J. Geogr. Syst. **10**(4), 317–344 (2008)
11. City of Chicago Data Potal, Crimes - 2001 to Present. https://data.cityofchicago.org/Public-Safety/Crimes-2001-to-present/ijzp-q8t2. Accessed Dec 2015
12. Current, J., Pirkul, H., Rolland, E.: Efficient algorithms for solving the shortest covering path problem. Transp. Sci. **28**(4), 317–327 (1994)
13. Current, J., et al.: The shortest covering path problem-an application of locational constraints to network design. J. Reg. Sci. **24**(2), 161–183 (1984)
14. Current, J.R., Schilling, D.A.: The median tour and maximal covering tour problems: formulations and heuristics. Eur. J. Oper. Res. **73**(1), 114–126 (1994)
15. Current, J.R., Velle, C.R., Cohon, J.L.: The maximum covering/shortest path problem: a multiobjective network design and routing formulation. Eur. J. Oper. Res. **21**(2), 189–199 (1985)
16. Downs, J., Horner, M.: Network-based kernel density estimation for home range analysis. In: Proceedings of 9th International Conference on Geocomputation, Maynooth, Ireland (2007)
17. Ester, M., et al.: A density-based algorithm for discovering clusters in large spatial databases with noise. In: KDD, vol. 96, pp. 226–231 (1996)
18. Fotakis, D., et al.: Space efficient hash tables with worst case constant access time. Theor. Comput. Syst. **38**(2), 229–248 (2005)
19. Gandhi, R., Khuller, S., Srinivasan, A.: Approximation algorithms for partial covering problems. J. Algorithms **53**(1), 55–84 (2004)
20. Ghasemalizadeh, H., Razzazi, M.: An improved approximation algorithm for the most points covering problem. Theor. Comput. Syst. **50**(3), 545–558 (2012)
21. Gonzalez, T.F.: Clustering to minimize the maximum intercluster distance. Theoret. Comput. Sci. **38**, 293–306 (1985)
22. Han, J., et al.: Data Mining: Concepts and Techniques. Elsevier, Amsterdam (2011)
23. Hifi, M., M'hallah, R.: A literature review on circle and sphere packing problems: models and methodologies. Adv. Oper. Res. (2009)
24. Hochbaum, D.S., Shmoys, D.B.: A best possible heuristic for the k-center problem. Math. Oper. Res. **10**(2), 180–184 (1985)
25. Hochbaum, D.S., Shmoys, D.B.: A unified approach to approximation algorithms for bottleneck problems. J. ACM (JACM) **33**(3), 533–550 (1986)
26. LAPD Crime and Collision Raw Data for 2015. https://data.lacity.org/A-Safe-City/LAPD-Crime-and-Collision-Raw-Data-for-2015/ttiz-7an8. Accessed Dec 2015
27. Li, X., Han, J., Lee, J.-G., Gonzalez, H.: Traffic density-based discovery of hot routes in road networks. In: Papadias, D., Zhang, D., Kollios, G. (eds.) SSTD 2007. LNCS, vol. 4605, pp. 441–459. Springer, Heidelberg (2007)
28. Marchette, D.: Scan statistics on graphs. Wiley Interdiscip. Rev.: Comput. Stat. **4**(5), 466–473 (2012)
29. Murawski, L., Church, R.L.: Improving accessibility to rural health services: the maximal covering network improvement problem. Soc.-Econ. Plan. Sci. **43**(2), 102–110 (2009)

30. Okabe, A., Sugihara, K.: Spatial Analysis Along Networks: Statistical and Computational Methods. Wiley, Hoboken (2012)
31. Okabe, A., Yamada, I.: The k-function method on a network and its computational implementation. Geogr. Anal. **33**(3), 271–290 (2001)
32. Oliver, D., et al.: A k-main routes approach to spatial network activity summarization. IEEE Trans. Knowl. Data Eng. **26**(6), 1464–1478 (2014)
33. OpenStreetMap. http://www.openstreetmap.org/. Accessed Sept 2015
34. O'Sullivan, D., et al.: Geographic Information Analysis. Wiley, Hoboken (2014)
35. Priebe, C.E., et al.: Scan statistics on enron graphs. Comput. Math. Organ. Theor. **11**(3), 229–247 (2005)
36. Samet, H.: Foundations of Multidimensional and Metric Data Structures. Morgan Kaufmann, San Francisco (2006)
37. Shekhar, S., Chawla, S.: Spatial Databases: A Tour. Prentice Hall, Upper Saddle River (2003)
38. Shiode, S., Shiode, N.: Detection of multi-scale clusters in network space. Int. J. Geogr. Inf. Sci. **23**(1), 75–92 (2009)
39. Sørensen, T.: A method of establishing groups of equal amplitude in plant sociology based on similarity of species content and its application to analyses of the vegetation on danish commons. Biologiske Skrifter **5**, 1–34 (1948)
40. Spooner, P.G., et al.: Spatial analysis of roadside acacia populations on a road network using the network k-function. Landsc. Ecol. **19**(5), 491–499 (2004)
41. Tan, P., et al.: Introduction to Data Mining. Addison-Wesley, Boston (2005)
42. Yamada, I., Thill, J.C.: Local indicators of network-constrained clusters in spatial point patterns. Geogr. Anal. **39**(3), 268–292 (2007)
43. Yang, K., et al.: Capacity-constrained network-Voronoi diagram. IEEE Trans. Knowl. Data Eng. **27**(11), 2919–2932 (2015)

GIScience Considerations
in Spatial Social Networks

Dipto Sarkar[1]([⊠]), Renee Sieber[1,2], and Raja Sengupta[1,2]

[1] Department of Geography, McGill University, Montreal, QC, Canada
dipto.sarkar@mail.mcgill.ca
[2] School of Environment, McGill University, Montreal, QC, Canada

Abstract. There has been a proliferation of literature that incorporates social network analysis (SNA) to study geographic phenomena. We argue that these incorporations have mostly been superficial. What is needed is a stronger interrogation of the challenges and possibilities of a tight coupling of spatial and social network concepts, which take advantage of the strengths of each methodology. In this paper, we create a typology of existing research focused on the integration of geography into SNA: nodal, topographic and spatial. We then describe three core concepts that co-exist in the two fields but are not necessarily complementary: distance, communities, and scale. We consider how they can be appropriated and how they can be more tightly coupled into spatial social networks. We argue that the only way we can move beyond a superficial integration is to holistically identify the challenges and consider new methods to address the complexities of integration.

Keywords: Spatial social networks · Geography · GIScience · Social networks

1 Introduction

In recent years, Social Network Analysis (SNA) has generated considerable attention due to the distinctive ways in which it characterizes and prioritizes the relationships among entities. The diagrammatic approach of social networks serves as a starting point for visual exploratory analysis. Social Network Analysis's foundation in graph theory provides a strong backbone for deriving metrics to analyze network patterns.

The basic premise of a social network is to define a society as a group of entities with persistent interactions and shared attributes. A society can refer to a group of people sharing the same territory, subject to the same laws, interested in same activities (forming clubs), and belonging to the same economic or social status. It is the common attribute(s) between the individuals that gives rise to social interactions. In geography, the concept of a society is formed on the common attribute, irrespective of interactions between the entities. In SNA, it is not presumed that similar entities will interact. Thus, the focus is on the explicit interactions. This provides an interesting avenue to uncover patterns, discover important individuals and reveal interesting facts about the society, which a procedure starting with the assumption that everyone with similar qualities interact with each other may not provide. Thus, SNA provides a complimentary approach, focusing on studying individuals, groups and ultimately the society by

© Springer International Publishing Switzerland 2016
J.A. Miller et al. (Eds.): GIScience 2016, LNCS 9927, pp. 85–98, 2016.
DOI: 10.1007/978-3-319-45738-3_6

concentrating on the known interactions that exist between the entities. Both methods of enquiry, spatial and social analysis, can however benefit each other by a coupling of knowledge.

In this paper, we explore some of the requirements for tighter integration of SNA with geography. Although [1] highlighted the long tradition of network analysis in Geography, whereby both human and physical phenomenon have been modeled as networks [2], the resurgence in SNA in various fields compels revisiting the tradition and offer new perspectives for understanding social networks in the context of GIScience. Despite a proliferation of research that integrates geographic aspects in SNA, the current literature lacks a framework for classifying the different methods by which the integrations have been accomplished. We introduce a typology of integrating geography and SNA in the current literature. We then highlight three concepts commonly used in SNA and geography but warrant deeper understanding of what they mean in either context, highlighting the problems as a way to form working definitions required in the realm of spatial social networks. We hope to offer additional interesting avenues for exploring interconnectivity and interactions between people and also between people and their surroundings. Referred to by various terms, such as location based social networks [3, 4], geo-social [5], and spatial social networks [6], we prefer the terms spatial social networks, as it highlights the social connections, recognises the embeddedness of the interactions in geographic space, not confining geographic notions to just a pair of coordinate locations.

2   Social Network Analysis (SNA)

Social Network Analysis represents relationships between connected entities such as individuals, organizations, and groups. In SNA, a social network is computationally represented as a collection of nodes and edges. Specifically, a network is usually expressed as a non-directed graph defined $G = (V, E)$ where $V = v_1, v_2, v_3,..., v_n$ represents the set of nodes and $E = e_1, e_2, e_3,..., e_n$ is the set of edges. Each edge e_k is associated with an unordered pair of vertices (i, j). In some applications, the edges between the nodes are non-reciprocate and hence the graph is directed, where each edge e_k is associated with an ordered pair of vertices (u, v). The main focus of the social network is on the edges (or ties), that is, the relationships that exist amongst the nodes [7]. Using graphs to represent social networks limits the possibility of self-loops, as in terms of social relationships, the concept of a person being a friend with themselves does not make sense.

Most social networks are unweighted graphs. The presence of an edge between two nodes is binary, indicating whether there exists a relationship amongst the two nodes or not. In unweighted graphs, the edges do not convey any other information besides connectivity of nodes. Hence, navigation in the network space is only possible by moving along existing edges from node to node (like navigation on a road network). A sociogram is a visualization of the social network. In the sociogram, the widths, or the lengths of the edges are arbitrary. The nodes in a sociogram are located with the attempt to show interconnected nodes close to each other [8]. The position of the nodes in the layout is not directly interpretable on its own on a Cartesian plane and only has meaning in relation to other nodes [9]. Any scaling or rotation of the sociogram does not change the underlying information [10].

Various graph layouts have been developed to produce aesthetically pleasing drawings by modifying the position of the nodes and edges and by changing the length of the edges [11]. The drawings should not be confused with the graph itself; very different layouts can correspond to the same graph [11]. Figure 1a and b shows the same graph represented with two different layout algorithms applied; the adjacency matrix of nodes and edges is shown in Fig. 1c. In the adjacency matrix (Fig. 1c), each non-diagonal entry, a_{ij}, is the existence of an edge connecting node i to node j. Usually the entries a_{ij} are binary and denote the existence or non-existence of an edge between the two nodes i and j. Unlike sociograms, there exists a unique adjacency matrix for each graph (up to permutations of rows and columns) [12].

Increasing amounts of data can now be used to nuance social relations. In many cases, an entity can be thought of as a cluster of structured or unstructured attributes, distinguished by a unique identifier. The emergence of big data has spawned new perspectives on social network structures [13–15], energized the development of new metrics [16, 17], and increased the availability of software and libraries [18–20]. Geography is increasingly playing a role in characterizing social networks. However, geography often tends to be treated similarly to other attributes and has, until recently, merited little critical attention about how it can be coupled with SNA to exploit spatial embeddedness of the network. In this paper, we use the term Geography rather broadly to refer to the field itself; a term synonymous with describing "the earth's surface from a standpoint of distributions and interactions" [21].

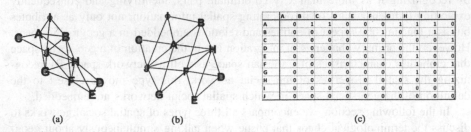

Fig. 1. Various representations of the same social network. (a) Sociogram with ForceAtlas2 layout [9] (b) sociogram with Fruchterman–Reingold layout [22] (c) adjacency matrix

3   Existing Methods of Coupling Geography and SNA

Numerous articles have discussed spatial social networks. We characterize the literature into three main types.

First, in its simplest form, articles treat geography as a nodal attribute [23–26]. This approach has location information of the entities stored in terms of nominal location (e.g., gazetted placenames). The location is often treated similarly to other nodal attributes (e.g., age, gender). The location provides information about similarity among different entities. This allows us to infer location based homophily (i.e. propinquity) or to consider places as promoters of social tie formation and maintenance. The main benefits of this type are that it is easier to treat one attribute like any other and does not demand the geometric transformation of a nominal location to a feature type. The analysis methods are usually rooted in SNA, as opposed to spatial analysis, and locational information merely adds more context to aid the SNA.

Second, researchers may treat location of the entities as a topographic attribute [27–30]. This is a more sophisticated way of integrating spatial information into social networks by associating x, y locations with nodes, edges, or both. In this type, edges can take on two meanings: the social connection between two entities or the physical path between the entities. Having x, y locations aids representation of the social network on a Cartesian space enabling use of visual as well as spatial analysis techniques to understand the spatial characteristics. This type of research is typically accompanied by basemaps to visualize the social network. More importantly, these methods fit social relationships on to a Cartesian space. The fitting of the social network to Cartesian space makes spatial analysis, such as kernel densities, a more vital component than in the nodal attribute treatment. Nonetheless, reducing geography to be a mere nodal attribute or simply to a x, y pair to be rendered on or analyzed against a map disregards the nuanced effects on actors and tie formation in the social network.

Third, is the treatment of geography as a spatial property of the network [6, 31–39]. This not only considers the geographic locations of the nodes and/or edges but also exploits spatial properties and patterns to infer spatio-temporal characteristics of the network. Common ways of considering spatial aspects include not only Euclidean distance but also social distance, contiguity in terms of geography and in terms of social relationships, to name a few. This alleviates the handling of spatial information by recognizing it as more than x, y co-ordinate pairs, identifying and consequently exploiting different means of incorporating spatial information, not only as attributes but as a fundamental aspect of entities and relations embedded in a geographic space. However, the primary challenge of integration lies in the creation of a geo-social space that embodies characteristics of Cartesian space as well as network space. By recognizing the spatial properties of the social network, this type moves closest to the definition of the geo-social space in which spatial social networks are embedded.

In the following section, we encompass all three types of spatial social networks to discuss the terminological chaos that ensue when talking simultaneously about geography and social network analysis. Creating a typology helps understand the various levels of sophistications of integration, creating a baseline for further deliberation.

4   Different Expressions and Challenges

Table 1 shows how specific terms compare between SNA and geography literature. We discuss the terms as they appear in the two different contexts and move on to highlighting the challenges as well as the importance of creating solidarity of the terms for spatial social networks.

Table 1. Parallel concepts of SNA and geography and examples of how they tend to be expressed

Concept	Expressed in SNA	Expressed in geography	Coupling problems
Distance	Counts of edges; connectivity; shortest path; degrees of centrality; weighting	Measures in Euclidean space; shortest path; homophily of non-geographic attributes in Cartesian space; impedances; distance decays	Incongruent spatial metaphors
Communities	Shared attribute; areas; number of social interactions; homophily	Static measures (jurisdictions); Dynamic measures (spatial distribution and clustering)	Semantics
Scale	Number of nodes and edges; completeness of capture; characteristic nodes	Resolution of collection and representation; spatial extent; edge effect	Reconciling and integrating the many interpretations

4.1 Distance

In social networks, distance is measured by movement from one node to another node, travelling along the edges. Nodes connected by an edge are said to be adjacent. Two nodes i and j are considered reachable, if there is a sequence of one or more edges that connects the said nodes. The sequence of edges between i and j is called a path. The number of links one traverses to reach another node equals the distance on the graph. More specifically, the geodesic distance, $d(i,j)$, between two nodes is defined as the shortest path between them [38]. It is also possible to create social networks with unconnected nodes, for example, separate groups of friends with no common friend between the groups. In SNA, the groups themselves are called connected components. If there is no path connecting the two nodes, that is, if they belong to different connected components then the distance between them is conventionally defined as infinite. Hence, two nodes belonging to two different connected components are unreachable from each other.

In SNA, the simplest way to characterise importance of nodes is by looking at the number of edges incident on it. Thus, an important node has many adjacent nodes by virtue of having a high *degree centrality* [38]. In a social network, being friends with an important person is always beneficial because one potentially becomes *closer* to many other people in the network. It is important to note that being 'close' considers the geodesic distance on the social network and not physical distance in a Cartesian Space. Thus, even without having a high degree centrality, by virtue of having a few important friends, a node may have highly efficient paths connecting it to most other nodes in the network. These nodes are said to have a high *closeness* [38].

In a weighted graph, one cannot just as easily traverse one path as another. In certain scenarios, edges may be associated with weights to represent factors like strength of a tie, probability of forming a tie, or in case of spatial social networks, geographic distance between the nodes. Weights add a new property coaxing geodesic distance calculations to account for the different weights of the edges. Links may also have directionality (non-reciprocal relations). If the graph is directional then distance d(i,j) and distance d(j,i) are not symmetrical. The concept of adding weights to the edges is similar to geographical impedances along a road or stream network.

Similar to associating weights with links, attribute information (e.g. age, gender) can be affixed to nodes. Each attribute can be considered a dimension and projected on to axes creating an n-dimensional attribute space[1]. Each node is represented as a point in this attribute space. The position is determined by the particular set of values of the node's attributes. The locations of the nodes are no longer arbitrary and the distance between them is interpretable [10]. This information in the attribute space determine the similarity of the nodes with respect to their attributes. For example, people with similar incomes and similar age are closer together in the attribute space.

If the geographic location of each node is stored as attribute information, then, longitude and latitude may be used to characterise the X and Y axes of the attribute space. The distance between the two nodes in the attribute space represents the distance in geographic space. Since, there may be a variety of attribute information collected about the nodes, the specific attributes selected to characterize the axes in the attribute space may produce different sociograms for the same social network.

Incorporating location into a node's attributes allows exploration into the geographic properties of networks. One way geographic effects may be considered is by calculating the distance between nodes with connections. Tobler's First Law of Geography states that near features are more alike than distant features [39]. Nodes located in proximity to one other in geographic space thus have properties that are similar to each other. In social networks, it is known that similar nodes tend to form connections (i.e., homophily) [40]. Thus, geographically closer nodes are more likely to have an edge than nodes that are further apart. Propinquity has been acknowledged to play a role in forming social relationships [41]. Many social processes are considered to be an outcome or affected by spatial proximity [42]. Milgram's [43] landmark work on "small world networks", which led to the famous concept of "Six degrees of separation", contained a geographic component as the letters were posted from the different cities to reach their final destination. This concept caught the attention of researchers in exploring the relationship between geographic distance and social ties.

Despite the telecommunication revolution and the fabled "death of distance" [44], researchers reiterate that relationships are often geographically local with the probability of forming long distance ties diminishing exponentially with increase in distance between the actors [45–48]. The dependence on geographic distance to form social ties

---

[1] Machine learning uses the term 'feature' to refer to each attribute used to characterize an entity. Consequently, the n-dimensional space where the features live is referred to as a feature space. Here we use the term attribute space to avoid confusion with geographic features.

can be exploited to form generative models for social networks, which mimic the properties found in real world networks [49, 50].

The geographic distance-friendship relationship is recognized as having important consequences on the structures and processes of the network [48–50]. While studying sparsely connected social networks, where despite the low density of links between the nodes, all nodes are reachable from each other via only a few steps (i.e. small world networks) [15, 49] concluded that if the probability of linking two individuals is inversely proportional to the geographic distance separating them, a simple greedy strategy (i.e., searching by making the locally optimal choice at each step) based on geography is able to find a short path to a target in $(\ln N)^2$ time. The author also pointed out that if a network is not structured like this, it is impossible to find the target using a simple greedy strategy in a poly-logarithmic time, making searches computationally expensive. A model proposed by [50] to explain the 'searchable' nature of small world networks considers individuals to belong to groups, which in turn are embedded hierarchically inside larger groups. The group can refer to any attribute, for example profession or geography. In this model too, searching using only local information (i.e., selecting a neighbouring node of the current node that has the same attribute as the target) was successful only when the probability of acquaintance between two individuals was inversely related to distance.

Distance metric is an abstract notion, appropriated by different fields in various ways to describe what 'near' and 'far' means in the subject's realm. Geography usually uses measures in Cartesian space. Distance in Cartesian space can be measured in different ways (e.g., Euclidean distance, Manhattan distance). Social networks on the other hand use network space where distance is measured as edge sequences between nodes. While studying social networks that are situated in geography, distance of nodes on the surface of the earth is a strong determinant of social relationships and hence affects geodesic distance in the social network. In spatial social networks, metrics can be developed to leverage the different distance conceptualizations to characterize the nodes as well as the entire network. However, the starting point for developing new distance measures will require a conceptualization of a geo-social space in which spatial social networks are embedded.

4.2 Communities

Sporadic debates in geography and SNA have recommended various definitions of communities. The simplest differences between communities and societies are in terms of size and interactions. In a social network, parts of the network may be highly connected to each other. These sub-structures inside the social network are referred to as clusters, communities, cohesive groups, or modules [51]. The principal elements defining a community in geography are usually identified as social ties, social interactions and area [52, 53]. However, this definition is not all encompassing, nor are all the elements described above a necessary condition for a society [54, 55]. Professional societies (e.g., the scientific community, community of lawyers) may not satisfy the requirement of sharing common geographic territory, or may not even interact with one another, yet form a community based on homogeneity of profession. Communities hence have three primary dimensions determining them, namely, shared area, social interactions, and homophily. The intersection of geography and social network helps explore communities as a function of both shared area and social connections, integrating social as well as spatial communities. Thus, a working definition for communities for spatial social networks adheres to the old-school definition encompassing shared area and based on social ties, yet is flexible to account for spatially discontinuous communities if the social ties between entities are known.

The mere fact that nodes are geographically co-located is insufficient to firmly combine spatial concepts of community to SNA ideas of community. Hence spatial cluster detection methods (e.g., Getis Ord, global Moran's I and Ripley's K) and even Tobler's law may not be informative in detecting social network communities. Even non-spatial topological community detection algorithms normally used in SNA, like clique or modularity based approaches, may be insufficient to find communities in spatial social networks. In the case of spatial SNA, the spatial arrangement of nodes as well as the nature of ties must be factored in to extract information from the network topology. Thus, for spatial social networks, it is important to consider not only the Euclidean distance between the nodes, but also the social distance between them, for most socio-spatial analyses including cluster and community detection [56].

Modularity [57] is a metric that provides a measure of the quality of graph partition. Modularity calculation for detection of social communities in spatial social networks must control for the spatiality of the network. Hence, community detection in spatial social networks needs to be perceptive of both spatial and network auto-correlation to distill social and geographic determinants of community formation. The modularity calculation can be modified to factor in the location of the node to find communities that are firmly determined by geographic factors [34]. However, researchers have argued that this form of approach to community detection provides little information about the underlying forces actually shaping the topology of the network, and have proposed a modularity measure that can factor out the effects of space, thus finding clusters of nodes that are similar but not just because of their location relative to one another [33]. Interesting community patterns can also be extracted from spatial social networks by applying standard modularity based community detection approaches coupled with innovative ways of visualizing the community. One such visualization approach plots the communities on maps and uses Kernel Density Estimate to characterize the relative occurrence of a user in any given community in any given location [28, 30].

Despite the nihility of a widely accepted unique definition of communities in geography and SNA, the existence of smaller connected structures within the larger society is a signature of the hierarchical nature of the complex social structure [51]. Identification of topological clusters moves the focus from quantifying the importance of individual nodes to identifying important sub-structures in the network, representing a jump in the entity of analysis, i.e., instead of just one node we look at a cluster of nodes. Coupling SNA and geography provides avenues for consolidating the various conceptualizations of communities, opening up opportunities to compare and contrast the various definitions of communities and their corresponding usefulness in revealing socio-spatial patterns and processes.

4.3   Scale

Scale has been a central notion in geography and also a particularly confusing one depending on the context [58–61]. Thus it is imperative to reconcile concepts of spatial social networks with different meanings of geographic scale.

The observation scale or the measurement unit [60] needs two specifications, one for the geography and another for social network elements. Whereas the specifications in geography include the smallest object discernable and the smallest measurable units, social networks need to include disclaimers about what resolution of data is collected about the nodes and the edges. Details about nodes include not only a list of the attributes collected, but also metadata about the attributes. For example, age is denoted as a specific number or as a range. In terms of geography, it is vital that location resolution of the node is known. Moreover, if the nodes denote people, it is important to know how they are located and assess the implications of the locations for the study. For example, is the location of the person's home recorded or is it the location of work. Moreover, a person, unlike a house, is not stationary in space. Thus, it is imperative to reflect on how the recorded location(s) affect the inferences from the spatial social networks and how qualifying the observation scale serves as the entry point for understanding the simplified model of reality at which the study is implemented [61].

The geographic scale or the spatial extent refers to the area on the surface of the earth spanned by the social network under study [60]. Thus, analysis on data from Facebook may have a geographic scale spanning the entire earth. However, depending on the phenomenon under study, only a subset of the network may be used. In terms of analysis boundaries, social networks pose a two-fold problem, finding the entire population and then determining the links between the entities. When resorting to sampling, decisions need to be made to limit the population, or the links, or both. If geographic constraints are used, then the geographic boundaries used to subset the network itself define the geographic scale. Conversely, the social network itself might dictate the geographic scale that needs to be considered. For example, an experiment that requires the creation of the social network by following the connections of a person, the geographic scale determined by how far from the original person their connections live. Someone residing in Montreal may have friends only in Montreal in which case the geographic scale will be small, or may have friends residing all over the world requiring a very large geographic scale of study. When resorting to sampling, all relationships that lie beyond the sampling boundaries are ignored. As network algorithms are fundamentally relational, the results obtained from these algorithms will be erroneous as a result of the edge effect [62].

When social networks are studied in the context of geography, the spatial extent at which the various social network metrics are reported may convey interesting information. For example, it is known that most of our social connections are local, with only a few long distance links. Thus, it may be interesting to classify the degree of a node with varying spatial extents. A person who has more long distance links than average may be of more interest in connecting disparate spatial locations. Similarly, when studying real world social networks, people living in small towns or villages often know each other, forming closely knit social networks. However, emergence of communities at a larger spatial extent with similar population density may be more interesting because of the lower probability of such an event.

Additionally, scale can be studied in terms of its phenomenology. The argument about Stommel diagrams [63, 64] in which geographic features only have meaning when observed in space-time (e.g., a flood, an oak forest), can be extended to social networks. The recognition of the fact that social networks are a spatio-temporal process is highlighted by the adoption of check-ins and timelines by almost all social media sites. Thus, in spatial social networks, a coupling of social networks requires recognition of the fact that social networks are not only contextually based on the conceptualization of edges, but also contextual in space-time.

Perhaps the most confusing use of the term 'scale' in social networks is when referring to the existence of characteristic nodes. In a uniform network, every node has an approximately equal number of edges. The degree distribution of a uniform network has a sharp peak with a very small standard deviation. Hence, a node with the mean number of edges is considered to be representative of all the nodes in the network. However, most social networks do not have an egalitarian degree distribution. Few nodes have disproportionately more edges compared to the majority of the nodes in the network. In terms of social networks, these nodes play a vital role in keeping the network connected. Thus, the degree distribution is skewed. These networks are considered to be scale-free because there is no characteristic node to represent all the other nodes [65]. The closest equivalence in geography is regarding aggregating and rescaling data [66]. Data is said to be rescaled to a lower resolution by combining smaller regions into larger ones, aggregating the values based on some central tendency. This process is often used to aggregate data from county level up to the provincial level. Thus, though the commonly used term for this operation is 'rescaling' or 'upscaling' as it involves a change in the observational spatial resolution, the idea is similar to 'scale' in networks where a large population is said to be represented by a single entity. The use of scale in social networks to refer to existence of a characteristic node is an incongruence of terminology between social networks and the different meanings of scale used in geography.

5   Conclusion

Spatial social networks have gained traction in the literature as a method of incorporating geographical information into SNA. In this paper, we have highlighted some of the inconsistencies that require closer deliberation for a tighter coupling of spatial information in SNA. We proposed a typology of the current literature of spatial SNA. We also highlighted some of the parallel concepts that exist in spatial and social network analysis. SNA provides an interesting perspective on explicitly studying relationships between entities. However, the incorporations of geography in social networks have been rudimentary thus far, and critical introspection is essential to incorporate concepts and concerns from the perspective of GIScience. We need to draw upon the long tradition of geography in working with non-Cartesian notions of space, moving towards a definition of geo-social space to succinctly reflect the subtleties of spatial social networks.

When using SNA, analysts must remain cognizant of the fact that a network is usually a snapshot of a social system in both time and conceptualization. Depending on the conceptualization of the relationships, multiple social networks can be created. For example, if explicitly declared friendships are used, it results in a particular social network that is different from the one when some common attribute between people is used to conceptualize the edges. Society is multi-faceted and the different conceptions highlight different aspects. Situating social networks in geography not only allows several new conceptualizations of relationships between entities, but has the potential to enrich analytic capabilities even when the network ties are based on non-geographic factors. In this paper, we have highlighted some of the considerations for progressing socio-spatial analytics utilizing spatial social networks. Investigators of new metrics, analysis techniques, and algorithms designed to leverage spatial and social networks should remain vigilant about unifying spatial and social concepts to reveal interesting phenomena, possible only through more deeply interrogating both spatiality and sociality together.

References

1. Barthélemy, M.: Spatial networks. Phys. Rep. **499**, 1–101 (2011)
2. Haggett, P., Chorley, R.J.: Network Analysis in Geography. Edward Arnold, London (1969)
3. Scellato, S., Noulas, A., Lambiotte, R., Mascolo, C.: Socio-spatial properties of online location-based social networks. ICWSM **11**, 329–336 (2011)
4. Ye, M., Yin, P., Lee, W.-C.: Location recommendation for location-based social networks. In: Proceedings of the 18th SIGSPATIAL International Conference on Advances in Geographic Information Systems - GIS 2010, p. 458. ACM Press, New York (2010)
5. Scellato, S., Mascolo, C., Musolesi, M., Latora, V.: Distance matters: geo-social metrics for online social networks. In: Proceedings of the 3rd Conference on Online Social Networks, vol. 8, p. 8 (2010)
6. Radil, S.M., Flint, C., Tita, G.E.: Spatializing social networks: using social network analysis to investigate geographies of gang rivalry, territoriality, and violence in Los Angeles. Ann. Assoc. Am. Geogr. **100**, 307–326 (2010)
7. Wasserman, S.: Social Network Analysis: Methods and Applications. Cambridge University Press, Cambridge (1994)
8. Krzywinski, M., Birol, I., Jones, S.J., Marra, M.A.: Hive plots–rational approach to visualizing networks. Brief Bioinform. **13**, 627–644 (2012)
9. Jacomy, M., Venturini, T., Heymann, S., Bastian, M.: ForceAtlas2, a continuous graph layout algorithm for handy network visualization designed for the Gephi software. PLoS ONE **9**, e98679 (2014)
10. Hanneman, R.A., Riddle, M.: Introduction to Social Network Methods. University of California, Riverside, Riverside (2005)
11. Di Battista, G., Eades, P., Tamassia, R., Tollis, I.G.: Algorithms for drawing graphs: an annotated bibliography. Comput. Geom. **4**, 235–282 (1994)
12. Garrido, A., De Ciencias, F., Uned, D.: Graph properties and invariants, by their associate matrices. Adv. Model. Optim. **11**, 337–348 (2009)
13. Albert, R., Barabasi, A.-L.: Statistical mechanics of complex networks. Rev. Mod. Phys. **74**, 54 (2001)

14. Barabasi, A.-L., Albert, R.: Emergence of scaling in random networks. Science **286**, 509–512 (1999)
15. Watts, D.J., Strogatz, S.H.: Collective dynamics of "small-world" networks. Nature **393**, 440–442 (1998)
16. Opsahl, T.: Structure and Evolution of Weighted Networks Diss. University of London, Queen Mary (2009)
17. Newman, M.E.J.: Assortative mixing in networks 5 (2002). doi:10.1103/PhysRevLett.89. 208701
18. Hagberg, A.A., Schult, D.A., Swart, P.J.: Exploring network structure, dynamics, and function using {NetworkX}. In: Proceedings of the 7th Python in Science Conference (SciPy2008), pp. 11–15 (2008)
19. Borgatti, S.P., Everett, M.G., Freeman, L.C.: UCINET 6 for Windows: Software for Social Network Analysis. http://www.analytictech.com/
20. Bastian, M., Heymann, S., Jacomy, M.: Gephi: an open source software for exploring and manipulating networks. In: International AAAI Conference on Weblogs and Social Media (2009)
21. De Geer, S.: On the definition, method and classification of geography. Geogr. Ann. **5**, 1–37 (1923)
22. Fruchterman, T.M.J., Reingold, E.M.: Graph drawing by force-directed placement. Softw. Pract. Exp. **21**, 1129–1164 (1991)
23. Pelechrinis, K., Krishnamurthy, P.: Socio-spatial affiliation networks. Comput. Commun. **73**, 251–262 (2015)
24. Cranshaw, J., Toch, E., Hong, J., Kittur, A., Sadeh, N.: Bridging the gap between physical location and online social networks. In: Proceedings of the 12th ACM International Conference on Ubiquitous Computing - Ubicomp 2010, p. 119. ACM Press, New York (2010)
25. Crandall, D.J., Backstrom, L., Cosley, D., Suri, S., Huttenlocher, D., Kleinberg, J.: Inferring social ties from geographic coincidences. Proc. Natl. Acad. Sci. U. S. A. **107**, 22436–22441 (2010)
26. Entwisle, B.: Putting people into place. Demography **44**, 687–703 (2007)
27. Batty, M., Hudson-Smith, A., Neuhaus, F., Gray, S.: Geographic analysis of social network data. In: Proceedings of the AGILE 2012 International Conference on Geographic Information Science, pp. 24–27 (2012)
28. Nag, M.: Mapping networks: a new method for integrating spatial and network data. Unpublished manuscript. Princeton University, Department of Sociology (2009)
29. Comber, A., Batty, M., Brunsdon, C.: Exploring the geography of communities in social networks. In: Rowlingson, B., Whyatt, D. (eds.) Proceedings of the GIS Research UK 20th Annual Conference, pp. 33–37. Lancaster (2012)
30. Koylu, C., Guo, D., Kasakoff, A., Adams, J.W.: Mapping family connectedness across space and time. Cartogr. Geogr. Inf. Sci. **41**, 14–26 (2014)
31. Butts, C.T., Acton, R.M., Hipp, J.R., Nagle, N.N.: Geographical variability and network structure. Soc. Netw. **34**, 82–100 (2012)
32. Daraganova, G., Pattison, P., Koskinen, J., Mitchell, B., Bill, A., Watts, M., Baum, S.: Networks and geography: modelling community network structures as the outcome of both spatial and network processes. Soc. Netw. **34**, 6–17 (2012)
33. Expert, P., Evans, T.S., Blondel, V.D., Lambiotte, R.: Uncovering space-independent communities in spatial networks. Proc. Natl. Acad. Sci. U. S. A. **108**, 7663–7668 (2011)
34. Onnela, J.-P., Arbesman, S., González, M.C., Barabási, A.-L., Christakis, N.P.: Geographic constraints on social network groups. PLoS ONE **6**, e16939 (2011)

35. Doreian, P., Conti, N.: Social context, spatial structure and social network structure. Soc. Netw. **34**, 32–46 (2012)
36. Luo, W., MacEachren, A.M.: Geo-social visual analytics. J. Spat. Inf. Sci. **8**, 27–66 (2014)
37. Kwan, M.P.: Mobile communications, social networks, and urban travel: hypertext as a new metaphor for conceptualizing spatial interaction. Prof. Geogr. **59**, 434–446 (2007)
38. Freeman, L.C.: Centrality in social networks conceptual clarification. Soc. Netw. **1**, 215–239 (1979)
39. Tobler, W.R.: A computer movie simulating urban growth in the detroit region. Econ. Geogr. **46**, 234 (1970)
40. McPherson, M., Smith-Lovin, L., Cook, J.M.: Birds of a Feather: Homophily in Social Networks (2001). doi:10.1146/annurev.soc.27.1.415
41. Festinger, L., Back, K.W., Schachter, S.: Social Pressures in Informal Groups: A Study of Human Factors in Housing. Stanford University Press, Palo Alto (1950)
42. Downey, L.: Using geographic information systems to reconceptualize spatial relationships and ecological context. Am. J. Sociol. **112**, 567–612 (2006)
43. Milgram, S.: The small world problem. Psychol. Today **2**, 60–67 (1967)
44. Sempsey, J.J.: The death of distance: how the communications revolution will change our lives. J. Am. Soc. Inf. Sci. **49**, 1041–1042 (1998)
45. Mok, D., Wellman, B., Carrasco, J.: Does distance matter in the age of the internet? Urban Stud. **47**, 2747–2783 (2010)
46. Preciado, P., Snijders, T.A.B., Burk, W.J., Stattin, H., Kerr, M.: Does proximity matter? Distance dependence of adolescent friendships. Soc. Netw. **34**, 18–31 (2011)
47. Liben-Nowell, D., Novak, J., Kumar, R., Raghavan, P., Tomkins, A.: Geographic routing in social networks. Proc. Natl. Acad. Sci. U. S. A. **102**, 11623–11628 (2005)
48. Wong, L.H., Pattison, P., Robins, G.: A spatial model for social networks. Phys. A: Stat. Mech. Appl. **360**, 99–120 (2006)
49. Kleinberg, J.: Navigation in a small world. Nature **406**, 845 (2000)
50. Watts, D.J., Dodds, P.S., Newman, M.E.J.: Identity and search in social networks. Science **296**, 1302–1305 (2002)
51. Palla, G., Derényi, I., Farkas, I., Vicsek, T.: Uncovering the overlapping community structure of complex networks in nature and society. Nature **435**, 814–818 (2005)
52. Hillery, G.A.: Definitions of community: areas of agreement. Rural Sociol. **20**, 111–123 (1955)
53. Clark, D.B.: The concept of community: a re-examination. Sociol. Rev. **21**, 397–416 (1973)
54. Smith, D.M.: Geography, community, and morality. Environ. Plan. A. **31**, 19–35 (1999)
55. Frug, J.: The geography of community. Stanford Law Rev. **48**, 1047–1108 (1996)
56. Crampton, J.W., Graham, M., Poorthuis, A., Shelton, T., Stephens, M., Wilson, M.W., Zook, M.: Beyond the geotag: situating "big data" and leveraging the potential of the geoweb. Cartogr. Geogr. Inf. Sci. **40**, 130–139 (2013)
57. Newman, M.E.J.: Modularity and community structure in networks. Proc. Natl. Acad. Sci. U. S. A. **103**, 8577–8582 (2006)
58. Quattrochi, D.A., Goodchild, M.F.: Scale in Remote Sensing and GIS. CRC Press, Boca Raton (1997)
59. Goodchild, M.F.: Scale in GIS: an overview. Geomorphology **130**, 5–9 (2011)
60. Wu, H., Li, Z.-L.: Scale issues in remote sensing: a review on analysis. Process. Model. Sens. **9**, 1768–1793 (2009)
61. Atkinson, P.M., Tate, N.J.: Spatial scale problems and geostatistical solutions: a review. Prof. Geogr. **52**, 607–623 (2000)

62. Gil, J.: Street network analysis "edge effects": examining the sensitivity of centrality measures to boundary conditions. Environ. Plan. B Plan. Des. 147:1–147:16 (2016). doi:10.1177/0265813516650678
63. Stommel, H.: Varieties of oceanographic experience: the ocean can be investigated as a hydrodynamical phenomenon as well as explored geographically. Science 80(139), 572–576 (1963)
64. Holling, C.S.: Cross-scale morphology, geometry, and dynamics of ecosystems. Ecol. Monogr. 62, 447 (1992)
65. Barabási, A.-L.: LINKED: The New Science of Networks. Perseus Books Group, New York (2002)
66. Openshaw, S., Rao, L.: Algorithms for reengineering 1991 Census geography. Environ. Plan. A 27, 425–446 (1995)

Spatial Analysis

On Distortion of Raster-Based Least-Cost Corridors

Takeshi Shirabe[(✉)]

School of Architecture and the Built Environment,
Royal Institute of Technology (KTH), Stockholm, Sweden
shirabe@kth.se

Abstract. Given a grid of cells each having a cost value, a variant of the least-cost path problem seeks a corridor—represented by a swath of cells rather than a sequence of cells—connecting two terminuses such that its total accumulated cost is minimized. While it is widely known that raster-based least-cost paths are subject to three types of distortion, i.e., deviation, distortion, and proximity, little is known about potential distortion of their corridor counterparts. This paper studies a raster model of the least-cost corridor problem and analyses its solution in terms of each type of distortion. It is found that raster-based least-cost corridors, too, are subject to all three types of distortion but in different ways: elongation distortion is always persistent, deviation distortion can be substantially reduced, and proximity distortion can be essentially eliminated.

1 Introduction

To store and process data pertaining to geographic phenomena in a digital device such as a geographic information system (GIS), a raster model discretizes geographic space into a grid of equally-sized square units referred to as 'cells' or 'pixels.' In this model, continuous surfaces of variables (e.g., elevation, vegetation, and temperature) are represented by sets of values assigned to those cells, and discrete features (e.g., junctions, roads, and districts) are represented by individual cells or groups of cells.

The least-cost path problem is a classic application of a raster model. Given a grid of cells each weighted by its associated cost, the problem is to find a cellular path, or a sequence of adjacent cells, between two specified terminal cells such that its cost does not exceed that of any other such sequence. The cost, $d(P)$, of a path, P, is typically defined by the following formula.

$$d(P) = \sum_{(i,j) \in P} \left(\frac{c(i) + c(j)}{2} \right) \cdot l(i,j) \tag{1}$$

where $c(i)$ and $l(i,j)$ denote the cost of cell i and the distance between cells i and j, respectively.

Note that this paper takes the 8-adjacency assumption under which two cells are regarded as 'orthogonally' or 'diagonally' adjacent if they share a cell side or a cell

© Springer International Publishing Switzerland 2016
J.A. Miller et al. (Eds.): GIScience 2016, LNCS 9927, pp. 101–113, 2016.
DOI: 10.1007/978-3-319-45738-3_7

corner, respectively, and that the distance between two orthogonally adjacent cells is equal to the length of a cell side and that between two diagonally adjacent cells is $\sqrt{2}$ times longer.

The least-cost path problem in a grid can be seen as an instance of the shortest path problem, and solved as such if the grid is, implicitly or explicitly, transformed into a graph by equating each cell with a vertex and connecting two adjacent cells with an edge weighted by the cost of transition between them. This is a highly efficient approach in terms of computation, but care must be taken because the raster solution may well be distorted compared to a more realistic, smoother path that might exist in Euclidean space.

The raster path distortion can be generally classified into three types: "deviation" [12], "elongation" [12], and "proximity" [10]. The first two types of distortion are related, and caused by restricting the alignment of a raster path to a finite number of directions, eight under the present 8-adjacency assumption. To see this, let us place two points in the Euclidean plane such that they are 10 and 4 units of distance apart horizontally and vertically, respectively (Fig. 1). Suppose that it costs 1 unit of cost to pass through any point in the plane. Then, there exists only one least-cost path between the two points, which is the straight line segment between them. Now let us impose a 5×11 grid on the plane and assign 1 unit of cost to each cell to simulate the uniform cost surface (Fig. 1). In this grid, more than one least-cost path can connect the two points and their cost is calculated as the sum of six cell sides and four cell diagonals, that is, $(1 \cdot 6) + (\sqrt{2} \cdot 4) \approx 11.657$. Significantly, all the paths *deviate* from the true least-cost path (which is the straight line segment) and their cost is *elongated* compared to the true least-cost value (which is $\sqrt{10^2 + 4^2} \approx 10.770$). In fact, they almost reach the largest possible degree of elongation, 8.24 % [16], which occurs when the true least-cost path follows a straight line at an angle of 22.5° from an orthogonal axis.

Many approaches have been proposed to resolve deviation and elongation distortion ([2] briefly considers some of these). Perhaps the most popular one is to extend the definition of adjacency by including more distant cells than the eight closest ones in the neighborhood of each cell [10, 17]. The "extended raster" [4] is another approach, which represents a path as a sequence of points that do not necessarily correspond to cell centers but to vertices on cell boundaries. Douglas [7] and Tomlin [16] took yet

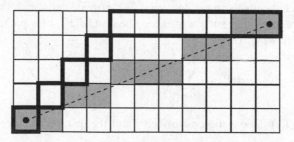

Fig. 1. A true least-cost path (dotted line) between two points over a uniform cost surface in the Euclidean plane and two least-cost raster paths (one shaded and the other enclosed by solid lines) between two cells in a uniform cost grid.

another alternative, which simulates refracting and diffracting waves, and Tomlin's algorithm in particular does this in a very efficient manner.

Proximity distortion generally refers to incorrect passing through costly location. A least-cost raster path experiences this type of distortion when it goes near high-cost cells even though it is supposed to influence (or to be influenced by) its nearby cells. In the most extreme case, a raster path falsely penetrates a linear barrier through an artificial gap between two diagonally adjacent cells (see Fig. 2).

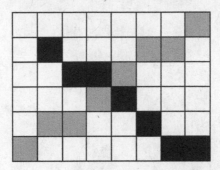

Fig. 2. False penetration of a raster path (lightly shaded) through an impenetrable linear barrier (darkly shaded).

Fortunately, all such "holes" [1] or "cracks" [13] can be filled by removing the corresponding edges from the grid-induced graph or heavily weighting them during or prior to the execution of a shortest path algorithm. However, a more general problem remains: if a raster path has an impact on a larger area beyond its (imaginary) centerline, how can its cost be correctly evaluated? This has an important implication: at geographic scales there are paths whose width cannot be assumed to be zero or negligible. Gonçalves [9] called such paths "wide paths" or "corridors" and formulated a problem of finding one with the least possible cost. The most innovative aspect of his formulation was that a corridor is seen as the area swept by a cellular line segment called a "path front." Gonçalves [9] proposed a solution method, too, which converts a cost grid to a graph in which each vertex represents a path front and each edge represents a possible transition from one path front to another. This method may not be most efficient—in fact its complexity is exponential in the number of cells comprising the corridor width, but guided Shirabe [15] to a highly efficient alternative. The key difference was to replace a path front with a set of cells resembling a 'marker pen tip' as a corridor sweeper (see Fig. 3 for an example).

Although it seems that least-cost raster *corridors* (rather than *paths*) by nature avoid proximity distortion, this has yet to be confirmed. Also it is still unknown whether least-cost raster corridors are subject to the other types of distortion and, if so, by how much. This papers aims to answer these questions. The remainder of the paper is organized as follows. Section 2 reviews the raster-based least-cost corridor algorithm. Section 3 analyzes the forms and costs of output corridors of varying widths and varying resolutions. Section 4 concludes the paper.

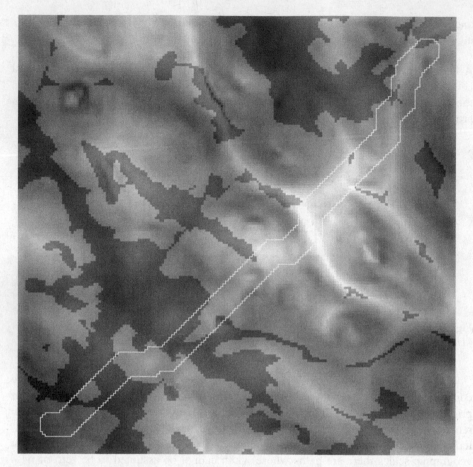

Fig. 3. A least-cost corridor (enclosed by solid lines) swept by a marker pen tip (a set of squares) over a hypothetical cost grid.

2   Model and Algorithm for the Least-Cost Corridor Problem

Assuming that a corridor having a constant width can be seen as an area swept by a disk of that width in the Euclidean plane, Shirabe [15] proposed a new raster model in which a corridor of a width of w cell sides is represented by a set of cells swept by a regular cellular 'octagon' with a width of w cell sides and its cost is defined as the sum of the costs of all the swept cells. Since it is not possible to compose a truly regular octagon from cells, it is approximated by a w-by-w block of cells with d diagonal arrays of cells removed from each corner, where d is the largest integer not greater than $\frac{2-\sqrt{2}}{2}w$ (see Fig. 4 for examples). The form keeps its orthogonal width equal to w cell sides and minimizes the difference of its diagonal width from its orthogonal width. More specifically, the diagonal width alternates around w between $\sqrt{2}(w-d-1)$ and

$\sqrt{2}(w - d)$. For instance, when $w = 1$, the diagonal width alternates between 0 and 1.41; when $w = 8$, it alternates between 7.07 and 8.49.

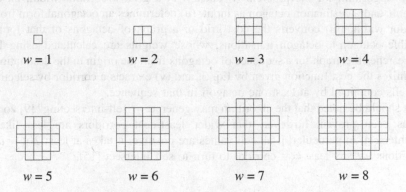

| $w = 1$ | $w = 2$ | $w = 3$ | $w = 4$ |

| $w = 5$ | $w = 6$ | $w = 7$ | $w = 8$ |

Fig. 4. Regular cellular octagons of different widths. Note that the smallest three octagons are actually squares as d is set to 0, but they are still called 'octagons' for ease of discussion.

In this model, the form/location of a corridor is incrementally determined by transitioning orthogonally or diagonally from one octagon to another. As illustrated in Fig. 5, each transition extends the corridor as much as the corresponding set difference and increases the cost of the corridor as much as the sum of the costs of all the cells in that difference.

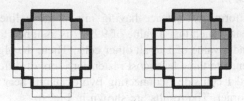

Fig. 5. Orthogonal (left) or diagonal (right) transition from one octagon (a set of white cells) to another (enclosed by a bold line). As the result of each transition, a corridor is extended as much as the corresponding set difference (shaded).

We assign each transition from an octagon, N_i, to an octagon, N_j, a weight, $w(N_i, N_j)$, as follows:

$$w(N_i, N_j) = \sum_{k \in N_j \setminus N_i} c(k) \tag{2}$$

Then, the cost, $a(P)$, of a corridor, P, is expressed by:

$$a(P) = \sum_{k \in N_0} c(k) + \sum_{(N_i, N_j) \in P} w(N_i, N_j) \tag{3}$$

where N_0 represents the first octagon in P. Accordingly, a raster version of the least-cost corridor problem is defined as: *Given a grid of cells each having a cost value, find a sequence of octagons of a specified form between two specified octagons, such that its cost does not exceed that of any other such sequence.*

An algorithm for the problem (i) takes a cost grid, a corridor width, an origin octagon, and a destination octagon as input, (ii) determines an octagonal form from the corridor width, (iii) converts the cost grid to a graph of octagons of that form and possible octagon-to-octagon transitions, whose weights are calculated using Eq. 2, (iv) searches the graph for a sequence of octagons from the origin to the destination that minimizes the cost function given by Eq. 3, and (v) extracts a corridor by selecting all the cells contained by at least one octagon in that sequence.

It should be noted that the algorithm may generate a "self-intersecting" [9] corridor. It has been proven, however, that wider least-cost corridors are less likely to self-intersect; in particular, if all cell values are positive, it takes at least $2(w - d + 1)$ transitions for any least-cost corridor to turn to self-intersect [15].

3   Analysis of Distortion

This section closely examines whether and how least-cost corridors represented by sequences of cellular octagons are subject to each of proximity, deviation, and elongation distortion. For numerical experiments, the algorithm reviewed in the previous section was implemented in Java adapting Dijkstra's shortest path algorithm [6].

3.1   Proximity Distortion

First, to model a uniform cost surface having impenetrable linear barriers on it, a 60×60 grid was created such that a value of 99999 is assigned to all cells where the barriers are located and a value of 1 to all other cells. Then, the algorithm was applied to this cost grid to generate three least-cost raster corridors of different widths—8 cell sides, 6 cell sides, and 1 cell side—connecting two octagons near the top-right corner and the bottom-left corner. The results are shown in Fig. 6.

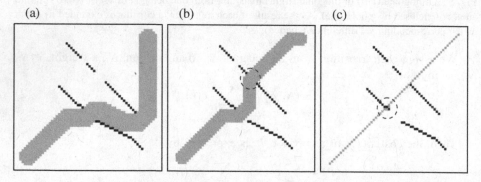

(a)   (b)   (c)

Fig. 6. Least-cost raster corridors of different widths (lightly shaded) in a uniform cost grid with linear barriers (darkly shaded). (a) The 8-cell-side-wide corridor turns most to avoid linear barriers and small gaps between them. (b) The 6-cell-side-wide corridor goes through a gap (encircled) that is too narrow for the 8-cell-side-wide corridor. (c) The 1-cell-side-wide corridor goes between two diagonally adjacent high-cost cells (encircled).

Overall, the 6-cell-side-wide corridor is more winding than the 1-cell-side-wide corridor but less so than the 8-cell-side-wide, because the corresponding octagon is too wide to squeeze into any artificial gap but narrow enough to go through some real gap (encircled in Fig. 6(b)) the 8-cell-side-wide corridor could not.

The 1-cell-side-wide corridor runs almost straight because the algorithm let it go between two diagonally adjacent high-cost cells (encircled in Fig. 6(c)), but not completely straight because it was not allowed to cross either of those high-cost cells. This rather awkward behavior is a consequence of the way the weight of each octagon-to-octagon transition is calculated (see Eq. 2 with Fig. 5). The weight of transition from one 1-cell-side-wide octagon (which is a single cell) to another depends solely on the cost of the latter, not on the cost of any other nearby cell.

The three instances highlight an important aspect of the algorithm: it does not distinguish groups of cells that form linear features from individual cells that happen to be adjacent or close to each other. Therefore, it may unexpectedly let corridors through artificial or real gaps, no matter how costly their nearby cells are. However, these errors can be eliminated simply by increasing the corridor width *not* by pre-processing (e.g., smoothing) the cost grid. This means that the algorithm can take any raster layer as input without any transformation. Moreover, notice that Eq. 2 does not assume the cost grid to have a uniform cost distribution. Thus, regardless of what values are assigned to individual cells, the algorithm selects a swath of cells that may contain high-cost cells but still minimize its overall cost. The corridor shown in Fig. 3 is one such example.

3.2 Deviation Distortion

True least-cost paths follow shortest paths over a uniform cost surface in the Euclidean plane. So do true least-cost corridors. For instance, a true least-cost corridor of a width of 1 unit of distance is drawn below under the same condition that the true least-cost path was drawn in Fig. 1.

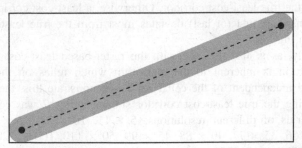

Fig. 7. A true least-cost corridor of a width of 1 unit of distance (shaded). It follows the line segment (dotted line) between two points (separated by a horizontal distance of 10 units and a vertical distance of 4 units) over a uniform cost surface in the Euclidean plane.

Now recall that the present algorithm discretizes a cost surface into a cost grid, represents a corridor as a sequence of regular cellular octagons, and restricts each octagon-to-octagon transition to an orthogonal or diagonal direction. According to Eq. 2, all orthogonal transitions have the same weight in a uniform cost grid. The same is true for all diagonal transitions. Therefore, if two raster corridors contain the same number of orthogonal transitions and the same number of diagonal transitions, they must have the same cost. Equation 2 also implies that one diagonal transition costs less than two orthogonal transitions in a uniform cost grid. This can be confirmed by counting the number of cells associated with each of orthogonal and diagonal transitions in Fig. 5. Hence, one can easily find at least one least-cost raster corridor between two terminal octagons in a uniform cost grid by starting with one terminal octagon and successively making diagonal transitions until reaching an octagon that aligns on the same orthogonal line as the other terminal octagon. Figure 8 shows two such corridors of different resolutions as examples (to be discussed later).

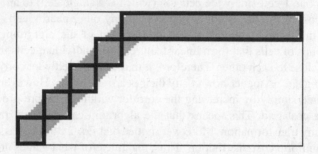

Fig. 8. Least-cost raster corridors of a width of 1 unit of distance (one enclosed by solid lines and the other shaded) in 5 × 11 and 50 × 110 uniform cost grids (gridlines are not shown).

The above two observations imply that given two terminal octagons, there may be no raster corridor coinciding with the true least-cost corridor connecting them and there may be more than one least-cost raster corridor connecting them. How much a least-cost raster corridor can deviate from the true least-cost corridor depends on the relative location of its terminal octagons. If they are aligned orthogonally or diagonally, there is only one least-cost raster corridor, which just runs in a single (orthogonal or diagonal) direction along the true least-cost corridor. Otherwise, a least-cost corridor that makes all diagonal transitions first (or last) deviates most from the true least-cost corridor, such as those in Fig. 8.

Unfortunately, as is also the case with the raster-based least-cost path problem, deviation distortion is inherent to the algorithm which relies on the 8-adjacency assumption, and independent of the cell size. To demonstrate this, the uniform cost surface underlying the true least-cost corridor shown in Fig. 7 was converted to 14 uniform cost grids of different resolutions: 5 × 11, 10 × 22, 15 × 33, 20 × 44, 25 × 55, 30 × 66, 35 × 77, 40 × 88, 45 × 99, 50 × 110, 100 × 220, 200 × 440, 400 × 880, and 800 × 1760. Then, the algorithm was applied to each grid to delineate a least-cost raster corridor of a width of 1 unit of distance from the top-right corner to the bottom-left corner. Note that the actual dimensions of these cost grids are the same but their resolutions vary, so that the number of cell sides to which 1 unit of distance corresponds increases inversely proportionally with the cell size. For example, 1 unit of distance is 1 cell side long in the 5 × 11 grid and 10 cell sides long in the 50 × 110 grid.

All the resulting 14 raster corridors (Fig. 8 shows two of them) similarly deviate from the true least-cost corridor (drawn in Fig. 7)—similarly, i.e., they are made of an orthogonal sequence of octagons and a diagonal sequence of octagons. This is a consequence of the way multiple shortest paths are resolved in the underlying shortest path algorithm, that is, of all shortest paths to each vertex, the one found first will be regarded as *the* shortest path to that vertex. In a graph induced by a uniform cost grid, if there are two shortest paths to a vertex such that one ends with an orthogonal edge and the other with a diagonal edge, the former will not be found earlier. As suggested by Goodchild [12], the distortion caused by "a systematic resolution of ties in the solution algorithm" should be reduced by replacing it with a random one. Statistically speaking, this heuristic is more effective for longer paths.

3.3 Elongation Distortion

The analysis of elongation distortion begins with the exact calculation of the cost of a true least-cost corridor in a uniform cost surface. If its two terminal disks are separated by a horizontal distance of x units and a vertical distance of y units, the cost, $a(x, y)$, of a true least-cost corridor of a width of 1 unit of distance between them in a uniform cost surface (which assigns a value of 1 to every location) is given by:

$$a(x, y) = \sqrt{x^2 + y^2} + \pi/4 \tag{4}$$

According to this formula, the cost of the true least-cost corridor illustrated in Fig. 7 is $\sqrt{10^2 + 4^2} + \pi/4 \approx 11.556$. We calculated the costs of the 14 least-cost raster corridors obtained earlier and compared them to this true least-cost value in Table 1.

Table 1. Costs of least-cost raster corridors in uniform cost grids

Grid	Cell size	Corridor width	Cost	Rel. Dif.
5 × 11	1	1	11.000	0.952
10 × 22	1/2	2	13.000	1.125
15 × 33	1/3	3	13.667	1.183
20 × 44	1/4	4	11.750	1.017
25 × 55	1/5	5	12.440	1.076
30 × 66	1/6	6	12.889	1.115
35 × 77	1/7	7	11.898	1.030
40 × 88	1/8	8	12.313	1.065
45 × 99	1/9	9	12.630	1.093
50 × 110	1/10	10	12.880	1.115
100 × 220	1/20	20	12.650	1.095
200 × 440	1/40	40	12.535	1.085
400 × 880	1/80	80	12.478	1.080
800 × 1760	1/160	160	12.506	1.082

Note: "Grid" represents the number of rows and columns of the cost grid from which each corridor is delineated, and "Cell Size" represents the cell size of that grid relative to that of the 5 × 11 grid. "Corridor Width" represents the width (in terms of numbers of cell sides) specified for each corridor. "Corridor cost" and "Rel. Dif." represent the cost of each corridor and its ratio to that of the true least-cost corridor (≈ 11.556), respectively.

With one exception, all the 14 least-cost raster corridors overestimate the true least-cost value. Their relative errors should, in theory, converge as the cell size becomes smaller, because cellular octagons become more similar to regular octagons while octagon-to-octagon transitions remain limited to four orthogonal and four diagonal directions. In the present example, those 14 raster corridors converge to an elongated polygon, referred to here as a 'decagon', of a width of 1 unit of distance, illustrated in Fig. 9.

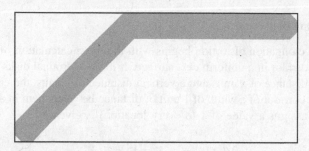

Fig. 9. A decagon to which least-cost raster corridors between the top-right and the bottom-left corners converge as the underlying grid becomes finer.

In general, the cost, $\hat{a}(x, y)$, of a decagon of a width of 1 unit of distance between two terminal octagons in a uniform cost surface (which assigns a value of 1 to every location) is calculated as:

$$\hat{a}(x,y) = x + \left(\sqrt{2} - 1\right)y + 2\left(\sqrt{2} - 1\right) \tag{5}$$

where x and y are defined in a similar way as in Eq. 4; that is, the decagon's two terminal octagons are separated by a horizontal distance of x units and a vertical distance of y units.

According to Eqs. 4 and 5, as the cell size decreases, the relative cost difference— or call it elongation—between a least-cost raster corridor and the corresponding true least-cost corridor converges to:

$$\hat{r}(y/x) = \frac{1 + (\sqrt{2} - 1)(y/x)}{\sqrt{1 + (y/x)^2}} \tag{6}$$

In the present example where $(x, y) = (10, 4)$, we have $\hat{r}(4/10) \approx 1.0823$. The worst case occurs when $y/x = \tan^{-1} 22.5$ and $\hat{r}(\tan 22.5) \approx 1.0824$. This implies that both least-cost raster corridors and paths suffer elongation distortion in the same way.

Finally, as pointed out earlier with reference to Table 1, one least-cost raster corridor (the one with a width of 1 cell side) *under*estimates the true least-cost value. This is explained by the fact that when a raster corridor is not very wide, its octagons are too coarse to fill the area of the corresponding decagon. The 1-cell-side-width least-cost corridor is the most extreme case, as there are two large triangular voids between every two diagonally adjacent cells in it (compare Figs. 8 and 9). Certainly, wider raster corridors have fewer such voids and are closer to decagons.

4   Conclusion

The paper reviewed a raster-based least-cost corridor algorithm originally proposed by Shirabe [15] and analyzed its solution in terms of three types of distortion: proximity, deviation, and elongation. The algorithm's uniqueness lies in its underlying corridor model, which (1) represents a corridor as a sequence of 'octagons'—i.e., sets of cells of an approximately octagonal form determined by a specified corridor width—and (2) gives each transition from one octagon to another a weight equal to the sum of the values of all cells that are included in the latter octagon but not in the former octagon. The model keeps the width of the corridor (approximately) constant throughout and simplifies the calculation of its cost to the sum of the weights of all its octagon-to-octagon transitions, which, in turn, enables the adaptation of Dijkstra's algorithm to the search for a least-cost corridor in a cost grid.

It was found, however, that raster corridors generated by the algorithm are subject to all three types of distortion but to different degrees of severity. A special kind of proximity distortion may occur when the corridor width is set to 1 cell side; that is, a 1-cell-side-wide least-cost corridor may incorrectly go through an artificial gap between two diagonally adjacent cells, no matter how high their costs are. However, if it is known that the influence of a corridor reaches farther than a single cell side from its centerline, this and more general proximity distortion can be eliminated by setting the corridor width sufficiently large.

Both deviation and elongation distortion are inevitable consequences of the algorithm's dependence on the 8-adjacency assumption and independent of the choice of corridor width or cell size. One might expect that the two related types of distortion could be reduced by extending the definition of adjacency, but doing so might cause other problems since the 8-adjacency assumption plays the key role in reducing the risk of self-intersection as well as linearizing the corridor cost function (see Eq. 3). Fortunately, deviation distortion can be reduced by employing Goodchild's random (rather than systematic) tiebreaker in the algorithm.

As for elongation, numerical experiments found some interesting pattern when the corridor width was relatively small (in terms of number of cell sides). That is, the cost of a not-very-wide least-cost raster corridor is significantly discounted by the void made by every diagonal octagon-to-octagon transition in that corridor and may actually underestimate the true least-cost value. In general, however, the relative difference in cost between a least-cost raster corridor and the corresponding true least-cost corridor converges to a positive theoretical bound as the corridor width increases (in terms of number of cell sides). Therefore, the effective and efficient resolution of elongation distortion should be left as an open question to future research.

Lastly, for its ability to take into account width, efficiency of computation, and ease of use, the least-cost *corridor* algorithm may be an attractive alternative to existing least-cost *path* algorithms in a wide range of applications where delineation of linear features from geographic space is a critical task. Examples include designation of wildlife corridors [1, 5] and alignment of highways [11, 14], power lines [3, 10], and pipelines [8]. Significantly, however, the findings of this paper imply that the algorithm still needs improvement and its users must be aware that the output corridors are not free of distortion and in particular their costs may be overestimated.

Acknowledgment. This work was partially supported by a grant (No. 942-2015-1513) from the Swedish Research Council Formas. The author would like to thank the three anonymous reviewers for their valuable comments and suggestions, which have improved the presentation of this paper. Still, any errors remain the author's own.

References

1. Adriaensen, F., Chardon, J.P., De Blust, G., Swinnen, E., Villalba, S., Gulinck, H., Matthysen, E.: The application of 'least-cost' modelling as a functional landscape model. Landsc. Urban Plan. **64**(4), 233–247 (2003)
2. Antikainen, H.: Comparison of different strategies for determining raster-based least-cost paths with a minimum amount of distortion. Trans. GIS **17**(1), 96–108 (2013)
3. Bagli, S., Geneletti, D., Orsi, F.: Routeing of power lines through least-cost path analysis and multicriteria evaluation to minimise environmental impacts. Environ. Impact Assess. Rev. **31**, 234–239 (2011)
4. van Bemmelen, J., Quak, W., van Hekken, M., van Oosterom, P.: Vector vs. raster-based algorithms for cross country movement planning. In: Proceedings of International Symposium on Computer-Assisted Cartography (Auto-Carto XI), Minneapolis, Minnesota, pp. 304–317 (1993)
5. Chetkiewicz, C.B., Boyce, M.S.: Use of resource selection functions to identify conservation corridors. J. Appl. Ecol. **46**, 1036–1047 (2009)
6. Dijkstra, E.W.: A note on two problems in connexion with graphs. Numer. Math. **1**, 269–271 (1959)
7. Douglas, D.H.: Least cost path in GIS using an accumulated cost surface and slope lines. Cartographica **31**(3), 37–51 (1994)
8. Feldman, S.C., Pelletier, R.E., Walser, E., Smoot, J.C., Ahl, D.: A prototype for pipeline routing using remotely sensed data and geographic information system analysis. Remote Sens. Environ. **53**, 123–131 (1995)
9. Gonçalves, A.B.: An extension of GIS-based least-cost path modelling to the location of wide paths. Int. J. Geogr. Inf. Sci. **24**(7), 983–996 (2010)
10. Huber, D.L., Church, R.L.: Transmission corridor location modeling. J. Transp. Eng. **111**(2), 114–130 (1985)
11. Lombard, K., Church, R.L.: The gateway shortest path problem: generating alternative routes for a corridor location problem. Geogr. Syst. **1**, 25–45 (1993)
12. Goodchild, M.F.: An evaluation of lattice solutions to the problem of corridor location. Environ. Plan. A **9**, 727–738 (1977)
13. Rothley, K.: Finding and filling the "cracks" in resistance surfaces for least-cost modeling. Ecol. Soc. **10**(1), 4 (2005)
14. Scaparra, M.P., Church, R.L., Medrano, F.A.: Corridor location: the multi-gateway shortest path model. J. Geogr. Syst. **16**(3), 287–309 (2014)

15. Shirabe, T.: A method for finding a least-cost wide path in raster space. Int. J. Geogr. Inf. Sci. **30**(8), 1469–1485 (2016)
16. Tomlin, C.D.: Propagating radial waves of travel cost in a grid. Int. J. Geogr. Inf. Sci. **24**(9), 1391–1413 (2010)
17. Xu, J., Lathrop, R.G.: Improving simulation accuracy of spread phenomena in a raster-based geographic information system. Int. J. Geogr. Inf. Sci. **17**(4), 153–168 (1995)

Model-Based Clustering of Social Vulnerability to Urban Extreme Heat Events

Joseph V. Tuccillo[✉] and Barbara P. Buttenfield

Department of Geography, University of Colorado-Boulder, Boulder, USA
{joseph.tuccillo,babs}@colorado.edu

Abstract. Geodemographic classification methods are applied to Denver Colorado to develop a typology of social vulnerability to heat exposure. Environmental hazards are known to exhibit biophysical variations (e.g., land cover and housing characteristics) and social variations (e.g., demographic and economic adaptations to heat mitigation). Geodemographic model-based classification permits a more extensive set of input variables, with richer attributions; and it can account for spatial context on variable interactions. Additionally, it generates comparative assessments of environmental stress on multiple demographic groups. The paper emphasizes performance of model-based clustering in geodemographic analysis, describing two stages of classification analysis. In so doing, this research examines ways in which high heat exposure intersects with socioecological variation to drive social vulnerability during extreme heat events. The first stage classifies tract-level variables for social and biophysical stressors. Membership probabilities from the initial (baseline) classification are then input to a second classification that integrates the biophysical and social domains within a membership probability space to form a final place typology. Final place categories are compared to three broad land surface temperature (LST) regimes derived from simple clustering of mean daytime and nighttime land surface temperatures. The results point to several broad considerations for heat mitigation planning that are aligned with extant research on urban heat vulnerability. However, the relative coarseness of the classification structure also reveals a need for further investigation of the internal structure of each class, as well as aggregation effects, in future studies.

Keywords: Geodemographic classification · Social vulnerability · Biophysical vulnerability · Urban heat exposure · Gini index

1 Introduction

Extreme heat events are a major cause of summertime mortality and health impacts in urban areas, whose consequences are expected to intensify given the onset of climate change [24]. To aid community and administrative decision support for extreme heat events, it remains critical to develop spatial metrics that identify different ways populations in an urban area may be vulnerable to

© Springer International Publishing Switzerland 2016
J.A. Miller et al. (Eds.): GIScience 2016, LNCS 9927, pp. 114–129, 2016.
DOI: 10.1007/978-3-319-45738-3_8

heat exposure. This study serves to develop place typologies of vulnerability to extreme heat events that can be scalable to a variety of cities and climatic regions prone to heat waves. The primary objective is to demonstrate an advanced clustering method to generate such typologies, using a case study of extreme heat vulnerability at the census tract scale in Denver on June 25, 2012, a day for which record-breaking temperatures upwards of 105 degrees Fahrenheit (40.6 degrees Celsius) were recorded.

Heat vulnerability metrics frequently integrate social, economic, physical and built environment characteristics. The literature on urban heat vulnerability metrics remains largely tied to index-based approaches [11,15,17,22,23,32]. Drawing from "Hazards of Place" models established by Cutter et al. [5] these studies utilize highly dimensional sets of socioeconomic and environmental variables to rank study units in terms of "high" or "low" vulnerability. Due to the reductionist nature of index-based techniques, their application to decision support frameworks may limit understanding of differential forms of socioeconomic and environmental risk occurring in different places. Vulnerability indices may limit abilities to link specific levels of vulnerability to variables of interest. Geodemographic classification techniques present a worthwhile alternative to index based approaches that enable direct targeting of resources and planning interventions to place-specific needs. These techniques' use of cluster analysis rather than weighted aggregation produce profiles of each place category that allow for the comparison of place-specific forms of risk to processes of hazard exposure [25,33].

For this analysis, Geodemographic classification is applied to consider two contexts of vulnerability, or domains: biophysical, including land cover and age, size, and improvements of housing stock; and socioeconomic, including income levels and resources for urban heat mitigation. This study examines the ways in which different populations experience extreme heat events in terms of assets and disparities related to socioeconomic status, physical environments, dwelling, and mobility. Model-based clustering techniques are applied to the data to provide new insight into the intersection of socioeconomic and biophysical domains of populations' vulnerability to extreme heat events. Model-based clustering assigns membership probabilities that quantify an observation's position in multivariate space for a domain of analysis [2,7]. This information identifies which observations, and where, closely correspond to defined place categories, versus those with mixed or unique multivariate characteristics. Domain-specific information linked to membership probabilities is then synthesized to produce a final classification. Being a first foray into this methodological approach, relatively coarse group numbers are examined in both the baseline and final classifications. The use of simpler classification structures provides insights about heat vulnerability, for example to examine information latent in the initial (baseline) classification that becomes apparent in the final typology, as well as assessing the complete model-based method by exploring what information in the variables remains latent at the end of both stages of clustering.

2   Background

Denver, Colorado was chosen as a test case for the analysis framework, due to its publicly available open land-use/land cover data, collected by the authors during a time period that matches the June 2012 heat wave. Denver is central to a major metropolitan region in the Western United States with a population of roughly 660,000 and a semi-arid climate. Days with temperatures in excess of 100° (37.8C) may occur periodically throughout the summertime. Since extreme heat days are uncommon for Denver, certain sub-populations may lack resources to handle sudden shocks of an extreme heat event, as related to household improvements for mitigating indoor heat exposure and daily activity patterns that may elevate health risks. This paper observes June 25, 2012, part of an extended heat event affecting Denver with maximum ambient temperatures observed on this day between 100–105° F (37.8C–40.6C).

2.1   Social Vulnerability and Study Variables

Social vulnerability to environmental hazards is generally discussed relative to limitations in populations' capacities to absorb impacts of environmental stress, given their levels of social and biophysical risk [1,8,27,30,31]. Environmental stressors like extreme heat are dynamic relative to extent, duration, and timing and intersect with spatial variation in social and biophysical risk to produce different place-specific vulnerabilities [1,8,27,30,31]. Place-specific impacts related to extreme heat exposure may include upswings in hospitalizations, heat-related mortality, reduced mobility, and strains on residential water and energy usage. Populations' varying degrees of *adaptive capacity*, reflected by social, financial, material, and natural resources for the absorption of and recovery from hazard impacts, are also considered for this study as a counterpoint to vulnerability [16].

The relationship between exposure of populations to environmental stress and resulting impacts is multifaceted [16,20]. This study considers separate socioeconomic and biophysical contexts, or "domains", of vulnerability to extreme heat exposure. Domains in this analysis consist of tract-level proxy variables for socioeconomic and biophysical stressors (Fig. 2). Variables were chosen from a combination of American Community Survey 5-Year Estimates (2008–2012) and data available in 2013 through the Denver Open Data Catalog. Parcel data consist of a combination of 2013 and 2015 datasets. Variables estimated from the 2015 parcel dataset excluded residential parcels with construction recorded after 2012. Impervious surface cover was derived from NLCD 2011.

Several drivers of vulnerability for the biophysical domain are considered. The first set includes housing stock size and type, age, and improvements. Given high heat exposure, vulnerabilities related to these variables extend to impacts including indoor and outdoor thermal comfort, diminished building mechanical functions, and strains upon affordability of energy and water given their increased usage [9]. Housing stock types include single-family residential detached dwellings, single-unit rowhouses/townhomes and condominiums. Large multi-unit structures include condominium and apartment complexes and senior

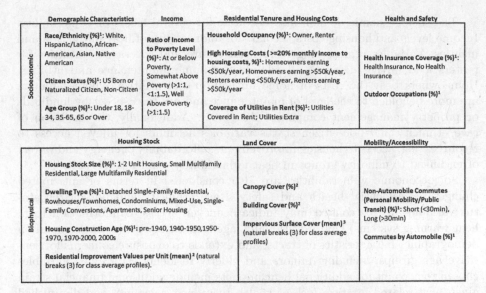

	Demographic Characteristics	Income	Residential Tenure and Housing Costs	Health and Safety
Socioeconomic	Race/Ethnicity (%)[1]: White, Hispanic/Latino, African-American, Asian, Native American Citizen Status (%)[1]: US Born or Naturalized Citizen, Non-Citizen Age Group (%)[1]: Under 18, 18-34, 35-65, 65 or Over	Ratio of Income to Poverty Level (%)[1]: At or Below Poverty, Somewhat Above Poverty (>1:1, <1:1.5), Well Above Poverty (>1:1.5)	Household Occupancy (%)[1]: Owner, Renter High Housing Costs (>=20% monthly income to housing costs, %)[1]: Homeowners earning <$50k/year, Homeowners earning >$50k/year, Renters earning <$50k/year, Renters earning >$50k/year Coverage of Utilities in Rent (%)[1]: Utilities Covered in Rent; Utilities Extra	Health Insurance Coverage (%)[1]: Health Insurance, No Health Insurance Outdoor Occupations (%)[1]

	Housing Stock	Land Cover	Mobility/Accessibility
Biophysical	Housing Stock Size (%)[1]: 1-2 Unit Housing, Small Multifamily Residential, Large Multifamily Residential Dwelling Type (%)[2]: Detached Single-Family Residential, Rowhouses/Townhomes, Condominiums, Mixed-Use, Single-Family Conversions, Apartments, Senior Housing Housing Construction Age (%)[1]: pre-1940, 1940-1950, 1950-1970, 1970-2000, 2000s Residential Improvement Values per Unit (mean)[2] (natural breaks (3) for class average profiles).	Canopy Cover (%)[2] Building Cover (%)[2] Impervious Surface Cover (mean)[3] (natural breaks (3) for class average profiles)	Non-Automobile Commutes (Personal Mobility/Public Transit) (%)[1]: Short (<30min), Long (>30min) Commutes by Automobile (%)[1]

Fig. 1. Study variables by socioeconomic and biophysical domains. Data sources: (1) American Community Survey 5-Year Estimates (2008–2012) (2) Denver Open Data Catalog (3) National Land Cover Database 2011.

housing. This differentiation is necessary since individually-owned housing units (whether owner-occupied or rented) may provide more direct access to upkeep of building structural and mechanical functions. In apartment housing, however, centralized operation and maintenance of utilities may be more common [10]. Greater need for building maintenance and improvements also extends to age of construction [10,24]. The mean value of home improvements per residential parcel for each tract is included to estimate the degree to which housing improvements are in place in addition to ages of housing construction in the tract.

The second set of vulnerability/adaptive capacity drivers considered for the biophysical domain consists of land cover categories including canopy, impervious surface, and building cover. These variables may alternatively contribute to or offset urban heat island effects. Places featuring sparse canopy cover may experience higher vegetative stress in periods of high heat, increasing the degree to which building and impervious cover may retain heat [4,6,10–13,34].

The third set of biophysical vulnerability drivers account for impacts to individual mobility and include diminished means of reaching services, employment, and amenities by regular commute patterns. These effects may be compounded by urban sprawl patterns and lack of a personal vehicle [29]. Chosen variables include personal vehicle ownership, and reliance upon automobiles versus public transit or personal mobility. For residents reliant on public transit commutes greater than 30 min, long wait times or walks to stops or destinations may exacerbate health risks for those with medical conditions.

The socioeconomic domain also accounts for several facets of vulnerability. Income levels and housing costs account for affordability of heat mitigation and utilities. Variables representing residential tenure and coverage of utilities for renters lend further detail to affordability of home maintenance and utilities. Homeowners bear the costs of home mitigation and utilities, whereas renters are more beholden to the level of maintenance and utility costs set by landlords or property management companies [10,18,21]. Additionally, the inclusion of race/ethnicity and non-citizen status variables accounts for uneven access to resources (e.g. access to housing improvements, affordability of energy and water) often linked to minority groups in heat vulnerability studies [10,11,24].

Socioeconomic vulnerabilities are also considered in terms of anticipated changes in residential health and safety. Race/ethnicity and non-citizen status variables in this context may indicate uneven access to information (e.g., heat warning systems) [26]. Additionally, age structure provides background on demographic characteristics of tracts, and extends to exposure concerns for sensitive age groups, including minors and elderly residents [11,23,24]. Variables chosen to account for additional heat impacts include additional financial strain given heat-related injuries (lack of health insurance coverage), and diminished occupational safety (residents with outdoor occupations).

2.2 Exposure Variables: Land Surface Temperatures

Land surface temperatures (henceforth "LSTs") serve as a proxy for exposure to high ambient temperatures during the study period. LST information was overlaid with the place-based vulnerability typology identified for Denver. Comparison of place categories to the ways in which they were exposed allows us to inspect differentiation of social and biophysical characteristics within LST regimes. Though warmer than ambient temperatures, the choice of LSTs is necessitated by the increased cost of monitoring ambient temperatures using temperature loggers and monitoring sites. MODIS 1 km thermal imagery available through USGS Global Visualization Viewer was used to estimate LSTs. Although sparse, this imagery allows for continuous representation of LSTs and precise targeting of dates featuring temperature extremes. Levels of relative humidity on this day were not considered because their influence on human thermal comfort was limited given semi-arid conditions (Fig. 2).

This analysis delineates groups of tracts in Denver by varying forms of social and built environment vulnerabilities present within three land-surface temperature (LST) regimes identified during June 25, 2012, a particularly hot day during a prolonged period of extreme heat. Mean land surface temperatures by tract climbed upwards of 130F (54.4C). Because dry conditions frequently prevail during Denver summers, a high differential often accompanies high temperatures and provides residents some relief in the form of strong nighttime cooling. Mean Nighttime LSTs for Denver cooled by a range of ˜63.6F–79F (17.6C–26.1C). However, the rate and duration of cooling varied considerably across the city from extreme daytime temperatures.

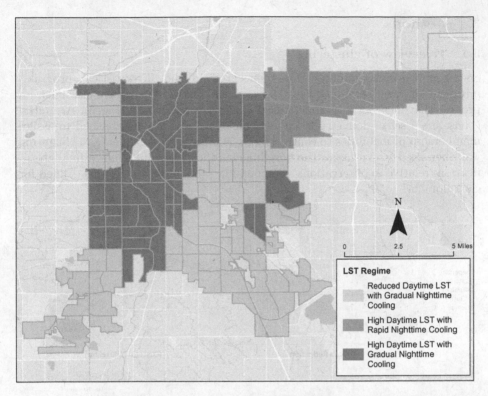

Fig. 2. Three Land Surface Temperature (LST) regimes estimated for June 25, 2012 in Denver from MODIS thermal imagery. Areal units shown on the map are census tracts.

In Fig. 1, these effects are generalized to tracts using three broad land surface temperature regimes derived from simple clustering of mean daytime and nighttime LSTs in ArcGIS. Aligned with urban heat island dynamics, many tracts in central and west Denver with high impervious surface and building cover reached extreme high daytime LSTs (121F–125F) and experienced limited cooling at night, forming the first "Reduced Daytime LST and Limited Nighttime Cooling" LST regime. Tracts closer to the SE and SW edges of Denver, and portions of NW Denver, with higher canopy cover, did not warm as much during the day; but their proximity to the urban heat island led to more limited cooling at night, forming the second "High Daytime LST with Gradual Nighttime Cooling" regime. Finally, tracts in extreme NE portions of Denver with recent residential development interspersed with open prairie, experienced the hottest LST's (125F–130F) during the observation period with strong nighttime cooling to less than ~73F, forming the third LST regime, "High Daytime LST with Strong Nighttime Cooling".

3   Methods

3.1   Overview of clustering

Social vulnerability profiles were generated in the R statistical programming environment (http://www.R-project.org/) using a two-stage classification process that incorporates model-based clustering techniques (Fig. 3). An initial ("baseline") set of domain-specific classifications was first established to assign membership probabilities to census tracts. Hierarchical agglomerative clustering was then used to synthesize the baseline classification results into a final classification, relative to observations' membership probabilities in groups defined for each domain.

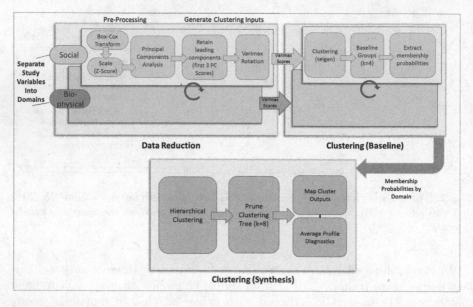

Fig. 3. Workflow showing processing for data reduction and two stages of clustering. Circular arrows indicate steps requiring iteration. Interested readers may contact the authors for additional detail on the data and code used for this analysis.

3.2   Data Reduction

Model-based clustering results are sensitive to high dimensional data, especially when there are a relatively small number of observations (n = 140). Data needs to be reduced into small sets of latent variables representative of composites of the variables of interest.

First, sparsely populated tracts were omitted, including Denver International Airport and the University of Colorado-Denver Auraria Campus. For the remaining 140 observations, a Box-Cox transformation corrected for skew in individual variables. Box-Cox transformed variables were standardized as z-scores) for consistency in value ranges.

Principal Components Analysis (PCA) reduced the dimensionality of data by domain. PCA scores were extracted for each run, retaining the leading three principal component scores. The retention of principal components (PCs) for the social domain is slightly more explanatory than for the built environment domain. For the social domain, the first three PCs explain 62 % of variance in the domain. For the built environment domain, the first three PCs explain 54 % of the variance. Although more variance within each domain may be explained by including more PCs, a tradeoff exists, to preserve most–but not all–of the explanation of variance for the domain while maintaining clarity in the cluster results.

Varimax rotation was then performed to generate composite input variables for cluster analysis that more clearly differentiate among the original variables of interest for each domain.

3.3   Baseline Clustering

Baseline clustering of Varimax scores for each domain was performed in R library "teigen" [2]. The clustering algorithm available in "teigen" employs the expectation-maximization (EM) method for grouping distributions and assumes models with a multivariate t-distribution [2]. Clustering was performed iteratively from $k = 4$ to $k = 9$ groups to optimize the number of groups for each domain's classification. The initial number of $k = 4$ groups was used to roughly approximate the information presented by the Varimax scores representing each domain. Varimax scores suggested continuums of two sets of highly influential variables per domain. For the biophysical domain, clear differentiation was observed between single and multifamily residential housing, as well as between old and new housing construction. Varimax scores for the socioeconomic domain encompassed high and low income levels, and tenure dynamics (renters and homeowners). For each domain, the clustering algorithm found an optimal number of $k = 4$ groups.

3.4   Final Clustering

Baseline cluster solutions were then synthesized into a final classification. Hierarchical agglomerative clustering was applied as a means of grouping observations in probability space by their relative proximity in membership probability space. Hierarchical agglomerative clustering was performed using the Ward.D2 method. The clustering tree was pruned at $k = 8$ groups to account for twice the number of baseline groups for each domain. The choice of $k = 8$ groups permits a unique pairing (overlay) of baseline groups from each domain (1:1 match). The degree to which the intersection of domains departs from a 1:1 match is considered, as well as the occurrence in final classification profiles of high levels of variables not strongly representative of baseline classes.

4   Results

Final classes are described by their positions in "membership space" relative to the baseline classification (Table 1). The goodness of variance fit for the final classification based upon standardized study variables was 0.39.

Table 1. Baseline classification groups and descriptions.

Class	Description
BP1	SFR and new construction
BP2	SFR, old construction, and housing risk
BP3	MFR and new construction
BP4	MFR and mixed housing ages
SES1	Homeowners and high adaptive capacity
SES2	Homeowners and limited adaptive capacity
SES3	Renters, young adults, and high adaptive capacity
SES4	Renter tendency and some adaptive limitations

To describe final classification, prominent variables of interest within each class (relative to global averages) are identified on the basis of even distribution (i.e., likely to occur in relatively consistent levels for every member tract) using a population-weighted Gini index [3,28]. Only variables within a threshold Gini score (0.0–0.3) throughout the class are described as potential forms of vulnerability for the class as a whole. Gini scores indicate the extent to which values of classified variables are evenly spread within each group, with scores of 0.0 indicating even spread, and 0.3 indicating a somewhat even spread. To account for data mismatch with Gini score profiles, residential improved values and impervious surface cover are represented in these results by the same variables binned into high, medium, and low breaks classes. The discussion of final categories below refers to values shown in Figs. 4 and 5.

For this data and case study area, the final classification structure, which used twice the number of baseline classes, primarily resembles an overlay analysis with most place categories represented by clear combinations of two baseline classes. The classification presents profiles of vulnerability on extreme heat days tied to housing stock age, type, size, and value; income levels and housing costs; medical risks (outdoor occupations and health insurance coverage); and sensitive age groups (minors and elderly). To a lesser degree, the results also differentiate land cover and commute types across place categories. Many low-density residential tracts with high daytime LST regimes belong to classes characterized by greater housing risk, more Hispanic/Latino residents, and more limited social and financial means of adaptation than other classes. Class 4 features above-average levels of older SFR, as well as increased levels of residents in poverty, non-citizen status, low household improvement values, and high homeowner costs regardless of

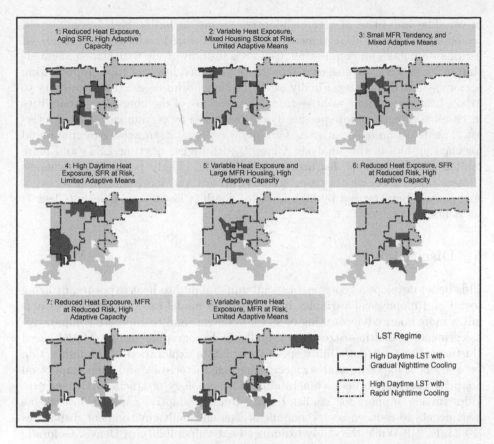

Fig. 4. Figure 4 Geographic distribution of the results of the final classification indicate eight classes. These are overlaid with boundaries of the LST regimes discussed earlier in the paper.

income level. Class 4 also features above-average levels of residents lacking health insurance, as well as those employed in outdoor occupations. Similar disparities in socioeconomic resources are visible for Class 8, despite somewhat more mixed income levels and newer housing stock compared to Class 4. Classification profiles further reveal that Class 2 tracts, also frequently occurring within high daytime LST regimes, are characterized more strongly by vulnerabilities associated with high levels of sensitive age groups (minors and elderly). By contrast, similarly exposed, more densely populated tracts around the CBD (Classes 3 and 5) are characterized by above-average levels of residents with greater adaptive means (higher incomes, newer housing, higher improvement values, inclusion of utilities in rent, and shorter commutes suggestive of more direct access to services and employment).

Conversely, tracts within reduced daytime and nighttime LST regimes (Classes 1, 6, 7) comprise place categories that feature above-average levels of variables indicative of greater adaptive means, including high income levels, increased residential improvement values, more insured residents, and increased use of automobiles for commuting. Class 8 is the only group with strong membership in cooler LST regimes to demonstrate increased levels of those representing socioeconomic disparities. Finally, several unique forms of social vulnerability to urban heat were found to arise from intersections of membership probabilities in baseline categories not specifically characterized by certain variables of interest. The detection of sensitive age groups was perhaps strongest example found for class profiles in this analysis. Class 2, with higher daytime LSTs and more limited adaptive means, features high youth and elderly populations, whereas Classes 6 and 7 — with greater adaptive means and more limited heat exposure — are characterized by high levels of youth (Class 6) and elderly (Class 7) residents.

5   Discussion

This paper develops a typology of social vulnerability to heat exposure, utilizing social and biophysical variables in a two-stage model-based clustering to permit a more nuanced accounting of spatial context, and to generate comparative assessments of multiple stressors on multiple demographic groups. Model-based clustering creates a probability space to facilitate exploration of the interaction between heat exposure and socioecological characteristics and their impact on adaptive capacity to heat vulnerability. The typology of tracts in Denver provides further support for studies linking limited adaptive capacity in extreme heat events to reduced socioeconomic status and built environment disparities [10,11,14,24]. While the study examines heat vulnerability in Denver Colorado for a single extreme heat event, the model-based clustering method demonstrated here can be readily applied to other urban areas using similar types of input data. Model-based clustering composites a number of statistical methods, which considered individually are not in themselves innovative, however the combination and application to the complex problem of heat vulnerability has not been reported elsewhere to the authors' knowledge. It is the integration that contributes to and advances knowledge in GIScience and spatial analysis.

The typology of tracts provides several examples of operational knowledge for decision support related to heat mitigation. First, disparities in exposure and risk characteristics are clearly interpretable for older single-family residential neighborhoods throughout Denver. High LST regimes are more highly comprised of classes with aging housing stock and/or reduced SES (Classes 2, 4). Class 4 features added concerns for heat mitigation in residential areas, particularly increased impervious surface cover. By contrast, older SFR neighborhoods in reduced LST regimes are more typically aligned with Class 1, with increased levels of financial and material resources available to residents for the mitigation of heat impacts. Where overlap between Class 1 and high daytime LST regimes occurs, resources like increased canopy cover may also be more frequently present in residential settings to offset effects of heat exposure. These effects concur with

Fig. 5. Selection of biophysical and socioeconomic input variables representing the most distinguishing and/or most uniform characteristics defining each of eight final classes. Space limitations for the conference paper prohibit including the full sets, which will be shown at the conference presentation. Numbers to the right of each variable name in parentheses show Gini scores within the group. Numbers in the righthand column indicate the "average profile", that is, the percentage difference between a variable's mean in the class compared to the overall mean. A score of 200 for example indicates that the level of that variable in that group is double the overall mean. Each group also shows the percentage of tracts situated in each LST regime.

Harlan et al.'s [10] findings of limited SES and high rentership in highly exposed inner-city neighborhoods, versus increased adaptive means in historic districts near the urban core.

Second, for gentrifying tracts around downtown Denver, evidenced through increased levels of SES and property improvements in Classes 3 and 5, a leading concern for mitigation may be physical risk related to mechanical functionality of cooling systems in multifamily residential structures and residents' indoor thermal comfort [19, 21, 24]. Additionally, surprising contrasts were found among variables occurring in increased for Class 3, particularly increased property improvement values versus residents at or below poverty and pre-1940 housing stock. These findings suggest vulnerabilities related to forms of socioeconomic and built environment risk that occur below the scale of tracts, and point to further community engagement as a potential solution for heat mitigation efforts in these portions of Denver.

Third, suburban portions of Denver represent the strongest align-ment between reduced LST regimes and increased socioeconomic/biophysical resources available to populations for heat mitigation. Yet another surprising contrast common among these groups is the prevalence of above-average levels of high housing costs for low-income renters. These findings again suggest the presence of processes that may occur at sub-tract scales (e.g. isolated blocks whose socioeconomic traits vary from their surroundings). Further, members of Classes 6 and 7 also manifest in peri-urban portions of northeast Denver, char-acterized by new housing development, high daytime LSTs and strong nighttime cooling. Also similar to findings by Harlan et al. [10], then, a contrast may exist in this portion of Denver between populations subject to higher heat exposure but largely retaining increased household-level assets for heat mitigation. How-ever, Class 8, which is also largely present in suburban and peri-urban portions of Denver, provides an interesting contrast to existing findings on urban heat, being characteristically more closely aligned with the increased levels of socioe-conomic disparities present in Class 4. These effects suggest mitigation priorities for heat events that may be more unique to growing semi-arid cities like Denver. Altogether, these types of rich insights demonstrate that clustering on mixed membership enables the definition of relatively unique, intuitive place typolo-gies of vulnerability to heat exposure across Denver, while keeping the number of classes relatively small.

5.1   Limitations and Future Work

Several notable shortcomings are present for this analysis. First, the coarseness of the classification was found to mask some of the more nuanced information critical to support heat mitigation decisions in Denver or a similar city. For example, some variables do not manifest in profiles of the final cluster solution. Variables that do not strongly manifest in any class include African-American and Asian-American residents at risk, Senior Housing, and residents with Long Public Transit commutes. Efforts to account for these effects in future work should consider the selection of more robust baseline classifications by domains, pruning the final hierarchical classification tree at a higher number of classes, and identifying of local multivariate outliers.

Additional shortcomings of this analysis are linked to internal validity of the methods. A major limitation of this analysis is related to uncertainty propa-gated by choice of analysis units and aggregation effects. First, the Modifiable Areal Unit Problem (MAUP) is propagated by bounding of study variables by administrative units (census tracts). It is highly likely that greater complexity of relationships among socioeconomic and environmental variables of interest exists beneath tract boundaries particularly because tracts may not effectively represent sociocultural features. MAUP effects propagate uncertainty in several notable ways in the results. For example, no common housing stock type can be discerned for Class 8, and no clear differentiation can be made for housing stock ages characterizing Class 4. Additionally, for Classes 3 and 5 characterizing downtown Denver, levels of multifamily residential housing well above-average renders it difficult to discern the presence of older housing stock, which appears to be represented less consistently in these classes. Further, for Class 3, it is difficult to discern how some variables of interest, since they often appear at odds (e.g. high residential improved values per unit and residents at or below poverty). Second, MAUP and scale effects also proved problematic for representing LST data used in this study. The design of this analysis bounded a continuous process (variation in temperature) and generalized it to tract boundaries, which may propagate uncertainty in our definition of LST regimes and how they align with different categories of places. This approach may be acceptable for this study, being a "first pass" that relies only upon a qualitative comparison (overlay) of place category characteristics with LST regimes. However, future iterations of this work should more directly account for levels of uncertainty in each tract relative to aggregation effects (inconsistencies between raw and aggregated LSTs).

Still, MAUP effects remain a critical consideration not only for this analysis, but for the wider development of vulnerability metrics, to consider problems related to grouping of social and environmental variables using administrative boundaries. Similarly, the effects of spatial autocorrelation among sets of variables of interest, as well as potential effects of spatial non-stationarity in variables' alignment, merit further consideration for future work.

Despite its limitations, the model-based clustering approach to Geodemographic classification overall provides a worthwhile direction for identifying patterns in social vulnerability to urban extreme heat events. In addition to methodological improvements, future applications of this framework should extend to comparison of cluster results outcome measures, particularly public health variables including morbidity and mortality. Pairing this approach with measures of disparities in access to services, public facilities, and employment centers also presents promise. Future studies should expand these methods to urban areas with differing of experiences of extreme heat events (e.g. high heat and relative humidity for New York City and Chicago, extreme high temperatures and greatly limited nighttime cooling for Phoenix).

References

1. Adger, W.N.: Vulnerability. Glob. Environ. Change **16**(3), 268–281 (2006)
2. Andrews, J.L., McNicholas, P.D.: tEIGEN: model-based clustering and classification with the multivariate t-distribution, 2015, r package version 2
3. Brown, M.C.: Using Gini-style indices to evaluate the spatial patterns of health practitioners: theoretical considerations and an application based on alberta data. Soc. Sci. Med. **38**(9), 1243–1256 (1994)
4. Buyantuyev, A., Wu, J.: Urban heat islands and landscape heterogeneity: linking spatiotemporal variations in surface temperatures to land-cover and socioeconomic patterns. Landscape Ecol. **25**(1), 17–33 (2010)

5. Cutter, S.L., Boruff, B.J., Shirley, W.L.: Social vulnerability to environmental hazards*. Soc. Sci. Q. **84**(2), 242–261 (2003)
6. Declet-Barreto, J., Brazel, A.J., Martin, C.A., Chow, W.T., Harlan, S.L.: Creating the park cool Island in an inner-city neighborhood: heat mitigation strategy for Phoenix, AZ. Urban Ecosyst. **16**(3), 617–635 (2013)
7. Fraley, C., Raftery, A.E.: Model-based clustering, discriminant analysis, and density estimation. J. Am. Stat. Assoc. **97**(458), 611–631 (2002)
8. Gallopín, G.C.: Linkages between vulnerability, resilience, and adaptive capacity. Glob. Environ. Change **16**(3), 293–303 (2006)
9. Guhathakurta, S., Gober, P.: The impact of the Phoenix urban heat Island on residential water use. J. Am. Plan. Assoc. **73**(3), 317–329 (2007)
10. Harlan, S.L., Brazel, A.J., Prashad, L., Stefanov, W.L., Larsen, L.: Neighborhood microclimates and vulnerability to heat stress. Soc. Sci. Med. **63**(11), 2847–2863 (2006)
11. Harlan, S.L., Declet-Barreto, J.H., Stefanov, W.L., Petitti, D.B.: Neighborhood effects on heat deaths: social and environmental predictors of vulnerability in Maricopa County, Arizona. Environ. Health Perspect. (Online) **121**(2), 197 (2013)
12. Jenerette, G.D., Harlan, S.L., Brazel, A., Jones, N., Larsen, L., Stefanov, W.L.: Regional relationships between surface temperature, vegetation, and human settlement in a rapidly urbanizing ecosystem. Landscape Ecol. **22**(3), 353–365 (2007)
13. Jenerette, G.D., Harlan, S.L., Stefanov, W.L., Martin, C.A.: Ecosystem services and urban heat riskscape moderation: water, green spaces, and social inequality in Phoenix, USA. Ecol. Appl. **21**(7), 2637–2651 (2011)
14. Jesdale, B.M., Morello-Frosch, R., Cushing, L.: The racial/ethnic distribution of heat risk-related land cover in relation to residential segregation. Environ. Health Perspect. (Online) **121**(7), 811 (2013)
15. Johnson, D.P., Stanforth, A., Lulla, V., Luber, G.: Developing an applied extreme heat vulnerability index utilizing socioeconomic and environmental data. Appl. Geogr. **35**(1), 23–31 (2012)
16. Lam, N.S., Reams, M., Li, K., Li, C., Mata, L.P.: Measuring community resilience to coastal hazards along the Northern Gulf of Mexico. Nat. Hazards Rev. **17**(1), 1–12 (2015). Article ID 04015013
17. Maier, G., Grundstein, A., Jang, W., Li, C., Naeher, L.P., Shepherd, M.: Assessing the performance of a vulnerability index during oppressive heat across Georgia, United States. Weather, Clim. Soc. **6**(2), 253–263 (2014)
18. Morrow, B.H.: Identifying and mapping community vulnerability. Disasters **23**(1), 1–18 (1999)
19. Nguyen, J.L., Schwartz, J., Dockery, D.W.: The relationship between indoor and outdoor temperature, apparent temperature, relative humidity, and absolute humidity. Indoor Air **24**(1), 103–112 (2014)
20. Polsky, C., Neff, R., Yarnal, B.: Building comparable global change vulnerability assessments: the vulnerability scoping diagram. Glob. Environ. Change **17**(3), 472–485 (2007)
21. Quinn, A., Tamerius, J.D., Perzanowski, M., Jacobson, J.S., Goldstein, I., Acosta, L., Shaman, J.: Predicting indoor heat exposure risk during extreme heat events. Sci. Total Environ. **490**, 686–693 (2014)
22. Reid, C.E., Mann, J.K., Alfasso, R., English, P.B., King, G.C., Lincoln, R.A., Margolis, H.G., Rubado, D.J., Sabato, J.E., West, N.L., et al.: Evaluation of a heat vulnerability index on abnormally hot days: an environmental public health tracking study. Environ. Health Perspect. **120**(5), 715 (2012)

23. Reid, C.E., O'Neill, M.S., Gronlund, C.J., Brines, S.J., Brown, D.G., Diez-Roux, A.V., Schwartz, J.: Mapping community determinants of heat vulnerability. Environ. Health Perspect. **117**(11), 1730 (2009)
24. Rosenthal, J.K., Kinney, P.L., Metzger, K.B.: Intra-urban vulnerability to heat-related mortality in New York City, 1997–2006. Health Place **30**, 45–60 (2014)
25. Rufat, S.: Spectroscopy of urban vulnerability. Ann. Assoc. Am. Geogr. **103**(3), 505–525 (2013)
26. Shiu-Thornton, S., Balabis, J., Senturia, K., Tamayo, A., Oberle, M.: Disaster preparedness for limited english proficient communities: medical interpreters as cultural brokers and gatekeepers. Publ. Health Rep. **122**(4), 466–471 (2007)
27. Smit, B., Wandel, J.: Adaptation, adaptive capacity and vulnerability. Glob. Environ. Change **16**(3), 282–292 (2006)
28. Spielman, S.E., Singleton, A.: Studying neighborhoods using uncertain data from the American community survey: a contextual approach. Ann. Assoc. Am. Geogr. **105**(5), 1003–1025 (2015)
29. Stone, B., Hess, J.J., Frumkin, H., et al.: Urban form and extreme heat events: are sprawling cities more vulnerable to climate change than compact cities. Environ. Health Perspect. **118**(10), 1425–1428 (2010)
30. Turner, B.L., Kasperson, R.E., Matson, P.A., McCarthy, J.J., Corell, R.W., Christensen, L., Eckley, N., Kasperson, J.X., Luers, A., Martello, M.L., et al.: A framework for vulnerability analysis in sustainability science. Proc. Natl. Acad. Sci. **100**(14), 8074–8079 (2003)
31. Wisner, B., Blaikie, P., Cannon, T., Davis, I.: At Risk: Natural Hazards, People's Vulnerability and Disasters. Routledge, Abingdon-on-Thames (2004)
32. Wolf, T., McGregor, G.: The development of a heat wave vulnerability index for London, United Kingdom. Weather Clim. Extremes **1**, 59–68 (2013)
33. Wood, N.J., Jones, J., Spielman, S., Schmidtlein, M.C.: Community clusters of Tsunami vulnerability in the US Pacific Northwest. Proc. Nat. Acad. Sci. **112**(17), 5354–5359 (2015)
34. Xu, Y., Dadvand, P., Barrera-Gómez, J., Sartini, C., Marí-Dell'Olmo, M., Borrell, C., Medina-Ramón, M., Sunyer, J., Basagaña, X.: Differences on the effect of heat waves on mortality by sociodemographic and urban landscape characteristics. J. Epidemiol. Commun. Health **67**(6), 519–525 (2013)

Representing the Spatial Extent of Places Based on Flickr Photos with a Representativeness-Weighted Kernel Density Estimation

Jiaoli Chen and Shih-Lung Shaw[(✉)]

Department of Geography, University of Tennessee, Knoxville, TN 37996, USA
{jchen42, sshaw}@utk.edu

Abstract. Geotagged photos have been applied by many researchers to explore the spatial extent of places. This paper addresses an important challenge of using geotagged Flickr photos to delineate the spatial extent of a vague place, which is defined as a place without a clearly defined boundary. We argue that the variation of location popularity has a great impact on the estimation of such vague spatial extent of a place. We propose an approach to model the representativeness of each geotagged photo point based on its location popularity. A modified kernel density estimation method incorporating the photo representativeness is developed and tested with eight places, which cover urban vs. non-urban areas, with vs. without an official boundary cases, and at various spatial scales of state, city and district levels. Our results indicate major improvements of the proposed representativeness-weighted kernel density estimation method over the traditional kernel density estimation method in estimating the spatial extent of vague places.

Keywords: Place · Geotagged photos · Flickr · Kernel density estimation

1 Introduction

Naïve geography in [1] envisions that the advanced geographic information systems (GIS) should "follow human intuition" (p. 1) and "support common-sense reasoning" (p. 5) so that ordinary people who do not need to know about GIS can use them easily. It is important for such GIS to understand and represent linguistic place names. Some places (e.g., administrative divisions) have a formally defined geographic extent to be represented in GIS, while many places (e.g., vernacular places) have no formally defined boundary but a vague geographic extent. Therefore, an effective and efficient representation of vague spatial extent of places is critical to GIS representation, query, analysis, and visualization of places.

Acquiring human knowledge of places is a traditional way to derive vague place extents. With the increasing popularity of geotagged social media (e.g., Flickr, Twitter, and Facebook), large numbers of place names exist in the contents from such platforms, carrying valuable information about people's perception of places. Among these social media, Flickr provides adequate and more direct associations between photo

© Springer International Publishing Switzerland 2016
J.A. Miller et al. (Eds.): GIScience 2016, LNCS 9927, pp. 130–144, 2016.
DOI: 10.1007/978-3-319-45738-3_9

geolocations and place name tags [2]. Thus, extracting the spatial extent of places from Flickr data[1] has been a GIS research topic (e.g., [3–6]), especially for purposes of enriching large-scale gazetteers and GIS services (e.g., [7, 8]). In the meantime, such crowd-sourced data also present challenges to place-related research, including to which degree the spatial extent of a place can be estimated from geotagged photos or other crowd-sourced social media data.

Past research has indicated the effectiveness of using Flickr data to identify major locations of vague places (e.g., [4, 9]). However, estimation of vague geographic extents still presents a major challenge. As suggested by Jones et al. [10], we argue that the underlying assumption of a random distribution of Flicker photos is incorrect when applying geotagged photos to estimate vague place extents. In general, there are fewer photos taken at locations with low accessibility and low popularity than those popular and easily accessible locations. The conventional approach of treating each geotagged photo with equal representativeness or importance, regardless of where it is located, is questionable. The representativeness of photo points located in unpopular areas could be under-weighted due to a low absolute number of photos taken in such areas, while the representativeness of photo points located in popular areas would be over-weighted due to a larger number of photos taken in these areas. As a result, unpopular locations (e.g., inaccessible parts of mountain areas) of a vague place (e.g., Rocky Mountains) would be significantly underestimated or even excluded in the derived geographic extent. On the other hand, popular locations could be overestimated and distort the boundary of a place. When the kernel density estimation (KDE) method is applied to delineate vague boundaries (e.g., [5]), the resulting surface usually looks like a hot spot map of the photos tagged with a target place name rather than a "probability field" (p. 205) [13] representing the place extent where a higher estimate indicates a higher probability of belonging to a target place. Thus, this study aims at improving the representation of vague place extents by adjusting photo point representativeness based on their location popularity. Note that we adopt the same term "target place" (p. 1047) from [10] to refer to a place whose extent needs to be estimated.

In the remainder of this paper, we start with a review of work related to the concept of place, georeferencing place names, and delineation of vague place extent from survey, web and social media data. In the methodology part, we discuss a proposed approach based on photo point representativeness and the representativeness-weighted KDE (RW-KDE) method. Next, we present the results of testing our proposed assumptions and method based on eight selected sample places, followed by comparisons with the results derived from the traditional KDE approach. We conclude this paper with contributions, limitations and future research directions.

2   Related Work

As an important concept in geography, place has been extensively studied, implying more than space by incorporating social, economic, cultural and political meanings through human experience [14–16]. Places are usually revealed in unstructured forms.

---

[1] Data source: https://www.flickr.com/services/api/, also the data source of this paper.

Their informality is also reflected by the variations in people's understanding of them [12]. Thus, they are often absent from current GIS in which geographic objects need to be disambiguated, abstracted and digitalized from the perspective of space. Great efforts have been made toward the convergence of place and GIS, such as theoretically modeling place in a computational environment (e.g., [17, 18]) and delineating the spatial extent of place names which is the focus of this paper.

There have been some place-friendly web applications using simple gazetteers to handle place names [10]. But retrieval of vague place extent is still beyond the ability of current gazetteers. Usually given an authority-recognized place name, a gazetteer provides information about its geographic location, feature type, and relationship to other places [19]. But it provides limited information on vernacular names. In most gazetteers, the location of a place with a large spatial extent is usually represented as a point, rectangle bounding box, or occasionally a polygon with crisp boundary [12]. Although an abstract and simple geometry benefits computational process in current information systems, it falls short in delivering information about the inherent vagueness of place boundary which often reflects human perception and cognition. Therefore, much research has focused on deriving vague place extents from a variety of data sources.

Montello et al. [20] conducted an empirical study in which human subjects were asked to draw shapes of the downtown Santa Barbara area based on their understanding of the place extent. The downtown shapes drawn by different participants were then aggregated to generate a probabilistic representation of the vague place extent. Montello et al. [21] interviewed another two groups of participants to unveil and measure the variation of vagueness of a place perceived by different people and at different locations. These survey-based approaches have an advantage that data structure and collection procedure can be designed to facilitate subsequent modeling and estimation of vague place extents as well as to answer specific research questions. However, the difficulty in collecting such empirical data obstructs their wide applications.

Some other research derived a place's extent using its topological relations to other clearly defined geographic objects. Based on a set of points covering a region, Parker and Downs [22] combined DBSCAN clustering technique and fuzzy set theory for the delineation of a vague extent. For another example, based on two sets of geographic points that fall inside and outside a place, Alani et al. [23] created a Voronoi diagram of these points and delineated the place extent from the Voronoi polygons. Their method only generated a crisp boundary. Taking advantage of a diversity of spatial information, Schockaert et al. [24] proposed a unique approach which could derive constraints from qualitative and quantitative spatial data and approximate a vague extent based on the derived constraints using techniques of genetic algorithm and ant colony optimization. All these approaches require that the geographic objects used as references are available and their geometries are already defined. They do not focus on how and where to collect these references and their spatial relations to the target place. For a large number of vernacular places, the practicality of these approaches depends on the availability of well-defined reference data.

Since a lot of place-related data can now be found from the web (e.g., web articles and documents, Internet yellow pages), much research used search engines to acquire references (e.g., hotels, cities) that are related to (e.g., inside, outside, covering, containing same words as) the target place name (e.g., [11, 25, 26]). After the geolocations of returned references were determined by geoparsing and georeferencing, the vague extents were then estimated from the reference locations using KDE approach [10, 12], fuzzy-set approach [25], or adapted α-shape and recoloring algorithms [26].

As online social media became popular, geotagged photos (often from Flickr) were widely used in recent research for place extent assessment and digital gazetteer enrichment. This is because, unlike other web-sourced data, they do not require the geoparsing and geocoding processes which could introduce unexpected errors [2, 3]. Li and Goodchild [5] applied KDE to Flickr geotagged photos, and found that the highest-density cells normally tell the major location of a place. They also pointed out a limitation of the data source being lack of sampling strategy. Martins [2] improved the KDE surface by removing overestimated locations based on land coverage information, assuming that a place boundary is usually related to a land cover change. But this study did not address underestimated locations. Instead of the local density perspective in traditional KDE, Grothe and Schaab [3] took a global perspective to generate the crisp boundary of places using a support vector machine (SVM) classification technique. Cunha and Martins [6] improved the SVM method by incorporating place semantics from Flickr photo tags, demographic characteristic from population dataset, and topographical characteristic from elevation and land cover datasets, considering that a place's boundary can be found along the line where these characteristics change. The boundary they generated was also crisp and did not handle vague place extent. There is limited research in the literature that both delineates the vagueness of place extents and deals with the variation of location popularity. Given this challenge, this study focuses on improving estimation of vague place extents based on geotagged Flickr photos.

KDE is a widely adopted method for estimating geographic extent in various domains such as animal home range [27]. It is relatively easy to use, fits well with data such as geotagged photos, and thus is frequently adopted by researchers. KDE generates a density surface through interpolation from the geographic points covering a place to reflect the inherent vagueness of place extent [10]. Unlike fuzzy-set methods, KDE avoids using a subjective fuzzy membership function to represent vagueness. In the following sections, a modified KDE approach along with its assumptions are discussed and it is evaluated with data of eight selected case studies.

3   Methodology

3.1   Data Acquisition and Preprocessing

In order to assess if the performance of our proposed method works for different place types and at various feature scales, we selected the following eight places as case studies: *Manhattan Chinatown* and *San Francisco Chinatown* that do not have an official boundary at the urban district scale; City of *Nashville* and City of *Philadelphia* with an official boundary at the urban city scale; *Rocky Mountain National Park* and *Great Smoky Mountains National Park* as non-urban features with an official boundary; State of *California* and State of *Utah* with an official boundary at the state scale. These places are denoted as the *target place*s.

Around each target place, a larger rectangle study area (about eight times larger in size) was defined and used to search for the data of all geotagged photos through the Flickr search API. Note that we downloaded the data of all geotagged photos, not just those tagged with the target place name. In the reminder of this paper, we use the term *target photos* to refer to the data of photos tagged with the target place name or its variants, to distinguish them from the data of the whole set of photos denoted as *all photos*. The time span used in our searches varies among the eight places. For Manhattan Chinatown and San Francisco Chinatown, we searched for photos between January 2013 and February 2015 since these two places did not have an official boundary and their geographic extents could change over time. For the validation purpose, we extended Liu's [30] approach of using the Street View imagery of Google Maps (https://www.google.com/maps) to manually delineate the boundary lines that separate locations with typical Chinese characteristics from those without Chinese themes. The derived boundary lines served as the benchmark reference to be compared with the geographic extents estimated by our proposed method. Most of the Chinatown Street View images were captured after 2013, which should be comparable with the time period of Flickr photos used in this study. For California, we downloaded Flickr photos posted in January and July of 2014 as our sample data to test if partial Flickr data could also give acceptable estimates. Similarly, for Utah, only photos between September and December of 2014 were used. The time span chosen for the other four selected case-study places was between February 2004 and February 2015 since they all had an official boundary that were relatively stable over time.

As suggested by Liu [30], redundant photos that were uploaded in bulk by the same user at the same location were removed to minimize the bias, with only one randomly chosen photo kept; place name variants (e.g., California abbreviated as CA; Nashville misspelled as Nashvile; Manhattan Chinatown shorten to Chinatown when the photo is posted on the Manhattan Island) were found by browsing through tags that occurred more than three times in the set of all photos. We then selected a set of target photos that were tagged with the place name or any acceptable variant from the set of all photos. Finally, based on the geolocations of target photos and all photos, two sets of geographic points for each place name were created respectively: *target points* and *all points*.

3.2   Representativeness of Geotagged Photos

The target points are those points assumed to fall inside the target place and used in many published work to generate the KDE surface. Here we denote the set of target points by T, and the set of all points by A. Their relation can be depicted by $T \subseteq A$. As discussed in the introduction section, unpopular locations tend to be underestimated, while popular locations tend to be overestimated. Intuitively, the contribution of a target point at a low popularity location should be increased to compensate for the disadvantage of photo availability at that location. Flickr data provide an opportunity to measure a location's popularity in terms of the photo volume. This makes it possible to model a target point's contribution level, namely representativeness, of the place in a study. Therefore, we assume that the location popularity is indicated by the volume of all points at a given location.

For implementation, we first discretize the earth surface in each study area into a regular grid with a cell size of $n \times n$, where n is the width of a cell. Each grid cell represents a location l. The popularity p_l of location l can be quantified by the count of all points falling in that cell. As stated above, when a location's popularity decreases, the representativeness of a target point at that location should increase. We assume that the representativeness r_t of target point t that falls in the cell of location k is inversely proportional to location k's popularity p_k. That is,

$$r_t = \frac{1}{p_k} \tag{1}$$

This function creates the value of representativeness falling within (0, 1]. The choice of grid cell size n is an important step. It can be observed from the set of all points in Fig. 1 that the sparsest points located in the most unpopular locations are usually isolated and distant from their nearest neighbor. The most isolated target point that has the largest nearest neighbor distance can be found and denoted as the *sparsest target point* which can reflect the most unpopular location near target points. Note that the nearest neighbor of a target point here is based on the set of all points. In order to maximize the possibility that the sparsest target point has the highest representativeness value 1 and to avoid assigning too many other target points with the value 1, we define n as the sparsest target point's nearest neighbor distance d divided by $\sqrt{2}$, which is:

$$d = max_{t \in T}(min_{a \in A-\{t\}} dist(t,a)); n = d/\sqrt{2} \tag{2}$$

where t represents a point in the set of target points T, and a represents a point in the set of all points A.

3.3 Outlier Removal

As shown in Fig. 1, the target points of a place with a continuous extent usually form an obvious cluster at that place with a few outliers surrounding it. There are several characteristics of the outliers that can help us remove them: (1) they lie remotely from the major cluster of target points; (2) they are thinly scattered around the major cluster;

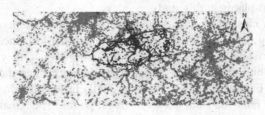

Fig. 1. Spatial distributions of target points (red) and all points (red and green) of the Great Smoky Mountains National Park. The official boundary is shown in black solid line. (Color figure online)

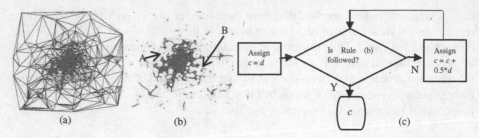

Fig. 2. DTC with the target points (red dots) of Nashville: (a) edges connecting neighbors constructed by Delaunay Triangulation; (b) resulting major cluster with its convex hull (in dash line). (c) Flow chart of the search procedure for cut-off distance c. (Color figure online)

and (3) their total count is relatively small. So our goal is to find the major cluster and separate those thinly scattered points that are distant from it.

The Delaunay Triangulation Clustering (DTC) can meet our goal of identifying the major cluster in the target points. According to [28], the basic idea is to first construct neighbors for each target point based on Delaunay Triangulation and use edges (black solid lines in Fig. 2(a)) to connect neighboring target points. For each pair of neighboring target points, if they are not close (i.e., their distance is beyond certain threshold, namely *cut-off distance c*), they should not be in the same cluster, and their connecting edge is removed [28]. Finally, sets of target points that are connected by remained edges (black solid lines in Fig. 2(b)) form clusters [28]. To choose a reasonable cut-off distance c for breaking edges, we assume the following two rules and the search procedure in Fig. 2(c).

Rule (a): c must be no less than the sparsest target point's nearest neighbor distance d in Eq. (2). This is because if c is less than d, the sparsest target point that is located in the most unpopular location will definitely be disconnected with any of its neighbors in target points, which can result in its isolation. This rule takes the location unpopularity issue into account when deciding the cut-off distance.

Rule (b): the largest cluster, denoted as the *major cluster*, in the clustering result under the cut-off distance c should contain more than 95 % of the target points. This is because the distribution pattern of target points shows that the majority of target points form a cluster at the target place with only a few outliers surrounding it. The major cluster should consist of the majority of target points, and here we assume the number of them to be more than 95 % of the target points. Empirically, this assumption works well across all eight places in this study.

When the major cluster is found, there are several isolated points and small clusters (e.g., A and B in Fig. 2(b)) that are quite near the major cluster. Since they do not have the outlier characteristic of being distant from the major cluster, they are not treated as outliers. A convex hull of the major cluster is created to capture these isolated points and small clusters. Finally, all other points or small clusters that do not intersect with the convex hull are removed from the set of target points. The remaining target points after the outlier removal step are denoted as the *cleaned points*.

3.4   Representativeness-Weighted KDE (RW-KDE)

We take the density surface approach to generate a raster surface representing vague place extents. The cleaned points provide limited observations about the location of a place. To measure the degree of other locations' inclusion in a place, interpolations can be made from the limited observations using the KDE. At each location to be estimated, the kernel estimator sums up the kernels centered at the cleaned points whose contributions decrease as distances increase [29]. The traditional KDE method considers all observations with equal importance. We incorporate photo point representativeness into the KDE defined by Silverman [29] using the following kernel estimator:

$$\hat{f}(X) = \frac{1}{h^2 \sum_{i=1}^{n} r_i} \sum_{i=1}^{n} r_i * K(\frac{X - X_i}{h}) \tag{3}$$

where n is the count of cleaned points; X_i is the coordinates of a cleaned point; X is the coordinates of a raster cell whose inclusion needs to be estimated; h is the smoothing bandwidth; r_i is the representativeness of cleaned point i calculated by Eq. (1); and K is the quadratic kernel function cited from Silverman [29] (p. 76):

$$K(X) = \begin{cases} 3\pi^{-1}(1 - X^T X)^2 & \text{if } X^T X < 1 \\ 0 & \text{otherwise} \end{cases} \tag{4}$$

For each set of cleaned points of a place, both KDE and RW-KDE are implemented using the same parameters. This ensures that their results are comparable. In order to both maintain a good resolution with details on the final raster surface and minimize the computational cost associated with high resolution, we choose about 1/300 of the width of a place study area to be the cell size of output surface. It is well-known that the smoothing bandwidth could have a significant impact on the output. For each place, we repeated KDE and RW-KDE using different bandwidths to evaluate the sensitivity of the improvements by RW-KDE method. We tested eleven bandwidths that are about 1/8, 3/16, 1/4, 5/16, 3/8, 7/16, 1/2, 9/16, 5/8, 11/16 and 3/4 of the width of the minimum bounding rectangle of the cleaned points. The reason for choosing this range is that almost all surfaces tend to be over smoothed when bandwidth approaches 3/4, and that for most places in this study, bandwidths smaller than 1/8 are too small for many locations within a place to find any cleaned point in its neighborhood when calculating kernel density estimate. Since the improvements can be observed across the eleven bandwidths, here we only present results based on three bandwidths: 1/4, 3/8, and 1/2.

4   Results

In this section, we use the official boundary and the surveyed boundary as references to compare the results derived from the RW-KDE method vs. the KDE method. All official boundaries come from the 2015 TIGER/Line® Shapefiles provided by the U.S.Census

Bureau[2] and the park maps on the website of U.S. National Park Service[3]. Chinatowns are usually recognized and known for its distinct Chinese characteristics in building styles, signs, decorations, and pedestrians. These characteristics change notably between adjacent streets that are respectively inside and outside Chinatown. Thus, it is reasonable to manually draw a relatively objective boundary separating Chinatown from the surrounding areas, based on the Street View images from Google Maps.

All eight places were estimated by the steps described in Sect. 3. Figure 3 shows the surfaces of vague extent generated by KDE vs. RW-KDE methods with the three selected bandwidths. All estimated values on the density surfaces are normalized to [0, 1] and linearly stretched for grey shades of [0, 255]. In the introduction section, we argue that location popularity could vary greatly across different areas of a place. To better know how strong the variation is, we take California as an example to estimate three popularity density surfaces (Fig. 4) based on all photo points in the study area using the KDE, each of which corresponds to one of the three estimated KDE surfaces and RW-KDE surfaces of vague extent in Fig. 3 under the same estimating parameters.

The popularity density surfaces show the advantage of San Francisco and Los Angeles areas which are two of the largest cities in the U.S. These surfaces are very similar to the estimated KDE surfaces of vague spatial extent in Fig. 3. To know if a location's membership of California estimated from the traditional KDE method is correlated to the popularity density of that location, we plot these two variables for each location within the official boundary and calculate a Pearson's r correlation coefficient. The results indicate a very strong correlation which supports our early argument that location popularity may impact the estimates by the traditional KDE method. Take the popularity density surface with the bandwidth of 755,000 feet (Fig. 4(c)) as an example, more than 65 % of the state area have a popularity density below 0.2 in a scale of [0,1], and only 4 % have a popularity density above 0.8. Among the low popularity-density locations (below 0.2), all of them have a membership value below 0.263 on the traditional KDE surface; 75 % have a value under 0.13; 50 % have a value under 0.075; and 25 % have a value under 0.033. This means that the majority of locations within the official boundary are much unpopular than San Francisco and Los Angeles areas, and they are estimated by the traditional KDE approach to have a much lower possibility of being in California than the San Francisco and Los Angeles areas which have an estimate above 0.42. This is a significant deviation from the ground truth. A qualitative observation of the large dark area inside the official boundary tells the same story.

In contrast, by incorporating photo representativeness, the estimated values of RW-KDE surface are no longer correlated to a location's popularity density (see blue dots in Fig. 4). Also among the same low popularity-density locations (below 0.2) within the official boundary, 75 % of them have an estimated value above 0.4 on the RW- KDE surface; 50 % have a value above 0.56; and 25 % have a value above 0.73. The RW- KDE has greatly increased the estimated values at unpopular locations, better representing their membership of California. These improvements support our argument that we need to treat each photo tagged with a place name differently according to its

---

[2] https://www.census.gov/cgi-bin/geo/shapefiles/index.php.

[3] https://www.nps.gov/.

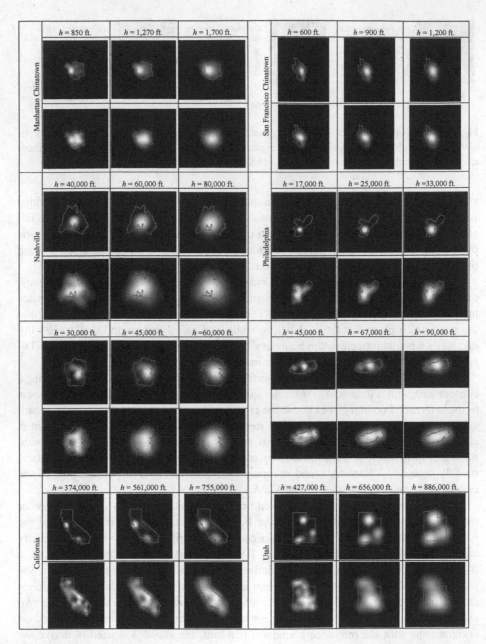

Fig. 3. Vague spatial extents represented by KDE (top row of each place) vs. RW-KDE surfaces (bottom row) under different bandwidths (*h*). Reference boundaries are in red line. (Color figure online)

location popularity; otherwise, less popular locations that obviously belong to a vague place would be underestimated in the derived geographic extent. Similar issues and improvements can be found in downtown vs. other areas of Philadelphia and Nashville,

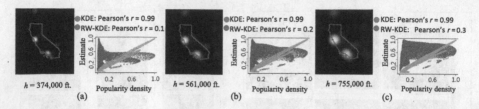

Fig. 4. In each pair of (a) to (c), (left) a popularity density surface based on all Flickr photos using KDE; (right) a scatter plot of the estimates from the resulting KDE and RW-KDE surfaces against the popularity densities within the official boundary of California.

low vs. high accessibility areas in national parks, and across different bandwidths. Although there exists linear distributions of photo points in national parks, RW-KDE can take advantages of the limited number of sparse points that are away from the roads by assigning them a higher weight. It better represents the less popular areas that are often underestimated by the KDE method.

Moreover, based on qualitative observations of Fig. 3, in the cases of Chinatowns, national parks, and Nashville, RW-KDE surfaces tend to produce the highest estimates near the center of reference boundaries. This is consistent with the common sense that the center of a place has the highest probability of membership, and that probability decreases when approaching the boundary line. However, the cores of traditional KDE surfaces tend to be distorted toward the most popular locations.

Both the KDE and the RW-KDE surfaces can produce a crisp boundary using a contour line with a threshold. To further quantify their performance, namely closeness to the reference boundary, we choose the commonly used measures of accuracy, recall, and precision. As defined in [24], recall measures how much of the reference extent A can be covered by the estimated extent A', namely area(A ∩ A')/area(A); precision measures how much of the estimated extent correctly falls in the reference boundary, namely area(A ∩ A')/area(A'); and accuracy measures the overall performance of the estimates, namely area(A ∩ A')/area(A ∪ A').

We use the rank of raster cells based on their density values in descending order rather than the value itself to derive the boundaries from the KDE surface and the RW-KDE surface, respectively. This is because the distributions of normalized density values from the KDE surface and the RW-KDE surface are quite different. If a threshold of density value is used to derive boundaries from these two surfaces, the returned boundaries could be very different in size. Since the size of derived boundaries can influence their comparisons with the reference boundary, we need to keep the two estimated boundaries comparable in size when examining their performance based on comparisons with the reference boundary. If a rank threshold β is used to derive the boundaries, the returned two sets of raster cells from the KDE surface and the RW-KDE surface will form two crisp boundaries with equal size but different shapes.

We calculate the recall, precision and accuracy measures for the boundaries generated with various rank thresholds from the KDE and the RW-KDE surfaces under the selected three bandwidths and plot them in Fig. 5. These plots also include additional

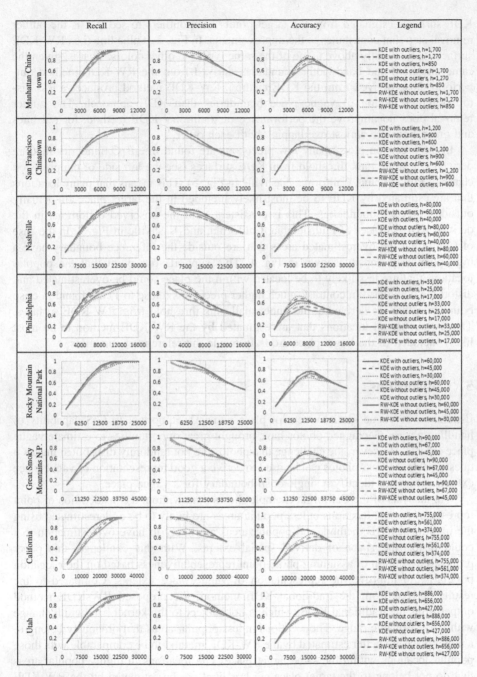

Fig. 5. Recall, precision, and accuracy of the boundaries derived from the KDE and RW-KDE surfaces. The x axis represents the rank threshold β used to derive crisp boundary. The y axis in the second, third, and fourth columns represent the recall, precision, and accuracy measures, respectively. (Color figure online)

KDE surfaces estimated from the target points with outliers to see if it is the outlier removal step that contributes to the improvements. In Fig. 5, a blue line (i.e., RW-KDE surface) is frequently above a line of the same type and other colors. That is, given the same threshold and bandwidth, boundaries derived from the RW-KDE method outperform those derived from the KDE method. RW-KDE method produces a less distorted representation of vague place boundaries than KDE approach. The outlier removal does not improve the performance of traditional KDE method, for red lines and green lines are almost overlapping with each other.

In the precision plots, many low precisions are found in crisp boundaries that are derived at a small rank threshold from KDE surfaces (see the left side of the precision charts of Philadelphia, California, Nashville in Fig. 5). This means that many locations that are estimated by KDE to be most likely included in the target place are actually outside reference boundaries. This is consistent with the observed distortions in the derived KDE surfaces of Philadelphia, California, and Nashville in Fig. 3. In the case of California, the overestimated areas that are outside the reference boundary have a high popularity density shown in Fig. 4. This supports our point that the popular locations that are less likely to be located within the place could be overestimated and distort the boundary. In contrast, much higher precisions (close to 1) are found in crisp boundaries that are derived at the same small rank threshold from RW-KDE surfaces. That is, almost all locations that are estimated by the RW-KDE method to be most likely included in the target place are truly inside the reference boundaries.

5   Conclusions

Estimation of vague spatial extent of place names is important to GIS capability of handling places. Geotagged photos have brought great opportunities to estimate vague place extents. However, the challenge that photos are not randomly distributed makes it questionable if Flickr photos can be used to derive a good representation of vague place extent, not just to identify a single crisp boundary for a vague place.

Our analysis of the California example shows that, without a consideration of point representativeness, locations in less popular areas are likely to be underestimated using the traditional KDE method, and popular areas are likely to have overestimation that could distort the shape of derived spatial extents. With this challenge, this paper proposes a solution of assigning photo point representativeness based on their location popularity to improve the representation of vague place extents. Compared to the results derived from the traditional KDE method, the proposed RW-KDE method outperforms the traditional KDE, which is not subject to the kernel bandwidth change within a reasonable range. The locations in less popular areas that obviously belong to a target place are better estimated by the RW-KDE method to be comparable with those popular locations within the same target place. The locations in highly popular areas that do not belong to the target place are less likely to be estimated by the RW-KDE method to be part of the target place. The RW-KDE method also derives crisp boundaries with higher recall, precision, and accuracy measures, which quantitatively suggest a less distorted representation of place boundaries.

The major contribution of this paper is two folds: First, it addresses and proposes a solution for the aforementioned important challenge that has been widely recognized in the literature but not fully explored. The proposed method has been tested with eight places and produced better representation of vague place extents. Second, the improvements show that it is feasible to use geotagged Flickr photos to construct a good representation of vague place extents where a higher estimate indicates a higher probability of belonging to the target place.

Good matches between the estimated vague extents and the reference boundaries indicate that Flickr users' perception of the eight place extents is close to the reference boundaries. However, as suggested by [6], there could be some places whose derived vague extents are different from their official boundaries. More places will be examined in the future to find out useful patterns about people's perception of place, such as in what situations a place's derived extent is quite different from the official boundary.

The method proposed in this paper may not be suitable for places having multiple parts. This is because the outlier removal process assumes that the photos tagged with a target place name usually form one major cluster at that place. For places containing disjoint parts or places sharing the same place name, additional considerations are needed in future research to detect the disjoint parts of a place. Then the proposed method in this study can be applied to generate a vague extent for each individual part.

References

1. Egenhofer, M.J., Mark, D.M.: Naïve geography. In: Frank, A.U., Kuhn, W. (eds.) Spatial Information Theory A Theoretical Basis for GIS. LNCS, vol. 988, pp. 1–15. Springer, Berlin, Heidelberg (1995)
2. Martins, B.: Delimiting imprecise regions with georeferenced photos and land coverage data. In: Kim, K.-S. (ed.) W2GIS 2011. LNCS, vol. 6574, pp. 219–229. Springer, Heidelberg (2011)
3. Grothe, C., Schaab, J.: Automated footprint generation from geotags with kernel density estimation and support vector machines. Spat. Cogn. Comput. **9**, 195–211 (2009)
4. Hollenstein, L., Purves, R.: Exploring place through user-generated content: using Flickr tags to describe city cores. JOSIS **1**, 21–48 (2010)
5. Li, L., Goodchild, M.F.: Constructing places from spatial footprints. In: Proceedings of the 1st ACM SIGSPATIAL International Workshop on Crowdsourced and Volunteered Geographic Information, pp. 15–21. ACM (2012)
6. Cunha, E., Martins, B.: Using one-class classifiers and multiple kernel learning for defining imprecise geographic regions. IJGIS **28**, 2220–2241 (2014)
7. Keßler, C., Janowicz, K., Bishr, M.: An agenda for the next generation gazetteer: geographic information contribution and retrieval. In: Proceedings of the 17th ACM SIGSPATIAL International Conference on Advances in Geographic Information Systems, pp. 91–100. ACM (2009)
8. Gao, S., Li, L., Li, W., Janowicz, K., Zhang, Y.: Constructing gazetteers from volunteered big geo-data based on Hadoop. Comput. Environ. Urban. Syst. (2014, in press)
9. Keßler, C., Maué, P., Heuer, J.T., Bartoschek, T.: Bottom-up gazetteers: learning from the implicit semantics of geotags. In: Janowicz, K., Raubal, M., Levashkin, S. (eds.) GeoS 2009. LNCS, vol. 5892, pp. 83–102. Springer, Heidelberg (2009)

10. Jones, C.B., Purves, R.S., Clough, P.D., Joho, H.: Modelling vague places with knowledge from the web. IJGIS **22**, 1045–1065 (2008)
11. Purves, R., Clough, P., Joho, H.: Identifying imprecise regions for geographic information retrieval using the web. In: Proceedings of the GIS Research UK 13th Annual Conference, pp. 313–318. University of Glasgow, Glasgow (2005)
12. Twaroch, F.A., Jones, C.B., Abdelmoty, A.I.: Acquisition of vernacular place names from web sources. In: King, I., Baeza-Yates, R. (eds.) Weaving Services and People on the World Wide Web, pp. 195–214. Springer, Heidelberg (2009)
13. Goodchild, M.F., Montello, D.R., Fohl, P., Gottsegen, J.: Fuzzy spatial queries in digital spatial data libraries. In: Proceedings of the IEEE World Congress on Computational Intelligence, pp. 205–210. IEEE (1998)
14. Relph, E.: Place and Placelessness. Pion, London (1976)
15. Tuan, Y.-F.: Space and Place: The Perspective of Experience. University of Minnesota Press, Minneapolis (1977)
16. Agnew, J.: Space and Place. In: Agnew, J., Livingstone, D. (eds.) The SAGE Handbook of Geographical Knowledge, pp. 316–330. SAGE, Thousand Oaks (2011)
17. Cohn, A., Gotts, N.: The 'Egg-Yolk' representation of regions with indeterminate boundaries. In: Burrough, P., Frank, A. (eds.) Geographic Objects with Indeterminate Boundaries, pp. 171–187. Taylor and Francis, Bristol (1996)
18. Dilo, A., De By, R.A., Stein, A.: A system of types and operators for handling vague spatial objects. IJGIS **21**, 397–426 (2007)
19. Goodchild, M.F., Hill, L.L.: Introduction to digital gazetteer research. IJGIS **22**, 1039–1044 (2008)
20. Montello, D.R., Goodchild, M.F., Gottsegen, J., Fohl, P.: Where's Downtown?: behavioral methods for determining referents of vague spatial queries. Spat. Cogn. Comput. **3**, 185–204 (2003)
21. Montello, D.R., Friedman, A., Phillips, D.W.: Vague cognitive regions in geography and geographic information science. IJGIS **28**, 1802–1820 (2014)
22. Parker, J.K., Downs, J.A.: Footprint generation using fuzzy-neighborhood clustering. Geoinformatica **17**, 285–299 (2013)
23. Alani, H., Jones, C.B., Tudhope, D.: Voronoi-based region approximation for geographical information retrieval with gazetteers. IJGIS **15**, 287–306 (2001)
24. Schockaert, S., Smart, P.D., Twaroch, F.A.: Generating approximate region boundaries from heterogeneous spatial information: an evolutionary approach. Inf. Sci. **181**, 257–283 (2011)
25. Schockaert, S., De Cock, M., Kerre, E.E.: Automatic acquisition of fuzzy footprints. In: Meersman, R., Tari, Z. (eds.) OTM-WS 2005. LNCS, vol. 3762, pp. 1077–1086. Springer, Heidelberg (2005)
26. Arampatzis, A., Van Kreveld, M., Reinbacher, I., Jones, C.B., Vaid, S., Clough, P., Joho, H., Sanderson, M.: Web-based delineation of imprecise regions. Comput. Environ. Urban Syst. **30**, 436–459 (2006)
27. Downs, J.A., Horner, M.W.: Analysing infrequently sampled animal tracking data by incorporating generalized movement trajectories with kernel density estimation. Comput. Environ. Urban Syst. **36**, 302–310 (2012)
28. Eldershaw, C., Hegland, M.: Cluster analysis using triangulation. In: Computational Techniques and Applications, CTAC 1997, pp. 201–208. World Scientific (1997)
29. Silverman, B.W.: Density Estimation for Statistics and Data Analysis. Chapman and Hall, London (1986)
30. Liu, Y.: A study of colloquial place names through geotagged social media data. Master thesis. University of Tennessee (2014)

Scaling Behavior of Human Mobility Distributions

Tuhin Paul[1]($\boxtimes$), Kevin Stanley[1], Nathaniel Osgood[1,2], Scott Bell[3], and Nazeem Muhajarine[2]

[1] Department of Computer Science, University of Saskatchewan, Saskatoon, SK, Canada
{tuhin.paul,kevin.stanley,nathaniel.osgood}@usask.ca
[2] Department of Community Health and Epidemiology, University of Saskatchewan, Saskatoon, SK, Canada
nazeem.muhajarine@usask.ca
[3] Department of Geography and Planning, University of Saskatchewan, Saskatoon, SK, Canada
scott.bell@usask.ca
http://www.usask.ca

Abstract. Recent technical advances have made high-fidelity tracking of populations possible. However, these datasets, such as GPS traces, can be comprised of millions of records, well beyond what even a skilled analyst can digest. To facilitate human analysis, these records are often expressed as aggregate distributions capturing behaviors of interest. While these aggregate distributions can provide substantial insight, the spatio-temporal resolution at which they are captured can impact the shape of the resulting distribution. We present an analysis of five spatial datasets, and codify the impact of rebinning the data at different spatio-temporal resolutions. We find that all aggregate metrics considered are affected by rebinning, but that some distributions do so regularly and predictably, while others do not. This work provides important insight into which metrics can be used to compare human behavior across datasets and the kinds of relationships between that can be expected.

Keywords: Spatial data · Mobility · GPS · Analytics

1 Introduction

Human spatial behavior underlies many disciplines, including geography, sociology, architecture, and many forms of engineering and research effort has been invested attempting to describe how people move through and use space. Through studies conducted with pen and paper through diaries, surveys, or ethnographies, researchers have made significant strides in codifying how people move through and utilize space. With the advent of mobile communications and location sensing technology, vast new repositories of spatio-temporal information on human mobility have become available. Voronoi diagram-based spatial

© Springer International Publishing Switzerland 2016
J.A. Miller et al. (Eds.): GIScience 2016, LNCS 9927, pp. 145–159, 2016.
DOI: 10.1007/978-3-319-45738-3_10

decompositions based on cell tower or WiFi router access logs, trajectory data from GPS logging, or interaction level data from RFID and Bluetooth (BT) beacons all provide previously unprecedented representations of a person's spatial trajectories [1,9,10,12,21,22,26,29,31]. However, all of these data sources have different characteristics: cell record and WiFi data are characterized by irregular spatial distributions contingent on inter-device spacing and only generate records when people connect, GPS logs are only reliable outdoors, and BT and RFID devices provide reliable measures of proximity but only in controlled settings. Even reliable measurements via GPS have variable accuracy depending on the device, atmospheric conditions, and built environment.

To cope with the large amounts of data generated by these new measurement techniques, researchers often employ aggregate metrics, which can be characterized as distributions over a single variable such as trip length, to help describe the data. The model parameters (e.g., mean and variance of a Gaussian) corresponding to these distributions can be used to describe the data concisely. For example, many human-centric statistics, such as visit frequency or interpersonal contact duration, are characterized by truncated power law distributions [15]. The power coefficient describing that distribution can inform an analyst about the relative behavior of two populations. However, because changing the spatial extent over which these data are collected can change the shape of the distribution, studies of the same populations at different spatio-temporal scales will be described by different model parameters, and by extension may generate erroneous conclusions. This hearkens back to the Modifiable Areal Unit Problem, a recurring challenge when working with data and variables that can be aggregated to different units of analysis [19,20]. Understanding to what extent these distributions are susceptible to the spatial and temporal resolution of collection, and to what extent these sizing and sampling effects are predictable based on underlying mathematical processes, would help human behavioral researchers make meaningful comparisons across datasets and between populations.

Employing five mobility datasets, recorded from either smartphone GPS or GPS logging devices, we analyzed sampling effects. To model spatial binning, an area of interest was binned into square sections of varying sizes. To model temporal granularity, we down-sampled the mobility traces at regular intervals. This selective and regular resampling allows us to examine the impact of spatio-temporal resolution on the resulting aggregate distributions. We find that some distributions have definitive scaling behaviors, indicating the possibility of meaningfully comparing datasets across resolutions. Other metrics do not vary as regularly under resampling, indicating that caution should be exercised when comparing results from different data sources using these techniques.

2   Related Literature

Human mobility is not random, but follows well defined patterns [15,22,24,25], sometimes characterized by aggregate statistics like: (1) Inter-contact time, (2) visit frequencies, (3) dwell time, (4) radius of gyration, (5) trip length, and

(6) trip duration [15, 24]. Because mobility is continuous in space and time, quantization (binning) is often applied [21, 23, 24]. While the transmission range of a GSM (Global System for Mobile Communications) base station is normally up to 35 km [13], Bluetooth and WLAN (Wireless Local Area Network) transmission ranges are limited to tens of meters to a few hundred meters [30]. A study found position errors of 2 m–15 m, on average, using GPS [17].

The examination of different units of analysis in geography has a long tradition [19, 20]. Persuasive arguments for considering these effects in GIScience are also well documented [8], including in work on the convergence of GIScience and Social Media [28]. Bell et al. examined similar sized units of different types (census vs neighbourhoods) and found similar patterns [4].

Eagle et al. used mobile phones to collect location history and behavioral data of people from multiple sources [6]. For better granularity of spatial data indoors, and social context, locations of surrounding Wi-Fi access points have been used by [9, 10, 27]. The resultant datasets provide valuable insight into the interwoven patterns in human movements. Research on intertwined patterns in human mobility lead the development of synthetic human mobility models, which emulate observed patterns in human mobility traces [3, 11, 15, 22].

Data driven geographic inquiry or algorithmic geographies [14], are increasingly important to understanding human behavior, complex systems, and our environment-behavior interactions [16]. Urban geographers, demographers, and behavioral geographers are using open, big, and real-time (or streamed) data in new ways. This includes health [18], networks and transportation [27], and behavior modelling [2]. In GIScience and its cognate geographic disciplines, the application of grid cells and varying spatial resolutions has primarily been in remote sensing and elevation modelling [5, 7].

3   Experimental Setup

We used five data sets: the Saskatchewan Human Ethology Datasets (SHED) 1, 2, and 5 [9, 10, 27], the open source dataset GeoLife [32], and GPS traces from the 'Seasonality and Active Saskatoon Kids' dataset (hereafter, the 'Kids' dataset) [18]. The SHED datasets are technical pilots for the ongoing development of iEpi [12], and contain detailed mobility, activity, and contact traces from graduate students and staff (SHED1 and SHED2) or undergraduate students (SHED5). GeoLife is an open source collection of mobility traces collected using GPS loggers by Microsoft Research [32] in China. The GPS traces in GeoLife correspond to self-identified trips taken by participants, and do not include stationary periods. The Kids study [18] used GPS loggers and wearable accelerometers to study a large number of elementary students from low income neighbourhoods over a week, to determine activity and mobility patterns.

Software glitches, hardware failure, and participant non-compliance lead to significant variance within the number of available records in each of the databases. Individual participants returned anywhere from negligible fractions, to almost complete records of possible data, but only a portion of the total number of records included GPS data (e.g., while at school or university, SHED or Kids participants might report accelerometer but not GPS records due to poor GPS reception indoors). Participants were classified into two groups, responders (at least 20 % of possible time slots or samples with GPS data over the data collection period) and non-responders, for all but Geolife, where compliance was difficult to assess because data corresponds to participant-identified trips. The threshold of 20 % was chosen arbitrarily based on inspection of trajectories. Participants whose GPS records were available for less than 20 % of the possible time slots were removed. GeoLife data were sampled at 1–5 s intervals [32], and participants were included in the analysis if they had recorded trips spanning at least two weeks. The number of participants and records before and after filtering are presented in Table 1.

Table 1. Dataset properties

	SHED1	SHED2	SHED5	Kids	GeoLife
Study duration	4 weeks	4 weeks	4 weeks	1 week	5+ years
#(participants)	38	37	29	745	182
#(used participants)	34	27	24	722	33
#(GPS records)	1.35e6	3.41e7	279,298	1.54e8	2.5e7
#(used records)	107,409	101,746	80,998	1.42e8	1.86e7

To determine the impact of the temporal sampling rate, we down-sampled the data (expressed by T), between subsequent measurements. A down-sampling period (T) is an integer multiple of the base period (T_0) at which GPS data are collected. Down-sampling at T is performed by taking every $(\frac{T}{T_0})^{th}$ sample from the base data. Because each dataset has a different minimum sampling time (between 1 s and 8 min), we standardized the minimum sampling duration to be 8 min for SHED5, and 10 min for others. For SHED5, $T \in \{8\min \times (1, 5, 10, 15, 30, 60)\}$ and for others, $T \in \{10\min \times (1, 3, 6, 12, 24, 48)\}$. The fastest sampling rate was chosen for consistency with SHED5, which had the slowest base rate; the slowest sampling rate was chosen to be 3 times per day, consistent with the minimum number of daily cellphone records required in [26]. This downsampled sequence was then sampled spatially using a regular square grid. If no location record existed at the downsampled timestep, then a special symbol for "unknown location" was used for the location of that participant at that timestep. These special symbols were ignored when creating the aggregate distributions, but they broke trips during trip length and duration calculations. Both choices were intentionally conservative; we assign no location if the location is unknown, and do not assume that a trip continues if data during a trip is missing. This will tend to make trips shorter, as potentially longer trips may be broken into a number of shorter sub-trips.

Location was binned with a maximum granularity of 4 km, consistent with a suburban cell tower area, with that granularity successively reduced by factors of 2 to a minimum of 15.625 m, consistent with the nominal accuracy of commodity GPS receivers common in typical smartphones. The spatial resolution is reported as the length of the square bins or grid cells, and given the symbol d. The coverage area of a dataset was gridded at the coarsest resolution (4 km edged squares), and increasingly finer resolution cells were created by subdividing these larger cells into 4, halving the edge dimension while conserving the topology of the spatial binning, until the finest resolution of 15.625 m was reached. Locations were taken to be the centers of the grid cell in subsequent calculations. Over short time scales and at fine resolutions, there was strong agreement between the recorded position and the binned locations; as temporal and spatial scales expanded, agreement between computed location and measured location began to diverge, as expected. Intra-step shifts in time or space (e.g., changing the start time or base grid locations) was not investigated, but would also be expected to have an impact, particularly at coarse spatial or temporal scales.

We computed five previously employed aggregate metrics [15, 24] for each dataset at each spatio-temporal resolution: visit frequency, dwell time, trip length, trip duration, and radius of gyration (RoG). All empirical distributions are aggregated across locations and participants through time.

Visit Frequency: The distribution of the count of participant samples in a given location. Remaining in a cell increases the count for that cell. This metric indicates overall place popularity.

Dwell Time: The distribution of the number of time steps participants spent in a cell without changing cells. This metric distinguishes between places visited often, for short duration, versus those visited occasionally for longer.

Trip Length: The distribution of contiguous trips, where a trip is defined as changing locations for at least three consecutive downsampled time steps. Distance is calculated as the Euclidean distance, which is an integer multiple of d, between cell centers for each stage of the trip. The trip length distribution specifies the probability of traveling a certain distance.

Trip Duration: The distribution of time spent in a trip (as defined above), with a resolution of the current sampling period. Trip duration describes how long participants are likely to remain in transit.

Radius of Gyration: This metric, represented as r_g, is defined in (1), where c is the center of the polygon bound by spatial resolution-dependent coordinates $\{r_i : i \in \mathcal{N}^+ \wedge i \leq N\}$ of trip samples. The RoG distribution describes how compact the areas traversed by participants are. We computed c as the centroid of the convex hull of the polygon defined by trip samples.

$$r_g = \sqrt{\frac{1}{N} \sum_i^N (r_i - c)^2} . \qquad (1)$$

Given the distributions of the above metrics at chosen spatio-temporal resolutions, we used regression for power law based fits of the distributions because the metrics have been reported to follow truncated power law distributions [15,24]. Under the power law model, each distribution has two parameters, a constant term and an exponent term, encoded as α and k, as shown in (2).

$$f(x) = \alpha x^k \ [x \geq x_0] \tag{2}$$

After determining the model parameters α and k of (2) from the distributions of each of the five metrics at different spatio-temporal resolutions, we determined how these model parameters varied with d and T using the following models on the basis of R^2-based goodness of fit:

Linear: $f(x) = c_1 + c_2 x$
Logarithmic: $f(x) = c_1 + c_2 \log x$
Exponential: $f(x) = c_1 c_2^x$
Power: $f(x) = \alpha x^k$

Data were stored as text files. Initial data exploration was done using Eureqa[1] from Nutonian, Inc. Our final fits were done using R statistical software[2] with R^2 as the goodness of fit metric. Calculations were carried out on a computer with four Core AMD processor and 8 GB memory running Ubuntu 15.10.

4 Results

As we are primarily interested in determining how aggregate distributions of mobility change under different spatio-temporal measurement regimes, we have plotted the distributions of aggregate metrics. Figures 1 and 2 show the variation of our key metrics at spatial dimensions of 31.25 m and 500 m, respectively. Each curve within each graph denotes a particular (dataset, sampling time) pair.

Several trends are notable within each graph. First, most curves show the characteristic forms for power law distributions, which is consistent with the literature [15,24]. All curves (with the exception of RoG) are characterized by linear descent on the log-log plots over large portions of their span, indicating heavy tailed power distributions.

Second, not all datasets are equal. The Kids dataset is characterized by longer dwell times than the other datasets. This is likely indicative of the relative difference between elementary school students' and university students' lifestyles. The GeoLife dataset, comprised exclusively of trips, has a lower dwell time, and higher visit frequency and RoG, which is as expected for participants who are always on the move.

Third, the RoG measure is noisy with respect to sampling regime. Given that the formulation for RoG implicitly depends on the sampling regime, this makes sense, as altering capture resolution alters the parameters of RoG. As a result, we conclude that RoG is a poor measure for inter-experiment meta analysis,

---

[1] http://www.nutonian.com/products/eureqa-server/.
[2] https://www.r-project.org/.

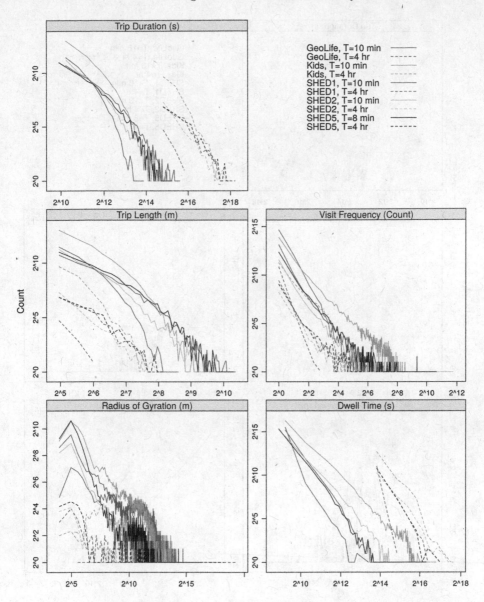

Fig. 1. Distribution of dataset features at $d = 31.25\,\mathrm{m}$

as significant variation in computing values will be expected due to the data capture resolution.

Fourth, dwell time, trip length, and trip duration are well characterized by power law distributions, as characterized by Fig. 3, where each box plot represents the distribution of R^2 values when fitting a power law to curves aggregated over participants, as seen in Figs. 1 and 2, for each d and T pair considered in

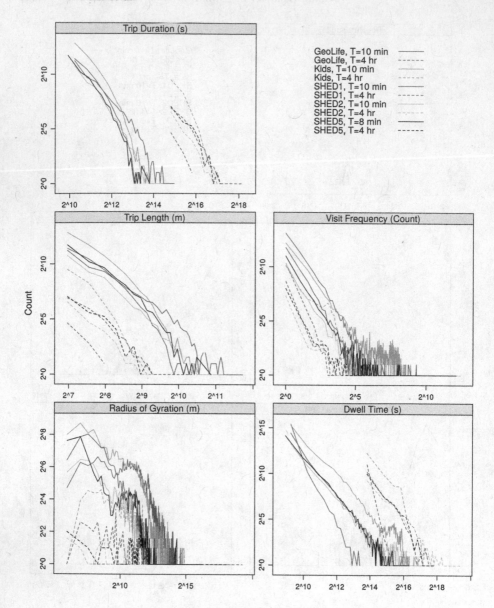

Fig. 2. Distribution of dataset features at $d = 500\,\mathrm{m}$

the expertiment. As expected from the noisy signal, RoG is poorly characterized by a power law. Visit frequency does not appear to be strongly power law
distributed, particularly near the tails. The noisy tails also make visit frequency
susceptible to changing fit quality with spatio-temporal resolution.

Fifth, there is apparent regularity in much of the variation in both Figs. 1 and 2, implying underlying mathematical relationships. To determine the regularity of effect, we further fit curves to the model parameters derived from the regression for power law distributions fits, to determine if the coefficients of the fit equations also vary regularly with resolution. That is, we wished to determine if the model parameters could be expressed as functions of the resolution.

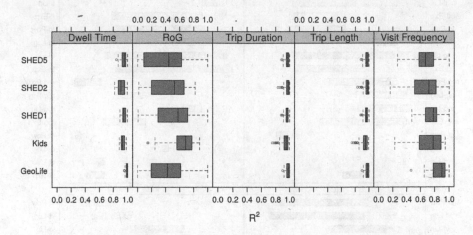

Fig. 3. R^2-based quality of power law fits of distributions of dataset features

Given the model parameters α and k, derived from power law based regressions of the distribution of key metrics, we tried to relate them to d and T. Figure 4 presents the R^2-based fit qualities, aggregated over all datasets, of exponential, linear, logarithmic, and power law regression models to establish relationships between model parameters (α and k) and spatio-temporal resolutions (d and T). Overall, power models explain the behavior of α and k with d and T best, exhibiting the largest mean R^2 values and smallest variances. However, values of k showed significant variance, and trip length and RoG had generally poor fits for all models tested.

Figure 5 presents the R^2 fit quality values of regression fits of power law model parameters (α and k), as d or T broken down by dataset. Each value in the boxplot is represented by a single spatio-temporal resolution (value of d and T), aggregated over all participants for a single dataset. Much of the variance in these fits can be ascribed to the power law only describing a region of variation, as would be expected from Figs. 1 and 2, where, for example, visit frequency becomes quite noisy with large T, or there is limited variation among datasets for trip duration at small T. Visit frequency and dwell time seem to have the strongest power dependence on both d and T. It is interesting to note, that while visit frequency did not consistently hue to a power law distribution, the variation of the model parameters did vary regularly. Trip length model parameters vary somewhat regularly with d, but are inconsistent across datasets with T.

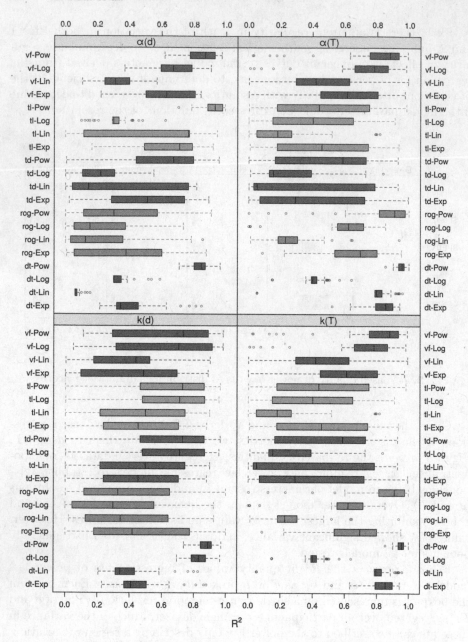

Fig. 4. Goodness of fit of $\alpha(d)$, $\alpha(T)$, $k(d)$, $k(T)$, for key metrics over all datasets, to exponential (Exp), linear (Lin), logarithmic (Log), and power law (Pow) models

RoG and trip duration do not exhibit strong fits. With RoG, this is expected, given the noisiness of the original signal, but with trip duration this is more likely due to the changing definition of a trip, as changing d and T changes possible trip lengths.

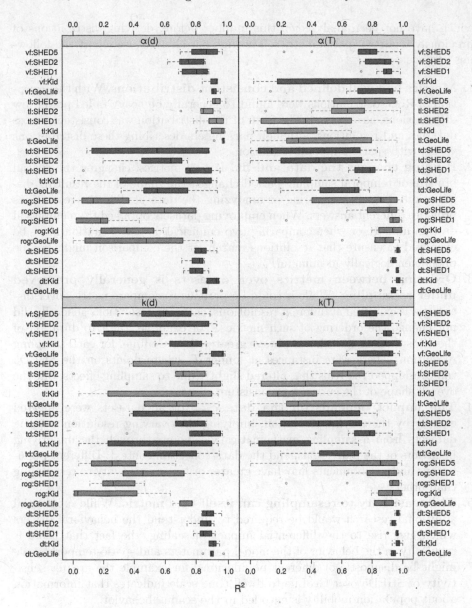

Fig. 5. Power function-based fit quality dependence of a and k on d and T

Discussion

Understanding human mobility and its measures is increasingly important for many fields. In this paper, we sought to examine the impact on aggregate metrics of spatial scale and temporal sampling period. We analyzed five spatial datasets, which have not been analyzed in this manner before, deriving distributions of previously reported spatial metrics. Through our analysis, we report the following findings:

1. **Metrics had well defined and consistent distributions.** With the exception of RoG, distributions were found to generally be heavy tailed power law distributions, as expected. The form of the distribution was consistent across datasets and resolutions, although parameters describing these distributions varied with spatio-temporal resolution.

2. **Binning changes the data and fit.** For all metrics, changing the spatial bin size or temporal sampling period changed the shape of the resulting distribution. That is, measuring or analyzing the data at different resolutions provides different answers. When employing datasets obtained from empirical data in models, or when comparing two empirical datasets, caution must be exercised to ensure that resolutions match, or the comparison might not be phenomenologically meaningful.

3. **Ordering between metrics over datasets is generally preserved under resampling.** While it could be perilous to compare metrics over distributions captured at different resolutions, changing resolutions generally did not change the ordering of such metrics. For example, the trip duration of the Kids dataset was almost always greater than GeoLife, for each sampling resolution. There were instances at longer T, where points on the SHED5 tail overlapped the Kids that altered slightly due to sampling effects, but the overall shape of the curves was consistent.

4. **The impact depends on the dataset.** Not all datasets were affected equally by the varying resolution, implying that varying resolution impacts datasets from a sampling mathematics viewpoint, through the underlying behaviors of the individuals, and the data collection context. Different populations and environments may have greater or lesser sensitivity to resampling than others.

5. **The sensitivity to resampling can itself be a metric.** While substantial additional research would be required to understand the behavioral drivers which give rise to the differential impact of scaling, the fact that there is regularity in the behavior of the model parameters and spatio-temporal scale might be diagnostic of different populations, for example the greater sensitivity of SHED5 over GeoLife to dwell time scale indicates that information about population mobility is encoded in the scaling behavior.

These findings have implications for how mobility data should be employed in research and practice. Finding 1 validates work from other researchers with new data [15,24]. Finding 2 cautions modelers and researchers employing this data. Because the distributions do not generalize across resolution, data from an empirical study conducted at one resolution cannot, with certainty, provide the underlying distributions for models with a different underlying spatial resolution.

Finding 3 indicates that derived metrics such as mobility entropy, which exhibit resolution sensitivity [21,23] may derive their resolution dependence from the variation described here. Finding 3 suggests that resampling within datasets will not compromise conclusions of an ordinal nature. Finding 4 indicates that the scaling effects are not entirely due to the mathematics of sampling: human behavior patterns in the data also contribute. Finding 5 hypothesizes that resampling behavior itself could be used as a metric of human mobility. These scaling metrics could also be used to evaluate agent-based models of human mobility used in simulation. Synthetic mobility models such as [11,15,22] should not only reproduce the distributions of key metrics at a given resolution, but the scaling behaviours noted here. Taken together, these findings provide a meaningful contribution to the study of human mobility metrics.

While we have made a significant contribution to the literature, several shortcomings of this study could be addressed in future work. First, while we used five datasets comprising millions of records, these datasets had a relatively small number of participants and durations measured in weeks. Further analysis of larger, longer duration, and more diverse datasets would help validate the work. Second, we employed GPS datasets, downsampled regularly in time and space. While this approach facilitated the analysis, location data sources such as WiFi and cell tower records have irregular shaped cells based on the Voronoi diagram of transmitter locations, and stochastic sampling patterns based on connectivity behavior. Understanding how irregularity in spatial and temporal sampling impacted these distributions would also be worthwhile. Finally, we have not attempted to employ these insights into building better models of human behavior. Further research into the application of these findings to building higher fidelity models of human behavior for simulation systems could have wide ranging impacts.

5   Conclusion

Spatio-temporal resolution changes the shape and model parameters of aggregate distributions used to describe human mobility. This variation appears to conserve, at least in ordering, the differences between datasets, implying that indications of the differences in human behavior being observed are also preserved. Because spatio-temporal resolution matters, making quantitative comparisons between datasets with different resolutions is potentially dangerous and should be avoided, at least until regularities in the scaling relationships can be better characterized. While significant research remains, this work represents an initial step in understanding how to properly employ newly available high-fidelity datasets in human mobility analysis.

Acknowledgments. We would like to acknowledge the Natural Sciences and Engineering Research Council of Canada for providing funding for this work.

References

1. Ahas, R., Aasa, A., Silm, S., Aunap, R., Kalle, H., Mark, Ü.: Mobile positioning in space-time behaviour studies: social positioning method experiments in Estonia. Cartography Geogr. Inf. Sci. **34**(4), 259–273 (2007)
2. Arribas-Bel, D.: Accidental, open and everywhere: emerging data sources for the understanding of cities. Appl. Geogr. **49**, 45–53 (2014)
3. Barrat, A., Fernandez, B., Lin, K.K., Young, L.S.: Modeling temporal networks using random itineraries. Phys. Rev. Lett. **110**(15), 158702 (2013)
4. Bell, S., Wilson, K., Bissonnette, L., Shah, T.: Access to primary health care: does neighborhood of residence matter? Annals Assoc. Am. Geogr. **103**(1), 85–105 (2013)
5. Dark, S.J., Bram, D.: The modifiable areal unit problem (MAUP) in physical geography. Prog. Phys. Geogr. **31**(5), 471–479 (2007)
6. Eagle, N., Pentland, A.: Reality mining: sensing complex social systems. Pers. Ubiquitous Comput. **10**(4), 255–268 (2006)
7. Gabriel, A.K., Goldstein, R.M., Zebker, H.A.: Mapping small elevation changes over large areas: differential radar interferometry. J. Geophys. Res.: Solid Earth **94**(B7), 9183–9191 (1989)
8. Goodchild, M.F.: Giscience, geography, form, and process. Annals Assoc. Am. Geogr. **94**(4), 709–714 (2004)
9. Hashemian, M., Knowles, D., Calver, J., Qian, W., Bullock, M.C., Bell, S., Mandryk, R.L., Osgood, N., Stanley, K.G.: iEpi: an end to end solution for collecting, conditioning and utilizing epidemiologically relevant data. In: Proceedings of the 2nd ACM International Workshop on Pervasive Wireless Healthcare, pp. 3–8. ACM (2012)
10. Hashemian, M.S., Stanley, K.G., Knowles, D.L., Calver, J., Osgood, N.D.: Human network data collection in the wild: the epidemiological utility of micro-contact and location data. In: Proceedings of the 2nd ACM SIGHIT International Health Informatics Symposium, pp. 255–264. ACM (2012)
11. Kim, M., Kotz, D., Kim, S.: Extracting a mobility model from real user traces. In: INFOCOM 2006, 25th IEEE International Conference on Computer Communications. Proceedings, pp. 1–13, April 2006
12. Knowles, D.L., Stanley, K.G., Osgood, N.D.: A field-validated architecture for the collection of health-relevant behavioural data. In: 2014 IEEE International Conference on Healthcare Informatics (ICHI), pp. 79–88. IEEE (2014)
13. Kos, T., Grgic, M., Sisul, G.: Mobile user positioning in GSM/UMTS cellular networks. In: 48th International Symposium ELMAR-2006 Focused on Multimedia Signal Processing and Communications, pp. 185–188. IEEE (2006)
14. Kwan, M.P.: Algorithmic geographies: big data, algorithmic uncertainty, and the production of geographic knowledge. Annals. Am. Assoc. Geogr. **106**(2), 274–282 (2016)
15. Lee, K., Hong, S., Kim, S.J., Rhee, I., Chong, S.: SLAW: a new mobility model for human walks. In: INFOCOM 2009, pp. 855–863. IEEE, April 2009
16. Miller, H.J., Goodchild, M.F.: Data-driven geography. GeoJournal **80**(4), 449–461 (2015)
17. Modsching, M., Kramer, R., ten Hagen, K.: Field trial on GPS accuracy in a medium size city: the influence of built-up. In: 3rd Workshop on Positioning, Navigation and Communication, pp. 209–218 (2006)

18. Muhajarine, N., Katapally, T.R., Fuller, D., Stanley, K.G., Rainham, D.: Longitu-
    dinal active living research to address physical inactivity and sedentary behaviour
    in children in transition from preadolescence to adolescence. BMC Public Health
    15(1), 1–9 (2015). http://dx.doi.org/10.1186/s12889-015-1822-2
19. Openshaw, S., Openshaw, S.: The Modifiable Areal Unit Problem. Geo Abstracts,
    University of East Anglia, Norwich (1984)
20. Openshaw, S., Taylor, P.J.: A million or so correlation coefficients: three exper-
    iments on the modifiable areal unit problem. Stat. Appl. Spat. Sci. **21**, 127–144
    (1979)
21. Qian, W., Stanley, K.G., Osgood, N.D.: The impact of spatial resolution and
    representation on human mobility predictability. In: Liang, S.H.L., Wang, X.,
    Claramunt, C. (eds.) W2GIS 2013. LNCS, vol. 7820, pp. 25–40. Springer,
    Heidelberg (2013)
22. Rhee, I., Shin, M., Hong, S., Lee, K., Kim, S.J., Chong, S.: On the levy-walk nature
    of human mobility. IEEE/ACM Trans. Networking **19**(3), 630–643 (2011)
23. Smith, G., Wieser, R., Goulding, J., Barrack, D.: A refined limit on the predictabil-
    ity of human mobility. In: 2014 IEEE International Conference on Pervasive Com-
    puting and Communications (PerCom), pp. 88–94. IEEE (2014)
24. Song, C., Koren, T., Wang, P., Barabási, A.L.: Modelling the scaling properties of
    human mobility. Nat. Phys. **6**(10), 818–823 (2010)
25. Song, C., Koren, T., Wang, P., Barabási, A.L.: Modelling the scaling properties of
    human mobility supplementary material. Nat. Phys. **6**(10), 1–20 (2010)
26. Song, C., Qu, Z., Blumm, N., Barabási, A.L.: Limits of predictability in human
    mobility. Science **327**(5968), 1018–1021 (2010)
27. Stanley, K., Bell, S., Kreuger, L.K., Bhowmik, P., Shojaati, N., Elliot, A., Osgood,
    N.D.: Opportunistic natural experiments using digital telemetry: a transit disrup-
    tion case study. Int. J. Geogr. Inf. Sci. (to appear)
28. Sui, D., Goodchild, M.: The convergence of GIS and social media: challenges for
    giscience. Int. J. Geogr. Inf. Sci. **25**(11), 1737–1748 (2011)
29. Versichele, M., Neutens, T., Claeys Bouuaert, M., Van de Weghe, N.: Time-
    geographic derivation of feasible co-presence opportunities from network-
    constrained episodic movement data. Trans. GIS **18**(5), 687–703 (2014)
30. Wu, X., Mazurowski, M., Chen, Z., Meratnia, N.: Emergency message dissemina-
    tion system for smartphones during natural disasters. In: 2011 11th International
    Conference on ITS Telecommunications (ITST), pp. 258–263. IEEE (2011)
31. Yuan, Y., Raubal, M., Liu, Y.: Correlating mobile phone usage and travel behavior-
    a case study of Harbin, China. Comput. Environ. Urban Syst. **36**(2), 118–130
    (2012)
32. Zheng, Y., Xie, X., Ma, W.Y.: Geolife: a collaborative social networking service
    among user, location and trajectory. IEEE Data Eng. Bull. **33**(2), 32–39 (2010)

Spatial Models

pFUTURES: A Parallel Framework for Cellular Automaton Based Urban Growth Models

Ashwin Shashidharan[✉], Derek B. van Berkel, Ranga Raju Vatsavai,
and Ross K. Meentemeyer

North Carolina State University, Raleigh, NC, USA
{ashdharan,dbvanber,rrvatsav,rkmeente}@ncsu.edu

Abstract. Simulating structural changes in landscape is a routine task
in computational geography. Owing to advances in sensing and data collection technologies, geospatial data is becoming available at finer spatial and temporal resolutions. However, in practice, these large datasets
impede land simulation based studies over large geographic regions due
to computational and I/O challenges. The memory overhead of sequential
implementations and long execution times further limit the possibilities
of simulating future urban scenarios. In this paper, we present a generic
framework for co-ordinating I/O and computation for geospatial simulations in a distributed computing environment. We present three parallel
approaches and demonstrate the performance and scalability benefits of
our parallel implementation pFUTURES, an extension of the FUTURES
open-source multi-level urban growth model. Our analysis shows that
although a time synchronous parallel approach obtains the same results
as a sequential model, an asynchronous parallel approach provides better
scaling due to reduced disk I/O and communication overheads.

1 Introduction

Urban Growth models (UGMs) serve as the foundation for analytical research
on complex urban growth phenomenon. These models capture the interactions
between various drivers of urbanization to explain changes in urban systems and
serve a common goal of predicting future scenarios based on past observations.
The functionality of an UGM also lies in its ability to simulate urbanization
scenarios under varying land use policies. However, these simulations are computationally expensive and run for long periods of time. Traditional sequential
approaches become too slow to allow scenario simulations on user desktop environments.

Large geographic extents and high spatial resolution input data also impacts
the simulation runtime. While the large geographic extents introduce more spatially related processes to the simulation, high resolution data increases the number of such underlying observable processes. Thus, owing to the size and spatial
details in the data, we see an increase in the spatio-temporal interactions which
add to the computational complexity of the model and simulation.

© Springer International Publishing Switzerland 2016
J.A. Miller et al. (Eds.): GIScience 2016, LNCS 9927, pp. 163–177, 2016.
DOI: 10.1007/978-3-319-45738-3_11

However, with advances in parallel computing, new techniques [12,14,17,31, 32] have emerged to scale geospatial simulations. New parallel execution strategies and algorithms have also been developed that provide the means to scaleup geospatial simulations. In this paper, we present pFUTURES, a large-scale simulation framework that we develop for cellular automaton based urban growth models. Further, we adopt the FUTURES UGM in this framework and demonstrate its scalability across different geographic regions.

Our specific technical contributions are as follows: (i) a parallel framework for executing cellular automaton based land-use simulation models, (ii) three parallel approaches to co-ordinate I/O and computation in a master-worker style configuration of a UGM (iii) flexible data partitioning technique at a granularity as defined by the UGM and, (iv) communication and I/O optimization strategies.

The rest of the paper is organized as follows: in Sect. 2, we summarize existing research in simulating urban geography. In Sect. 3, we provide an overview of the FUTURES simulation model. In Sect. 4, we describe our parallel system architecture, design and challenges, and how we adapt the FUTURES model within a master-worker style configuration. In Sect. 5, we describe our experimental setup and results from executing pFUTURES over three different geographic regions. Finally, we conclude in Sect. 6.

2   Related Work

Since Wolfram proposed the use of Cellular Automatons (CAs) [35] for the study of systems in nature, CAs have found wide adoption in land use and land change models. While a few of them feature as multi-criteria GIS analysis tools [10,22,30], most models have been implemented as standalone sequential systems. SLEUTH [9], FORE-SCE [23], CLUE-S [28] and FUTURES [20] are a few popular land use simulation models that are CA based systems. The SLEUTH urban growth model has been around for twenty years and modified for varying geographic extents [7]. FORE-SCE and CLUE-S are land use models. While FORE-SCE examines urbanization as one of the outcomes of land use change, CLUE-S is an instance of a land use model that analyzes land-use systems by considering hierarchy of land use at different spatial scales. However, these models define land change events as sequential CA transitions and do not perform them in parallel.

An overview of the parallel capabilities of CAs and their widespread applications can be found in [3]. In pRPL [12], the authors implemented a few of these capabilities as a parallel Raster Processing Library designed for scientific simulations. pRPL exposes an API that allows programmers to define transition rules, neighborhood configurations and raster layer specific routines. Internally, pRPL implements data decomposition, boundary communication and neighbor cell transitions based on a user-defined specification. The library has been written for geographers to develop models without any parallel programming experience. pSLEUTH [12] is a parallel version of the SLEUTH model built using pRPL. Extending pSLEUTH, [13,14] implemented a hybrid parallel cellular automata model for urban growth simulation using GPU/CPU heterogeneous architectures. A complete list of technical modifications and evolution of SLEUTH has been documented in [7].

Neighbourhood modeling is an inherent aspect in urban growth models. In [10], the authors proposed a new proximal space approach with related geo-algebraic operations to model neighborhood and neighborhood operations in CAs. Parallel data decomposition and execution strategies [11,18,32] have also been developed that preserve neighborhood relationships in resulting spatial structures. The evolution and adoption from SIMD to MIMD architectures in the field of geoprocessing has been described in [11].

Distributed computing based computational geometry advancements [2,4,17, 21,31,32] also feature in parallel geosimulations. Geospatial libraries like GDAL [32] have been modified [36] to parallelize common GIS vector and raster operations. The libraries exploit advancements in parallel computing to implement I/O operations in parallel. However, existing parallel approaches [12,14] do not intrinsically handle data decomposition of irregular boundaries. Duplication of data and redundant computation along boundary regions become necessary. To overcome these limitations, the pFUTURES framework implements a flexible data partitioning scheme as defined by the underlying UGM.

3 FUTURES: FUTure Urban-Regional Environment Simulation

FUTURES is an urban growth model that simulates land change events based on historical land growth patterns. It simulates emerging urban landscape patterns under varying environmental, infrastructural, socioeconomic and demographic factors.

Figure 1 provides an overview of the FUTURES simulation framework and its interacting components, namely, (i) DEMAND sub-model, (ii) POTENTIAL sub-model and, (iii) Patch Growing algorithm (PGA).

3.1 Development Potential Sub-model (POTENTIAL)

The POTENTIAL sub-model implements a site suitability modeling technique that formalizes the relationship between urban development and environmental, infrastructural, and socioeconomic changes over time in a region. The model considers a number of predictor variables as input to a multi-level logistic regression model. Each predictor variable accounts for a spatial or temporal aspect of land cover change and is used to define a *suitability score* for a site in a region. Finally, the output from the POTENTIAL model is normalized to produce a map of development probability values for all sites in a region.

The probability that an undeveloped cell becomes developed is defined as:

$$p_i = \frac{e^{s_i}}{1 + e^{s_i}} \tag{1}$$

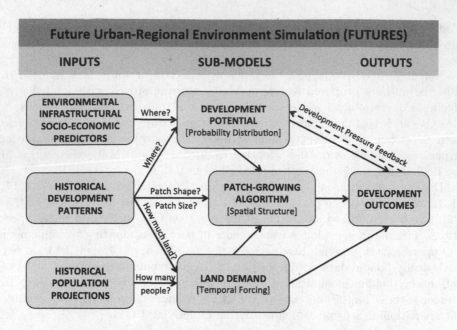

Fig. 1. The FUTURES land change modeling framework [20]

where, s_i is the composite development potential for a cell i. The development potential s_i is defined as a function of environmental, infrastructural, and socioeconomic predictor variables of site suitability as following:

$$s_i = a_{ji} + \sum_{h=1}^{n} \beta_{jih} * x_{ih} + \beta_{jih} * p_i{}' \tag{2}$$

where, for the ith undeveloped cell and varying across j groups (i.e., the level), a_{ji} is the intercept, β_{ji} is the regression coefficient, h is a predictor variable representing conditions at the start of a chosen simulation year, n is the number of predictor variables, x_{ih} is the value of h at i, and $p_i{}'$ is the dynamic development pressure variable due to neighboring developed sites.

$$p_i{}' = \sum_{k=1}^{n_i} \frac{State_k}{d_{ik}{}^\gamma} \tag{3}$$

where, $State_k$ is the current state (0/1) of the kth neighbor of cell i, d is its distance from the kth neighbor in its list of n neighbors and γ is a coefficient that controls the influence of distance between cell i and its neighboring cells.

3.2   Land Demand Sub-model (DEMAND)

The DEMAND sub-model establishes the relationship between historical land consumption and population growth under different development scenarios.

This relationship is established using an ordinary least squares regression technique. The regression model considers two parameters in estimating future land use for urbanization: (i) population growth and (ii) land consumption over time. The generated per capita demand projections from the model drive the PGA in the FUTURES UGM.

3.3 Patch Growing Algorithm (PGA)

The Patch Growing Algorithm implements the mechanism to simulate historically observed urbanization patterns based on the above two sub-models. Patch growth is defined as a 3-step process: (i) Monte Carlo based seed selection using the site development probability, (ii) patch size selection from a library of patch sizes, a weighted distribution of historically observed patch sizes, and (iii) patch growth through a neighbor discovery process.

A neighborhood configuration specified at the start of the simulation defines the neighbor discovery process. Further, a site suitability metric (Eq. 4) for newly discovered cells determines the most suitable cells that go towards patch growth.

$$s_i = s_i{'} * d^{-\alpha} \tag{4}$$

where, $s_i{'}$ is the underlying development potential of a cell i, d is the distance of cell i to a seed, and α is a patch compactness factor. The PGA continues the neighbor discovery process by using the newly added neighbors as potential seed cells. Thus, the patch growing algorithm continues till the value of patch size is met or else terminates when no more suitable sites for patch growth can be found.

4 pFUTURES System Architecture

pFUTURES adopts a centralized system architecture based on a master-worker model for parallel computing. In a large-scale urban simulation, this model allows a distributed execution of UGMs on smaller sub-regions in the landscape. The pFUTURES framework consists of a single master with multiple workers and relies on a messaging framework for communication. Figure 2 illustrates our master-worker model with multiple workers, centrally controlled by a single master.

4.1 Master Services

In this section, we describe the responsibilities of the master in pFUTURES, namely, (i) Data Partitioning, (ii) Task Scheduling and (iii) Task Synchronization.

Data Partitioning decomposes input data to create small distinct partitions of data for workers. This allows workers to execute independently with minimal task dependencies during parallel execution. In pFUTURES, data partitioning creates smaller data partitions for sub-regions in the landscape. The granularity

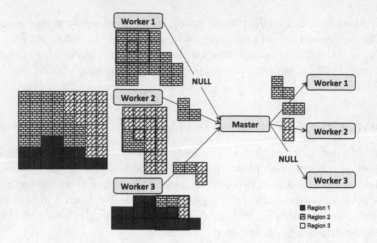

Fig. 2. Data processing in a master-worker style architecture

of this decomposition and the number of sub-regions can be created at "predictor levels" defined in the underlying UGM. Figure 3 illustrates a data partition created at a predictor level as defined by a UGM.

Task Scheduling decides the order in which tasks are executed by workers in the system. In the context of large scale urban simulations, task scheduling specifies a sequence in which sub-regions in the landscape are processed by workers. The task order is based on a dynamically generated or, a statically provided schedule at the master.

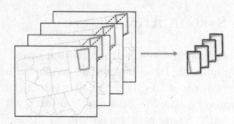

Fig. 3. Example of a data partition created from multiple data layers

Task Synchronization resolves data dependencies during task execution. Data dependencies arise due to (i) independent execution of a task at a worker and, (ii) task scheduling as enforced by the master. The master periodically synchronizes the UGM execution by communicating with the workers. In pFUTURES, task synchronization enables interactions among geographically distant regions, which results in similar outcomes to a sequential UGM simulation. In Sect. 4.4, we describe three synchronization strategies to co-ordinate worker tasks in pFUTURES.

4.2   Worker Services

In the master-worker parallel model, a worker executes a set of fixed computation steps on data partitions assigned by the master. In pFUTURES, a worker task entails the execution of a UGM on a region as assigned by the master. The responsibilities of the worker can be described in terms of *Task Execution* which deals with an individual worker running a UGM for a fixed number of time-steps on a data partition and *Task Updates*, which are carried out in response to task synchronization enforced by the master.

4.3   System Workflow

The pFUTURES system combines the master and worker services in a five-step workflow as shown in Fig. 4. The master begins by parsing a system configuration file which contains the global parameters that control the UGM simulation. The master then creates regional datasets based on regional configuration files and the input data layers. The configuration file for a region contains the affine geometry transforms, their inverse and the number of simulation time-steps. In the next step, the master creates a schedule and decides an order in which regions will be processed by workers in the system. The master sets up one task per worker and sends them a region and a time-step in the simulation to advance. A worker on receiving these values, parses the region specific configuration file and reads all input data for the simulation into memory. The workers update the development pressure values received due to development in adjacent regions and run the simulation. In every time-step, a worker also aggregates the development pressure updates for cells to be transmitted to adjacent regions. The master asynchronously tracks the status of the workers to receive these development pressure updates for all regions in the landscape. In the next time-step, the master iterates over these received development pressure updates, identifying and generating per-region lists with updates to transmit to the workers. On task completion, the task is invalidated from the schedule and new tasks are assigned to the workers. Thus, workers in pFUTURES execute the UGM over the landscape with synchronization support from the master (Fig. 2).

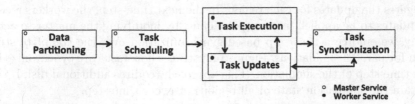

Fig. 4. pFUTURES system workflow diagram

4.4   Synchronization Strategies

In this section, we present three synchronization strategies which achieve varying level of interaction between the workers within pFUTURES, namely, (i) no border interactions, (ii) all border interactions and (iii) selective border interactions.

Approach 1: Data Parallel Approach. In the first approach, distinct partitions of data are assigned to the workers for UGM execution. Each worker runs the model independently and for all the simulation time-steps. There is no communication or synchronization with the master or other workers. On task completion, a worker writes the result to disk and signals the master to receive a new data partition, if available. In this approach, the UGM simulation can scale for any geographic extent without the need to accommodate all data in main memory at once. It also achieves speedup by executing the model simulation over multiple regions in parallel. However, this approach lacks the ability to capture interactions due to development among geographically neighboring regions, especially along the region boundaries. We propose Approach 2 to overcome this problem.

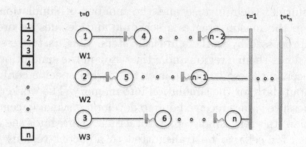

Fig. 5. Approach 2: time synchronous parallel approach

Approach 2: Time Synchronous Parallel Approach. In this approach, the master and worker functionalities are augmented to preserve all cross-border region interactions. In every time-step, a worker builds a list of development pressure updates for cells not belonging in its region and sends them to the master. The master waits for all regions in the landscape to be processed by workers and aggregates the updates for each region. In the next time-step, the workers receive the updates to be applied in its region. Thus, by modifying the master to act as a relay, we enforce a time-step based synchronization over the complete study region for parallel simulations. Figure 5 illustrates how tasks are synchronized at every time-step of the simulation. This approach requires additional disk I/O to maintain the simulation state of all regions at every time-step.

Approach 3: Time Asynchronous Parallel Approach. To reduce the additional disk I/O imposed by Approach 2, we propose an alternate approach based on visual analysis of the simulation results obtained using Approach 2. We make two observations: (i) spatial structural changes along the boundary line due to development in adjacent counties is limited and sparse and, (ii) development along

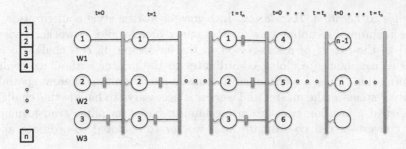

Fig. 6. Approach 3: time asynchronous parallel approach

county boundaries is not applicable to all regions in the study extent. Based on these observations, we modify our previous approach to support communication only among smaller conterminous regions of interest. Thus, we transform our parallel framework as a batch of in-memory simulations for conterminous regions that run on workers till completion. Figure 6 illustrates this idea of batch processing on the workers.

4.5   Challenges

The design of pFUTURES was primarily guided by a requirement for a parallel and scalable framework to augment the existing sequential version of FUTURES UGM. In this section, we describe the practical challenges and solutions in our adoption of the FUTURES UGM within the pFUTURES framework.

Challenge 1: Stateless Configuration of the Worker. The first challenge arises from the way workers are programmed for task execution in a master-worker configuration. At every time-step, a worker assigned to a region must use the most recent simulation state of the region. However, this state is not stored at the worker. This challenge is addressed by serializing the state of urban simulation at a worker in every time-step. This is achieved by: (i) maintaining the next simulation time-step for all regions at the master and (ii) storing onto disk, the state of the urban landscape for all regions. Thus, at the beginning of every new time-step, a worker is capable of advancing the simulation by querying the master and reading necessary state information from the disk.

Challenge 2: Isolated Execution of the Worker. In a master-worker style configuration, a worker is designed to have access only to specific data partitions as assigned by a master. A worker is unaware of other executing workers and the spatial configuration of other regions. Each worker is further limited by its ability to only communicate with the master. Thus, in adoption of a master-worker style configuration, it is a challenge to communicate development pressure due to sites developed along regional boundary lines across different workers. In Sect. 4.4, we have described our propagation mechanism for our three synchronization approaches.

Challenge 3: Location Translation. In a master-worker style configuration, the master maintains a global view of the study area, while, a worker's view is limited to the extent of a sub-region in the landscape. It is a challenge for a worker to communicate global co-ordinates to the master without co-ordinate transformation. Thus, a mechanism to translate between the different spatial co-ordinate systems at the master and worker is necessary. To handle this challenge, we designed a location translation mechanism based on affine transformation, which converts a cell co-ordinate at a worker to a global co-ordinate at the master.

Challenge 4: Communication and Memory Overhead. The overhead of centralized communication at the master due to development pressure updates relayed by the workers is unavoidable. However, we can reduce the memory overhead at the master to support communication in large study areas. To achieve this, we implement a disk-based query mechanism like *gdallocation* [33], to identify updates for specific regions. This approach reduces the memory consumption at the master by eliminating the need to store full-sized rasters of the landscape. It makes available more memory to store the communication updates received from the workers. Additionally, we reduce the communication between the master and a worker, by implementing a filtering sequence that eliminates updates for already developed and invalid cells in the landscape. Thus, the pFUTURES framework is scalable and supports parallel simulation of sub-regions with a low memory and communication overhead.

5   Experimental Evaluation

In this section, we explain our experimental methodology and discuss the results from executing the FUTURES UGM in pFUTURES over three different geographic regions.

5.1   Experimental Setup

To evaluate our parallel architecture, we carried out our experiments on a single node Linux based system with the MPI-2.1 message passing interface. The node was configured with two Intel(R) Xeon(R) CPU E5-2660 v3 @ 2.60 GHz processors, 256 GB of total main memory and a 20 TB disk with a SGI XFS parallel file system to store all input and output data.

5.2   Experiment Results

We evaluate the performance of our framework by conducting experiments over three study regions, namely, (i) Mountain region of North Carolina, (ii) State of North Carolina and (iii) South Atlantic States. We compare the execution times

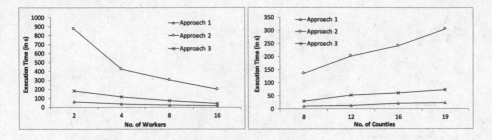

(a) Variable number of workers   (b) Variable number of counties

Fig. 7. Simulation time (Mountain region of North Carolina)

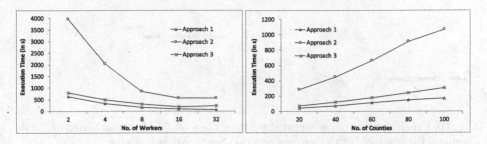

(a) Variable number of workers   (b) Variable number of counties

Fig. 8. Simulation time (State of North Carolina)

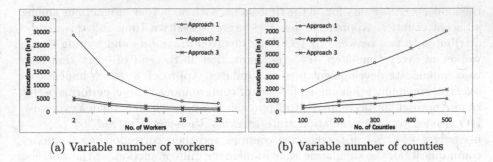

(a) Variable number of workers   (b) Variable number of counties

Fig. 9. Simulation time (South Atlantic States)

in our three approaches by: (i) increasing the number of workers under a fixed workload and, (ii) increasing the workload under a fixed number of workers. Figures 7, 8 and 9 summarize the results from our experiments. Figure 10 shows the output maps of simulation in our three study regions.

In terms of execution time performance, in all three approaches as shown in Figs. 7(a), 8(a) and 9(a), increasing the number of workers reduces the execution time of the simulation. We observe that Approach 1 requires the least amount of

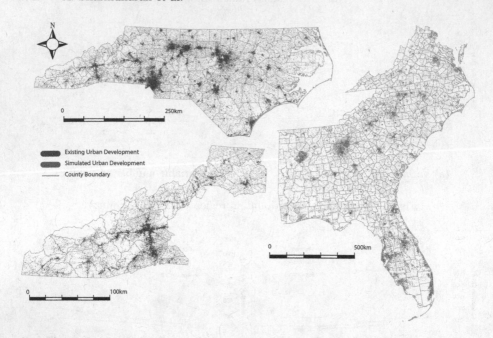

Fig. 10. Urban-regional environment simulation maps for (i) Mountain region of North Carolina, (ii) State of North Carolina and (iii) South Atlantic States

time to execute. This is expected as there is no synchronization or communication between the workers in the simulation. Each worker independently executes the model over a county for all the time-steps, with no border interactions among adjacent counties. Approach 2 has the largest execution time and this can be attributed to two reasons: (i) I/O costs involved in reading and writing at the worker at every time step, (ii) synchronization at the end of every time-step to communicate development pressure updates. Approach 3, which implements selective communication among batches of conterminous regions performs better than Approach 2. As compared to Approach 2, Approach 3 eliminates expensive I/O needed to save simulation state at every time-step. Instead, by processing only a select set of neighboring counties and their corresponding boundary communication, the simulation runs in-memory, till completion, at the workers. However, the approach still incurs synchronization costs associated with communicating development pressure updates which results in longer execution time as compared to Approach 1.

In our second set of experiments, as shown in Figs. 7(b), 8(b) and 9(b), we compare scalability with respect to the size of the study extents. The number of workers is set to eight and the number of counties is varied in each study region. It is observed that increasing the number of counties increases the amount of time taken to execute the simulation. This is consistent in all three approaches. Once again, Approach 1 performs the best as there is no communication and synchronization needed during the simulation. Approach 3 performs better than

Approach 2 but worse than Approach 1 due to reasons discusseds above. However, we notice that the performance of Approach 3, when compared to Approach 1, deteriorates with increase in workload. This is caused due to increased communication among neighboring counties as the study extent grows. We conclude that Approach 1 and Approach 3 scale well for increasing number of counties and, hence is our recommended approach.

We use a randomly generated schedule in our simulation experiments with the FUTURES UGM. This is justified as the underling UGM is insensitive to task scheduling. Data dependencies among workers executing this model are resolved at the start of every time-step and do not arise during execution of the time-step. Moreover, by design, Approach 1 and Approach 2 are agnostic to task scheduling. However, in Approach 3, the number of adjoining regions that can be processed in a time-step is dependent on the number of workers in the system. Thus, the impact of scheduling as described in [6] along with support for a variable grid approach [34] remains to be evaluated.

6   Conclusion

In this paper, we present a parallel framework, pFUTURES, for fast execution of cellular automaton based urban simulation models. We present three parallel approaches to coordinate I/O and computation in a master-worker style configuration for such models. We adopt the FUTURES UGM in this parallel framework and evaluate its scalability with varying geographic study extents. We conclude that the practical benefits from the flexible data partitioning scheme, the reduced communication and I/O costs from the parallel optimization strategies, and the overall simulation speedup from the framework make it suitable for geospatial simulations using high resolution data and large datasets.

References

1. Message Passing Interface Forum: MPI: Message-Passing Interface Standard Version 2.1, June 2008
2. Armstrong, M.P.: Geography and computational science. Ann. Assoc. Am. Geogr. **90**(1), 146–156 (2000)
3. Bandini, S., Mauri, G., Serra, R.: Cellular automata: from a theoretical parallel computational model to its application to complex systems. Parallel Comput. **27**(5), 539–553 (2001)
4. Batty, M.: Geocomputation using cellular automata. In: Abrahart, R.J., Openshaw, S., See, L.M. (eds.) Geocomputation, Routledge (2000). http://dx.doi.org/10.4324/9780203305805
5. Benenson, I., Torrens, P.M.: Geosimulation: Automata-Based Modeling of Urban Phenomena. Wiley, Hoboken (2004)
6. Bonnell, T.R., Chapman, C.A., Sengupta, R.: Interaction between scale and scheduling choices in simulations of spatial agents. Int. J. Geogr. Inf. Sci. **30**(10), 2075–2088 (2016). http://dx.doi.org/10.1080/13658816.2016.1158822

7. Chaudhuri, G., Clarke, K.C.: The SLEUTH land use change model: a review. Int. J. Environ. Resour. Res. **1**(1), 88–104 (2013)
8. Cheng, G., Liu, L., Jing, N., Chen, L., Xiong, W.: General-purpose optimization methods for parallelization of digital terrain analysis based on cellular automata. Comput. Geosci. **45**, 57–67 (2012)
9. Clarke, K.C., Hoppen, S., Gaydos, L.: A self-modifying cellular automaton model of historical urbanization in the San Francisco Bay area. Environ. Plan. B: Plan. Des. **24**(2), 247–261 (1997)
10. Couclelis, H.: From cellular automata to urban models: new principles for model development and implementation. Environ. Plan. B: Plan. Des. **24**(2), 165–174 (1997)
11. Ding, Y., Densham, P.J.: Spatial strategies for parallel spatial modelling. Int. J. Geogr. Inf. Syst. **10**(6), 669–698 (1996)
12. Guan, Q.: pRPL: an open-source general-purpose parallel Raster Processing programming Library. Sigspat. Spec. **1**(1), 57–62 (2009)
13. Guan, Q., Shi, X.: Opportunities and challenges for urban land-use change modeling using high-performance computing. In: Shi, X., Kindratenko, V., Yang, C. (eds.) Modern Accelerator Technologies for Geographic Information Science, pp. 227–236. Springer, Boston (2013)
14. Guan, Q., Shi, X., Huang, M., Lai, C.: A hybrid parallel cellular automata model for urban growth simulation over GPU/CPU heterogeneous architectures. Int. J. Geogr. Inf. Sci. **30**(3), 494–514 (2016)
15. Guan, Q., Wang, L., Clarke, K.C.: An artificial-neural-network-based constrained CA model for simulating urban growth and its application. Cartogr. Geogr. Inf. Sci. **32**(4), 369–380 (2005)
16. Guan, Q., Zeng, W., Gong, J., Yun, S.: pRPL 2.0: Improving the parallel Raster Processing Library. Trans. GIS **18**(S1), 25–52 (2014)
17. Hawick, K.A., Coddington, P.D., James, H.A.: Distributed frameworks and parallel algorithms for processing large-scale geographic data. Parallel Comput. **29**(10), 1297–1333 (2003)
18. Hutchinson, D., Lanthier, M., Maheshwari, A., Nussbaum, D., Roytenberg, D., Sack, J.R.: Parallel neighbourhood modelling. In: Proceedings of 4th ACM International Workshop on Advances in Geographic Information Systems, pp. 25–34. ACM (1996)
19. Liu, Y.: Modelling Urban Development with Geographical Information Systems and Cellular Automata. CRC Press, Boca Raton (2008)
20. Meentemeyer, R.K., Tang, W., Dorning, M.A., Vogler, J.B., Cunniffe, N.J., Shoemaker, D.A.: FUTURES: multilevel simulations of emerging urban-rural landscape structure using a stochastic patch-growing algorithm. Ann. Assoc. Am. Geogr. **103**(4), 785–807 (2013)
21. Murayama, Y., Thapa, R.B.: Spatial analysis: evolution, methods, and applications. In: Murayama, Y., Thapa, R.B. (eds.) Spatial Analysis and Modeling in Geographical Transformation Process: GIS-based Applications, pp. 1–26. Springer, Dordrecht (2011)
22. Park, S., Wagner, D.F.: Incorporating cellular automata simulators as analytical engines in GIS. Trans. GIS **2**(3), 213–231 (1997)
23. Sohl, T.L., Sayler, K.L., Drummond, M.A., Loveland, T.R.: The FORE-SCE model: a practical approach for projecting land cover change using scenario-based modeling. J. Land Use Sci. **2**(2), 103–126 (2007)

24. Terando, A.J., Costanza, J., Belyea, C., Dunn, R.R., McKerrow, A., Collazo, J.A.:
The southern megalopolis: using the past to predict the future of urban sprawl in
the Southeast U.S. PloS ONE **9**(7), e102261 (2014)
25. Torrens, P.M.: SprawlSim: modeling sprawling urban growth using automata-
based models. In: Agent-Based Models of Land-Use/Land-Cover Change, pp. 69–76
(2002)
26. Torrens, P.M.: Geosimulation and its application to urban growth modeling. In:
Portugali, J. (ed.) Complex Artificial Environments: Simulation, Cognition and
VR in the Study and Planning of Cities, pp. 119–136. Springer, Heidelberg (2006)
27. Torrens, P.M.: Calibrating and validating cellular automata models of urbanization. In: Urban Remote Sensing, pp. 335–345. Wiley (2011). http://dx.doi.org/10.
1002/9780470979563.ch23
28. Verburg, P.H., Soepboer, W., Veldkamp, A., Limpiada, R., Espaldon, V., Mas-
tura, S.S.: Modeling the spatial dynamics of regional land use: the CLUE-S model.
Environ. Manag. **30**(3), 391–405 (2002)
29. Waddell, P.: UrbanSim: modeling urban development for land use, transportation,
and environmental planning. J. Am. Plan. Assoc. **68**(3), 297–314 (2002)
30. Wagner, D.F.: Cellular automata and geographic information systems. Environ.
Plan. B: Plan. Des. **24**(2), 219–234 (1997)
31. Wang, F.: A parallel intersection algorithm for vector polygon overlay. IEEE Comput. Graph. Appl. **2**, 74–81 (1993)
32. Wang, S., Armstrong, M.P.: A quadtree approach to domain decomposition for
spatial interpolation in grid computing environments. Parallel Comput. **29**(10),
1481–1504 (2003)
33. Warmerdam, F.: The geospatial data abstraction library. In: Hall, G.B., Leahy,
M.G. (eds.) Open Source Approaches in Spatial Data Handling, pp. 87–104.
Springer, Heidelberg (2008)
34. White, R., Engelen, G., Uljee, I.: Modeling Cities and Regions as Complex Systems:
From Theory to Planning Applications. MIT Press, Cambridge (2015)
35. Wolfram, S.: Cellular automata as models of complexity. Nature **311**(5985), 419–
424 (1984)
36. Zhan, L.J., Qin, C.Z.: Parallel Geospatial Raster Processing by Geospatial Data
Abstraction Library (GDAL) - Applicability and Defects

From Data Streams to Fields: Extending Stream Data Models with Field Data Types

Qinghan Liang, Silvia Nittel[(✉)], and Torsten Hahmann

Spatial Informatics, School of Computing and Information Science,
University of Maine, Orono, USA
{qinghan.liang,torsten.hahmann}@maine.edu, nittel@spatial.maine.edu

Abstract. With ubiquitous live sensors and sensor networks, increasingly large numbers of individual sensors are deployed in physical space. Sensor data *streams* are a fundamentally novel mechanism to create and deliver observations to information systems, enabling us to represent spatio-temporal continuous phenomena such as radiation accidents, pollen distributions, or toxic plumes almost as instantaneously as they happen in the real world. While data stream engines (DSE) are available to process high-throughput updates, DSE support for phenomena that are continuous in both space and time is not available. This places the burden of handling any tasks related to the integration of potentially very large sets of concurrent sensor streams into higher-level abstractions on the user. In this paper, we propose a formal extension to stream data model languages based on the concept of fields to support high-level abstractions of continuous ST phenomena that are known to the DSE, and therefore, can be supported through queries and processing optimization. The proposed field data types are formalized in a data model language independent way using second order signatures. We formalize both the set of supported field types are as well as the embedding into stream data model languages.

Keywords: Data streams · Sensor data streams · Data stream engines · Fields · Field data types

1 Introduction

Motivation. With ubiquitous live sensors and wireless sensor networks, increasingly large numbers of individual sensors are deployed in physical space such as urban environments [25], forests [10], for earthquake monitoring [12], or precision agriculture. Such large numbers of live streaming sensors enable us to collect observations that are sufficiently *dense* in both space and time to now represent *continuous* change in space *and* time in near real-time. Examples for spatio-temporal continuous phenomena are, for instance, pollen distributions, toxic plumes, radiation accidents, or soil moisture distributions.

Sensor data *streams* are a fundamentally novel mechanism to create and deliver observations to information systems, enabling us to represent entities,

© Springer International Publishing Switzerland 2016
J.A. Miller et al. (Eds.): GIScience 2016, LNCS 9927, pp. 178–194, 2016.
DOI: 10.1007/978-3-319-45738-3_12

processes and events in information systems almost instantaneously as they happen in the real world [5,24]. While a stream seems similar to a time series, that is, it consists of an ordered set of time-stamp records, a stream is a significantly different *programming* abstraction: it is an *unbounded* multiset of elements, continuously producing records as time advances. New updates are constantly pushed into queries during query execution, resulting in real-time query answers. *Data stream engines* (DSEs) have been designed as a high-throughput alternative to database management systems (DBMS). Today, open source and commercial DSE such as Apache Spark [1] achieve query performance over streams with a throughput of >1 Million updates/s; in comparison, DBSs are limited to 500 updates/s [26]. Similar to DBSs, DSEs provide data model and query languages, which make it easier for users to program applications, enabling them to define data schemata and SQL-type queries over data streams. The core concept of stream data models is that of a *stream* (as opposed to a relation). Stream query languages contain *continuous queries* as well as *query windows* as evaluation contexts over streams. While DBS and DSE are separate technologies, stream data models and query languages are formally integrated with the relational algebra to guarantee compatibility between both technologies [4].

DSEs make it feasible to monitor and analyze phenomena that are continuous in space and time while delivering real-time answers to queries. However, today's stream data model languages provide concepts to represent individual sensor data streams. For instance, point geometry types are available to create stream tuples with sensor location attributes [3]. Such low-level support enables only modeling individual sensor data streams, and it is the responsibility of the user and application code to handle any tasks related to the *integration* of potentially very large sets of concurrent sensor streams into *higher-level abstractions* such as spatio-temporally continuous phenomena. Currently, the bulk of programming and understanding of continuous phenomena is pushed into the application code, and this code has to be re-implemented for each application again and again. We believe that DSEs should provide a generic, flexible and high-level abstraction for continuous spatio-temporal (ST) phenomena. The complexity of integrating individual sensor streams into a higher-level representation on-the-fly should be hidden from users while still allowing them to configure the mapping between sensor streams and a continuous phenomenon. Spatio-temporal continuous phenomena need to be supported on the level of both data model language and query execution support in DSEs.

Transforming large sets of individual sensor data streams into the high-level representation of a ST continuous phenomenon is not an easy feat. A representation of a continuous phenomenon must always be an approximation based on captured samples. When dealing with up to 1,000 concurrent streams, it is unlikely that their updates and sampling rates are synchronized. Instead, each stream might have its own sampling frequency. To create a snapshot of a spatially continuous phenomenon at a desired time stamp across all sensors requires resampling or interpolating existing streams. Further, the types of analyses of such phenomena will vary but still run concurrently. Thus, locking the representation of a continuous phenomenon into a single, fixed resolution of cells, triangulations or contour lines is severely limiting.

Contributions. In this paper, we propose an extension to stream data models that is based on *field* data types, which directly tap into streamed data rather than accessing stored data. Fields have long been proposed as a unifying information system abstraction for continuous phenomena [8,21] and are well understood on an ontological level [15,16], but are still uncommon as actual information system interfaces and implementations [7,13,22]. We believe that a field data type is the most promising approach to handling the complexities of sensor stream processing for continuous phenomena.

While the extension of DSEs with fields on the data model, query language and processing level is a complex task, we have investigated feasibility aspects of processing fields on-the-fly based on massive sensor data streams in previous work [27,28]. In this paper, we focus solely on introducing the *formal framework* of extending data stream models with field types. We introduce our proposed field data types for stream data models on an abstract level using second order signatures [17], which allow us to define the field types independently from specific data models and programming languages. The *abstract model* is continuous in space and time and used to formalize the universe of field types that can be constructed. The abstract model is complemented by a *discrete model* that is amendable to direct implementation. It grounds the purely continuous view of the abstract model in the realities of discrete computer systems. It does so by relating the abstract field types to computational concepts such as tuples, streams, windows, and interpolation functions. Our field stream data model with its dynamic spatial, temporal, and spatio-temporal fields serves as the foundation for sophisticated spatio-temporal data analyses, bridging the gap between raw point-based sensor streams and the detection of trends and events.

The following two Sects. 2 and 3 present the background of our work and the related work. The abstract model for the field types is presented in Sect. 4, while Sect. 5 describes the discrete model suitable for embedding in a data model language. Section 6 offers our conclusions and identifies future work.

2   Background

2.1   Data Streams

The core concept of relational database systems (DBS) is the *relation*, which is a set of persistently stored data tuples. Each relational query is performed over a stored relation in its *entirety*. On the other hand, data stream engines (DSEs) are concerned with frequently updated data and, thus, have *streams* as their core concept. A stream is an unbounded sequence of tuples that arrive in some temporal order, with multiple tuples likely arriving out of order or simultaneously. More formally, streams are defined as "unbounded, append-only multisets of time-stamped tuples" [4]. Unbounded means in this context that we cannot predict how many tuples will arrive at any point of time that is in the future; at any time point in the past up until now, this multiset is finite. Tuples have either explicit timestamps, which are added at the data source and denote the real-world event time, or implicit timestamps, which are added to tuples when they arrive at the DSE [5]. Stream data models support linear, dense and discrete time models [26].

While each tuple in a stream is assigned a timestamp, not every tuple necessarily has a spatial attribute. For that reason, we introduce a special type of stream – a *spatio-temporal stream* – as a stream wherein every tuple also has a spatial attribute. It is based on the following concept of a spatio-temporal relation.

Definition 1. *Let S be a discrete set of point locations, T be a discrete bounded set of timestamps, and $V_1,\ldots,V_n$ be value domains. Then a ST-Relation is a relation $\Re_{ST} \subseteq S \times T \times V_1 \times \cdots \times V_n$ such that for every $(s_i,t_i) \in S \times T$ there exists at most one $(v_1 \times \cdots \times v_n) \in V_1 \times \cdots \times V_n$ such that $(s_i, t_i, v_1, \ldots, v_n) \in \Re_{ST}$.*

Definition 2. *Let S be a discrete set of point locations, T be a discrete bounded set of timestamps, and $V_1,\ldots,V_n$ be value domains (e.g., values of measurements). Let further T_S be a set of timestamps with initial and last timepoints $t_{S1}, t_{Sf} \in T_S$ such that for every $t_{Si} \in T_S$, $t_{S1} \leq t_{Si} \leq t_{Sf}$. Then, an ST-Stream is a function $S_{ST} : T_S \to \Re$ such that for a fixed spatial domain S, a fixed temporal domain T, and a fixed value domain $V_1 \times \cdots \times V_n$, all $\mathcal{R} \in \Re$ are ST-Relations with $\mathcal{R} \subseteq S \times T \times V_1 \times \cdots \times V_n$.*

A ST-Stream can be the time series of observations from a single sensor or from an entire geosensor network. Viewing an ST-Stream as all streamed updates of geographically and thematically related sensors is a powerful abstraction that allows us to reason about complex spatio-temporal events that take place within the space and time observed by the streams. In our model, ST-streams encompass streams that consists of the updates from only a single sensor as well as the aggregation of streams that form continuous dynamic ST-Fields.

2.2 Fields as Formal Foundation for Streaming ST Continuous Phenomena

The term *field* (more precisely *geo-spatial fields*) is widely used to describe entities in physical space that are continuous in space and time and lack boundaries. A field implies a continuous quality of an observed phenomenon in the real-world, such as temperature, that is present at every point in time and space on Earth (and beyond). A plethora of different computer representations have been developed for continuous phenomena, and it is common practice to pick a representation that matches best the data capture method in order to represent a particular phenomenon. For instance, temperature may be represented by measurements at irregularly distributed sample points, foliage as regular grid cells, or pollen density as isolines, with more complex representations possible. While implementations of these computer representations enable many specialized analytical operations in geographic information systems, the diversity and lack of coherence of representation seriously impedes integration and cross-cutting analyses [8,21]. This problem is exacerbated if spatio-temporal fields are considered, and multi-faceted integration over space, time, multiple parameters, and various sensor platforms is the objective.

Over the last two decades, geo-spatial fields have been mathematically formalized, e.g., in [9,16]. In the context of our field stream data model, we are mainly concerned with fields that are approximated based on observations, so-called *sampled fields* (in contrast to equation fields). We present a mathematical definition of the following types of fields that are the underlying basic components of the field model: a *spatial field*, which addresses continuity over space, a *temporal field*, representing continuity over time, and a *spatio-temporal field*, representing continuity over both space and time. We are particularly interested in the important notion of a *sampled field* because it addresses the spatial domain of fields, that is, it represents either sensor locations or the continuous quality of the phenomenon.

A **spatial field** is defined as follows: Given a spatial domain S and an attribute domain V, a spatial field F_S over S is a computable, possibly partial, function $f : S \rightarrow V$ from spatial locations in S to attribute values in V. The spatial locations in S are points, and a subset of S are sensor locations. More details on spatial domains for fields are discussed in [15]. The attribute domain V can be finite or infinite, discrete or continuous, numeric or symbolic.

A **temporal field** F_t is a function $f : T \rightarrow V$ from the time points in T to attribute values in V. A temporal field represents the change of an attribute over time. The attribute could be a location (e.g., a trajectory of a moving object), a sensor measurement, or a stock price. For observed temporal fields, the time domain is both linear and dense, that is, time advances linearly. For a temporal field based on a sensor stream, the time domain is discrete and isomorphic to $\mathbb{N}$; each natural number corresponds to a non-decomposable unit of time which is the sampling time. For a continuous temporal field, the time domain is isomorphic to $\mathbb{R}$ since the real-world phenomenon exists without temporal 'gaps'.

Spatio-temporal fields have both spatial dimensions as well as a temporal dimension. For example, in a spatio-temporal temperature field over a lake, each value $f(s,t)$ identifies the temperature at location s and time t. Galton [16] defines a spatio-temporal field as a function $f : S \times T \rightarrow V$ that assigns each pair of a spatial location in S and a time point in T an attribute value in V. Again, the temporal domain T is linear and dense, and can be discrete or continuous. Choosing a snapshot point of view and a discrete temporal domain, the spatio-temporal field is equivalently defined as a function $f : T \rightarrow (S \rightarrow V)$ mapping time points to a spatial field $S \rightarrow V$.

In the next section, we discuss current support for continuous ST phenomena based on observation streams in information systems.

3 Related Work

Fields as data types are not commonly available in information systems, mostly due wide-spread use of established software, implementations and tools that are based on different types of representations. The OGC coverage interface specification [6] assumes a field-type representation for space-time varying phenomena, and has been an industry standard for more than a decade. However, its emphasis is on standardizing coverage operators, not data representations. Also, observation streams can not be accommodated directly with this interface specification. Work more closely related to ours is [7], which proposes a field data type as a generic data type to represent time series, trajectories and coverages. The idea of a single, generic data type to express different specialized fields is similar to our objectives. In this work, the field types have been prototypically implemented on top of an array database system, but, it is unclear if a formal, generic and reusable formal embedding has been attempted. [7] focuses on generic data types that are used via a library in a programming language, and the implemented data types use DBS technology for storage. Our work is different in two regards: first, our fields types are designed to extend stream data models (instead of subsuming them). Secondly, our work focuses on streams while the work in Camara et al. [7] addresses persistently stored and long-term collected data.

Similar to us, Ferreira et al. [13] are motivated by the increasing sampling density, and recognize the need to support how objects and fields evolve over time in a more flexible way so that integrated spatial analysis is simplified. An algebra for spatio-temporal data is proposed. This approach is less generic in its data types than [7] and our work since concrete types for time series, trajectories, coverages, as well as objects and events are proposed. Further, this work focuses on both fields and objects and their respective relationships, while our work aims at providing a flexible type systems for generic, composable and potentially complex continuous ST fields. Also, an embedding into a data model or query language is not addressed in [13].

The existing extensions to DSE data models and query languages for spatio-temporal streams are limited today. Therefore, fields can only be support by processing individual sensor data streams and integrating them in application code. Beside naively supporting points in stream data models, the work in [14, 20] focuses on spatio-temporal objects such as moving points and moving regions, and extend Gueting's work on spatio-temporal objects [18] for streams. Our work is complementary in addressing fields and streams. [2] introduces Nile-PDT, which correlates multiple concurrent streams; Nile-PDT is unaware of the spatial dimension, and is useful for extracting features rather than representing fields. GeoStreams [19] explores using DSE for processing large raster data streams, i.e. the input of DSE queries are entire rasters, not point observation streams.

In summary, extensions to stream data model language and query language to support spatio-temporal continuous phenomena that are based on point-based observation streams do not exist today.

4   Abstract Model of Continuous Spatio-Temporal Field Data Types

Our objective is to design a data model and integrate the types *streams*, *relations* and *fields* that we defined above seamlessly. When designing the new data model, we distinguish the *abstract (data) model* and the *discrete (data) model* as different levels of abstraction for the sensor data stream model [11]. In our terminology, the abstract model defines data types in an implementation-independent, high-level and formal way. Its definition is driven by semantic understanding of the concepts. The abstract model allows us to use infinite sets in the concept definitions, without worrying about the finite (computer) representations of these sets. Thus, we can define our field data types with an infinite time domain as well as an infinite space domain regardless of the finite data structures and corresponding algorithms at the stage of the discrete data model design. For example, there is no need to worry about whether a trajectory of a moving object shall be represented as a curve or as a polyline in a two-dimensional space, while it is defined as an infinite set of points in the plane. The discrete model is defined as a data model that serves as the basis for implementing the abstract data model; this model addresses the issue of finite computer representations in the context of handling and processing sensor data streams in DSE.

4.1   Second Order Signatures

Second order signatures [17], introduced by Güting in 1993, have been widely used in the database literature to formalize relations, spatial and spatio-temporal data types [14, 18, 20]. Second order signatures allow formalizing both the syntax and semantics of data types and defining operators for those data types. Furthermore, defining our proposed field types in second order signature provides a natural interface with the spatio-temporal type hierarchies mentioned above.

The basic idea of a second-order signature is using two coupled extended signatures to describe a data model: the first signature defines a *type system* and the second signature uses the types of the first signature as sorts and defines *operators* over these types. Since we only focus on defining data types in this paper, we primarily utilize the first signature as a tool. A *signature* is a pair (S, Σ), where S is a set whose elements are called *sorts* and Σ is a set whose elements are called *operators* (note, that this operators are not the same operators defined for the data types but type constructors). In addition, a signature has an associated set of *terms* defined. In a multi-sorted signature, if $t_1, \ldots, t_n$ are terms of sorts $s_1, \ldots, s_n$ and $\omega : s_1 \times \cdots \times s_n \rightarrow s$ is n-ary operator, then $\omega(t_1, \ldots, t_n)$ is a term of sort s. In a single-sorted signature, n is equal to 1. A 0-ary operator is called a *constant*.

The basic concept of a signature for a given set of sorts is then extended to introduce automatically list sorts, product sorts, union sorts, and function sorts [17]. Based on this concept extension, the first signature of a second-order signature defines a *type system*; the sorts of the signature describe so-called *kinds* and its operators are the *type constructors*. The *terms* of this signature introduce the available types of this type system.

4.2   Abstract Model: Continuous Spatio-Temporal Field Data Types

As said a (first-order) signature consists of two sets of symbols, i.e. sorts/kinds and operators, and defines a type system. First, we introduce the kinds of our data model in Table 1. The **type constructors** column lists the operators of a signature. Using the **argument sorts**, the *terms* of the result sort (**kind**) are the possible *new data types* we can construct in the proposed data model. Each kind describes a certain set of types; for example, BASE stands for the types int, float, string, and bool. The type constructors show the signature for constructing terms of each type. In this terminology, the symbol $(.)^+$ denotes a list of one or more operands of certain sorts.

Table 1. Abstract model of continuous spatio-temporal field types.

Argument sorts	Kind	Type constructor
	$\rightarrow$ BASE	integer
		float
		string
		bool
	$\rightarrow$ SPATIAL	point
	$\rightarrow$ TIME	instant
BASE$^+$	$\rightarrow$ SPATIALFIELD	simpleSfield
		simpleSfieldvector
BASE$^+$	$\rightarrow$ TEMPORALFIELD	simpleTfield
		simpleTfieldvector
BASE$^+$	$\rightarrow$ SPATIALTEMPORALFIELD	simpleSTfield
		simpleSTfieldvector
(BASE $\cup$ SPATIAL $\cup$ TIME $\cup$ SPATIALFIELD $\cup$ TEMPORALFIELD $\cup$ SPATIALTEMPORALFIELD)$^+$	$\rightarrow$ CSTFIELD	complexSfield,
		complexTfield
		complexSTfield
		complexSfieldvector
		complexTfieldvector
		complexSTfieldvector

The kinds BASE, SPATIAL and TIME are similar to the spatio-temporal data types as defined in [18]. In our model, SPATIAL emphasizes a particular spatial type, i.e. a point location, which is used to model geographic locations of a phenomenon; a subset of the point location are the locations of sensor devices. With regard to TIME, we only consider the type instant. In general, the data types in kind BASE refer to the measurements taken by sensor nodes, which can be represented as integer, float, string, and bool values. Abstract semantics of BASE, SPATIAL, TIME have been defined in [18].

The kinds SPATIALFIELD, TEMPORALFIELD and SPATIALTEMPO-RALFIELD represent definitions of field types that are a mapping from S, T or

$S \times T$ to a single attribute (e.g. temperature). These kinds represent spatial field data types, e.g., the temperature distribution over a specific geographic region (*simpleSfield* data type), temporal field data types, e.g., temperature readings from a single sensor over a time range (*simpleTfield*), and spatio-temporal field data types, e.g., temperature distribution over a specific area and time range (*simpleSTfield*), respectively. The fields have been described in more detail in Sect. 2. In addition, we introduce the new field data types *simpleSfieldvector*, *simpleTfieldvector* and *simpleSTfieldvector*, correspondingly, to support each space-time location being mapped to a vector of values (of potentially different types). For example, an instance of data type *simpleSTfieldvector* is a mapping from a spatio-temporal location to a vector consisting of a temperature value, wind speed measurement and so on. This extension provides the capability that multiple measurements that are associated with a single spatio-temporal location can be queried and analyzed at the same time.

The data types introduced so far are the basis for our further extensions. We introduce a new kind CSTFIELD that denotes Complex Spatio-Temporal Fields. Taking the *complexSfield* (complex spatial field) data type as an example, the value domain is not limited to basic measurements as in all other types defined before; instead, the value domain can be a combination of any of the data types defined in the data model so far. Therefore, an instance of *complexSfield* can be a spatial field, in which each location is mapped to a spatial object (e.g., each spatial location is mapped to a view shed region), which is similar to the idea of an object field as defined in [9]. Furthermore, a simple spatial/temporal field can also be a valid domain value for complex spatial/temporal fields. For example, a *complexTfield* (complex temporal field) can be a mapping from a temporal instant to a spatial field (as we discussed before, a snapshot point of view $f : T \rightarrow (S \rightarrow V)$ which the spatio-temporal field can be equivalently defined). Similarly, we introduce *complexSfieldvector*, *complexTfieldvector*, *complexSTfieldvector* data types for supporting the representation of multiple values/objects/fields under one spatio-temporal framework. If end users need multiple spatio-temporal fields to be correlated, the *complexSTfieldvector* data type is necessary. The data types of kind CSTFIELD are designed to add more query capability and representational flexibility for end users.

We use the notation A_α to denote the carrier set for each data type, where α is the data type. Each carrier set is extended with the null value $\perp$ that denotes a missing or undefined value. For convenience, we define $\bar{A}_\alpha = A_\alpha \setminus \{\perp\}$.

4.3   Semantics of the Type System

The type *simpleSfield* is similar to the commonly used spatial fields in a GIS.

Definition 3. *A simpleSfield(α) is a data type with a carrier set*

$$A_{simpleSfield(\alpha)} \equiv \{f \mid f : \bar{A}_{point} \rightarrow A_\alpha \cup \{\perp\}\}$$

where α is a data type (integer, float, string, or boolean) applicable to the type constructor simpleSfield and the carrier set A_α denoting any possible data type of BASE kind.

Next, we define a set *ValueVector* of BASE data types and then use that to define the data type *simpleSfieldvector*.

Definition 4. *A ValueVector is a list of basic data types BASE$^+$ with a carrier set*

$$A_{ValueVector} \equiv \{\{s_1, \ldots, s_n\} \mid \forall i\, [s_i \in A_{integer} \cup A_{real} \cup A_{string} \cup A_{bool} \cup \{\bot\}]\}$$

for some $n \geq 2$.

Definition 5. *A simpleSfieldvector(α) is a data type with carrier set*

$$A_{simpleSfieldvector(\alpha)} \equiv \{f \mid f : \bar{A}_{point} \to A_\alpha\} \cup \{\bot\}$$

where α is a data type in sort BASE$^+$, with carrier set A_α denoting ValueVector.

Similarly, *simpleTfieldvector* and *simpleSTfieldvector* can be defined as follows.

Definition 6. *simpleTfieldvector(α) and simpleSTfieldvector(α) are data types with respective carrier sets*

$$A_{simpleTfieldvector(\alpha)} \equiv \{f \mid f : \bar{A}_{instant} \to A_\alpha\} \cup \{\bot\},$$
$$A_{simpleSTfieldvector(\alpha)} \equiv \{f \mid f : \bar{A}_{point} \times \bar{A}_{instant} \to A_\alpha\} \cup \{\bot\}$$

where α is a data type applicable to the type constructor simpleTfieldvector and simpleSTfieldvector, respectively, with the carrier set A_α denoting ValueVector.

Next, we define *simpleTfield* and *simpleSTfield* analogous to the earlier definition of *simpleSfield*.

Definition 7. *simpleTfield(α) and simpleSTfield(α) are data types with respective carrier sets*

$$A_{simpleTfield(\alpha)} \equiv \{f \mid f : \bar{A}_{instant} \to A_\alpha\} \cup \{\bot\},$$
$$A_{simpleSTfield(\alpha)} \equiv \{f \mid f : \bar{A}_{point} \times \bar{A}_{instant} \to A_\alpha\} \cup \{\bot\}$$

where α is a data type applicable to the type constructor simpleTfield and simpleSTfield, respectively, with the carrier set A_α denoting any possible data type of BASE kind.

The final types – the *complexSfield*, *complexTfield*, and *complexSTfield* type constructors and their vector analogues – are high-level abstractions that hide individual sensors, their locations, and measurement values and provide an integrated field view.

Definition 8. *complexSfield*(α), *complexTfield*(α), *and complexSTfield*(α) *are data types with the respective carrier sets*

$$A_{complexSfield(\alpha)} \equiv \{f \mid f : \bar{A}_{point} \to A_\alpha\} \cup \{\bot\}$$
$$A_{complexTfield(\alpha)} \equiv \{f \mid f : A_{instant} \to A_\alpha\} \cup \{\bot\}$$
$$A_{complexSTfield(\alpha)} \equiv \{f \mid f : \bar{A}_{point} \times \bar{A}_{instant} \to A_\alpha\} \cup \{\bot\}$$

where α *is a data type applicable to the type constructor complexSfield, complexTfield and complexSTfield, respectively, with the carrier set* A_α *denoting any possible data type in* $A_{BASE} \cup A_{point} \cup A_{instant} \cup A_{simpleSfield} \cup A_{simpleTfield} \cup A_{simpleSTfield} \cup A_{simpleSfieldvector} \cup A_{simpleTfieldvector} \cup A_{simpleSTfieldvector}$.

Definition 9. *complexSfieldvector*(α), *complexTfieldvector*(α) *and complexSTfieldvector*(α) *are data types with the respective carrier sets*

$$A_{complexSfieldvector(\alpha)} \equiv \{f \mid f : \bar{A}_{point} \to A_\alpha\} \cup \{\bot\}$$
$$A_{complexTfieldvector(\alpha)} \equiv \{f \mid f : A_{instant} \to A_\alpha\} \cup \{\bot\}$$
$$A_{complexSTfieldvector(\alpha)} \equiv \{f \mid f : \bar{A}_{point} \times \bar{A}_{instant} \to A_\alpha\} \cup \{\bot\}$$

where α *is a data type applicable to the type constructor complexSfieldvector, complexTfieldvector and complexSTfieldvector, respectively, with the carrier set* A_α *denoting any possible data type in*

$$\{ \{s_1, \ldots, s_n\} \mid \forall i \; [s_i \in \; A_{BASE} \cup A_{point} \cup A_{instant} \cup$$
$$A_{simpleSfield} \cup A_{simpleTfield} \cup A_{simpleSTfield} \cup$$
$$A_{simpleSfieldvector} \cup A_{simpleTfieldvector} \cup A_{simpleSTfieldvector} \cup \{\bot\}]\}$$

We now have specified the data model from an abstract perspective by defining the range of spatio-temporal fields we support in our stream data model. We have defined which field types are possible, and how they are constructed.

5   Extending Stream Data Models with Field Data Types

The previous section formalized the spatio-temporal fields data types that our data model supports on an abstract level that assumes continuity in space and time. This section presents the discrete versions of these field data types that are necessary to implement the data types on the basis of streams of *discrete observations* and *discretized views of space and time*.

5.1 From Spatio-Temporal Streams to Continuous Spatio-Temporal Fields

As mentioned in Sect. 2, this paper addresses fields that are approximated based on observations, so-called *sampled fields* (in contrast to equation fields that are purely defined via equations). In particular, we introduce fields types that are canonically constructed from spatio-temporal streams. Our approach starts with the definition of a *spatio-temporal relation*, which is a finite set of tuples, and each tuple is an observation. Using a spatio-temporal relation that contains tuples with timestamps within a well-defined interval and spatial locations within a well-defined spatial region, we can construct an *observation field*. An observation field is simply a collection of raw sensor samples. Each sample represents a point in a spatio-temporal volume; other values are not available. Once we add an *interpolator* to an observation field, we can create a *continuous spatiotemporal field*. On behalf of the raw samples of the observation field and the interpolator, the continuous spatio-temporal field can produce *all* values within its continuous spatial and temporal domain. Some of these values correspond to actual samples; others are interpolated based on the samples. Below, we introduce streaming versions of the discussed types; this includes *sdstream*, *observationfield*, and *continuousSTfield*. The mapping between relational types and streaming types is bidirectional. Due to space constraints, we refer the reader to more details of this discussion in [23] (Table 2).

Table 2. Discrete model of spatio-temporal field types extending stream data models.

Argument sorts	Kind	Type constructor
	→ BASE	integer
		float
		string
		bool
	→ SPATIAL	point,
		geometry
	→ TIME	instant
BASE⁺	→ SDATA	sensordata
SDATA	→SDSTREAM	sdstream
	→ WINDOW	slidingwindow
SDSTREAM×WINDOW	→ STREL	observationfield
	→ INTERPOL	interpolator
observationfield×INTERPOL	→ CSTFIELD	continuousSTfield
		continuousSfield
		continuousTfield
(BASE∪SPATIAL∪TIME∪ CSTFIELD)⁺	→ CCSTFIELD	complexcontinuousSTfield

5.2   Discrete Data Model for Continuous Spatio-Temporal Field Data Types

The kinds BASE and TIME are maintained from the abstract data model. For the SPATIAL kind we add a *geometry* type that can represent various geometric objects such as points, lines, 2D regions, and aggregations thereof. This geometric type is introduced to better represent the results of operators over fields, whose discussion is beyond the scope of this paper.

Data Type for Sensor Tuples: This data type is defined to represent an individual sensor sample using the corresponding basic data types and the type constructor *sensordata*. A *sensor data tuple* represents a single update from a sensor, while the time series of a sensor's updates is a stream. The *sensordata*'s constructor uses BASE$^+$ as input sorts. The rationale behind this choice is that a sensor node can combine all or a subset of its sensor samples from different attached sensors that are taken at one time instant and forward them compactly as a single message, creating an update tuple. Sensors without an update at that time report a NULL value as part of the update tuple. We assume that a sensor data tuple will always contain at least one time stamp of the kind TIME and a location value of type *point* since a data value without any time or any location information is meaningless. A second time stamp might be created at the DSE to represent the arrival time of the update tuple.

Definition 10. *A sensordata(α) is a data type with a carrier set*

$$A_{sensordata}(\alpha) \equiv \bar{A}_{point} \times \bar{A}_{instant} \times A_{integer} \times A_{\alpha}$$

where α is a data type in sort BASE$^+$ that is applicable to the type constructor sensordata and produces an output of the kind SDATA.

The carrier set of data type α is a *Value Vector*, defined in Definition 4. The three other parameters A_{point}, $A_{instant}$, and $A_{integer}$ represent the carrier sets for location, explicit time stamp (time of observation), and implicit time stamp (arrival time at the DSE), respectively. The explicit time stamp is required whereas the implicit time stamp is an optional parameter, as indicated by the missing bar across $A_{integer}$.

Sensor Data Stream Type: Next, we define the data type for representing streams of sensor data tuples. Each value from the carrier set of this new data type *sdstream(α)* is a function that maps each implicit time stamp to a finite number (possibly zero) of sensor data tuples [4].

Definition 11. *A sdstream(α) is a data type with a carrier set*

$$A_{sdstream(\alpha)} \equiv \{f \mid A_{instant} \rightarrow \{S \subseteq A_{\alpha} \mid |S| < \infty\}\}$$

where α is a data type of sort SDATA that is applicable to the type constructor sdstream.

Sliding Window Type: For kind WINDOW, the data type constructor *slidingwindow* represents the concept of sliding windows in DSE. We adopt the two commonly used parameters to specify sliding windows: window size *ws* and update interval *ui*. We only consider tuple-based (count-based) or time-based sliding windows.

Definition 12. *A slidingwindow is a data type with a carrier set*

$$A_{slidingwindow} \equiv \Big\{ f \mid f : T_{eval} \to T_S \times T_E \quad \text{such that } \forall (t_{start_i}, t_{end_i}) \in T_{eval}$$
$$[t_{start_i} \leq t_{end_i} \text{ and } t_{end_i} - t_{start_i} = ws \text{ and}$$
$$t_{start_i} - t_{start_{i-1}} = ui] \Big\}$$

where ws is the sliding window size, ui is the sliding window update interval, T_{eval} is the carrier set of time stamps when a sliding window will be evaluated, and $T_S \times T_E$ is the carrier set that indicates the start and end time of a specific sliding window.

For a tuple-based (count-based) window $T_{eval}, T_S, T_E \subseteq \bar{A}_{integer}$ holds, while for a time-based window $T_{eval}, T_S, T_E \subseteq \bar{A}_{instant}$ holds. This is due to the fact, that we consider two different types of windows semantics: count-based window change their designated time interval with the arrival of a new streaming tuple, while time-based windows change their time interval with the advancement of time.

Observation Field Type: Using sliding windows, the most recent portion of a data stream can be regarded as a temporalized relation [4]. However, we prefer to define the more meaningful type *observationfield* that captures the finite set of sensor data points – each containing a spatial location, a timestamp, and a measurement value – from a window and defined them as a single field. An *observationfield* represents raw sensor data measurement streams as a ***discrete spatio-temporal field*** where values are only valid at the spatial/temporal locations where actual sensor measurement are available.

Definition 13. *An observationfield is a data type with carrier set*

$$A_{observationfield(sdstream(\alpha),\omega)} \equiv \Big\{ f \mid f : T_{eval} \to S_{eval} \quad \text{such that } \forall f(t_i) \in S_{eval_i}$$
$$[f(t_i).timestamp \in R(\omega, t_i)] \Big\}$$

where $S_{eval} = \{S \subseteq A_\alpha \mid |S| < \infty\}$, α is a data type of sort SDATA, $sdstream(\alpha)$ is a data type of sort SDSTREAM, ω is a data type of sort WINDOW, T_{eval} is the carrier set of timestamps when a sliding window ω is evaluated, and $R(\omega, t_{eval})$ is the time range at each evaluation timestamp $t_{eval} \in T_{eval}$ of the sliding window ω.

Interpolator Type: Now, we introduce the interpolator type required to convert streams of discrete sensor measurements into continuous spatio-temporal fields. More precisely, it is intended to estimate any values at arbitrary spatio-temporal locations within a specific instance of type *observationfield*, even if no sensor measurements are available at those precise spatio-temporal locations.

Definition 14. *An interpolator is a data type with a carrier set*

$$A_{interpolator} \equiv \{f \mid f : A_{observationfield(sdstream(\alpha),\omega)} \times \bar{A}_{point} \times \bar{A}_{instant} \rightarrow A_{\sigma}\}$$

where α is a data type of sort SDATA, sdstream(α) is a data type of sort SDSTREAM, ω is a data type of sort WINDOW, observationfield(sdstream(α), ω) is of data type observationfield, and σ is a data type in sort BASE$^+$.

Continuous Spatial Temporal Field Type: Since we define the continuous spatio-temporal field type in the context of handling discrete streaming sensor data, the new type *continuousSTfield* is explicitly based on the observation field and the corresponding interpolator function. The resulting *continuousSTfield* is continuously updated with the windows parameters from the underlying observation field.

Definition 15. *Assume a given interpolator with output of type σ and an observationfield denoted as observationfield(sdstream((α), ω). Then a continuousSTfield is a data type with a carrier set*

$$A_{continuousSTfield(observationfield(sdstream(\alpha),\omega))} \equiv$$
$$\left\{f \mid f : T_{eval} \rightarrow \{f^c \mid f^c : \bar{A}_{point} \times \bar{A}_{instant} \rightarrow A_{\beta} \text{ such that } A_{\beta} \subseteq A_{\sigma}\}\right\}$$

where α is a data type of sort SDATA, sdstream(α) is a data type of sort SDSTREAM, ω is a data type of sort WINDOW, T_{eval} is the carrier set of timestamps when a sliding window ω is evaluated, and β is a data type in sort BASE$^+$.

Other continuous spatial/temporal field types *continuousSfield* and *continuousTfield* and the complex type *complexcontinuousSTfield* closely follow the definitions from the abstract data model, except that they are now streaming versions correspondingly to the definition of a *continuousSTfield*.

6   Conclusions and Future Work

In this paper, we presented our formal extension of stream data models with field data types. Using the field data types, users can define high-level abstractions of continuous ST phenomena based on large numbers of concurrent, bursty and unpredictable sensor data streams that are now known to a DSE. Therefore, they can be supported through queries and processing optimization, thus, unburdening the user and application code. We introduced field types specifically for sampled fields, and their streaming counterparts. Our stream model extension formally integrates spatio-temporal streams, spatio-temporal relations and field types. We formalized the proposed types using second order signature to achieve independence from the details of a specific data model language implementation, and formalized the syntax as well as semantics of the proposed types. As for future work, several aspects closely related to this work require more research. For instance, the challenging question of generic operators over the proposed fields types that can be integrated into stream query languages requires further investigation. Similarly, a prototypical implementation of the type system as part of various actual data model languages is of significant interest.

Acknowledgement. The authors would like to thank Mark Plummer for many, fruitful discussions and the National Science Foundation for supporting this work via Award No. 1527504.

References

1. Apache Spark (2016). http://spark.apache.org
2. Ali, M.H., Aref, W.G., Bose, R., Elmagarmid, A.K., Helal, A., Kamel, I., Mokbel, M.F.: NILE-PDT: a phenomenon detection and tracking framework for data stream management systems. In: VLDB 2005, pp. 1295–1298. VLDB Endowment (2005)
3. Ali, M., Chandramouli, B., Raman, B., Katibah, E.: Spatio-temporal stream processing in Microsoft StreamInsight. IEEE Data Eng. Bull. **33**(2), 69–74 (2010)
4. Arasu, A., Babu, S., Widom, J.: The CQL continuous query language: semantic foundations and query execution. VLDB J. **15**(2), 121–142 (2005)
5. Babu, S., Widom, J.: Continuous queries over data streams. ACM SIGMOD Rec. **30**(3), 109 (2001)
6. Baumann, P.: The OGC web coverage processing service (WCPS) standard. Geoinformatica **14**(4), 447–479 (2010)
7. Camara, G., Egenhofer, M.J., Ferreira, K., Andrade, P., Queiroz, G., Sanchez, A., Jones, J., Vinhas, L.: Fields as a generic data type for big spatial data. In: Duckham, M., Pebesma, E., Stewart, K., Frank, A.U. (eds.) GIScience 2014. LNCS, vol. 8728, pp. 159–172. Springer, Heidelberg (2014)
8. Couclelis, H.: People manipulate objects (but cultivate fields): beyond the raster-vector debate in GIS. In: Frank, A.U., Campari, I., Formentini, U. (eds.) Theories and Methods of Spatio-Temporal Reasoning in Geographic Space. LNCS, vol. 639, pp. 65–77. Springer, Berlin (1992)
9. Cova, T., Goodchild, M.: Extending geographical representation to include fields of spatial objects. Int. J. Geogr. Inf. Sci. **16**(6), 509–532 (2002)
10. Duckham, M., Zhong, X., Toohey, K.: Challenges to using decentralized spatial algorithms in the field: the risernet geosensor network case study. SIGSPATIAL Newlett. Spec. Issue "Geosens. Netw." **7**(2), 14–21 (2015)
11. Erwig, M., Güting, R.H., Schneider, M., Vazirgiannis, M.: Spatio-temporal data types: an approach to modeling and querying moving objects in databases. GeoInformatica **3**(3), 269–296 (1999)
12. Faulkner, M., Olson, M., Chandy, R., Krause, J., Chandy, K., Krause, A.: The next big one: detecting earthquakes and other rare events from community-based sensors. In: International Conference on Information Processing in Sensor Networks (IPSN), pp. 13–24 (2011)

13. Ferreira, K.R., Camara, G., Monteiro, A.M.V.: An algebra for spatiotemporal data: from observations to events. Trans. GIS **18**(2), 253–269 (2014)
14. Galić, Z., Baranović, M., Križanović, K., Mešković, E.: Geospatial data streams: formal framework and implementation. Data Knowl. Eng. **91**, 1–16 (2014)
15. Galton, A.: A formal theory of objects and fields. In: Montello, D.R. (ed.) COSIT 2001. LNCS, vol. 2205, pp. 458–473. Springer, Heidelberg (2001)
16. Galton, A.: Fields and objects in space, time, and space-time. Spat. Cogn. Comput. **1**, 39–68 (2004)
17. Güting, R.: Second-order signature: a tool for specifying data models, query processing, and optimization. In: SIGMOD 1993, pp. 277–286. ACM, New York (1993)
18. Güting, R., Michael, H., Erwig, M., Jensen, C., Lorentzos, N., Schneider, M., Vazirgiannis, M.: A foundation for representing and querying moving objects. 1(212), 1–37 (2000)
19. Hart, Q., Gertz, M.: Querying streaming geospatial image data. In: SSDBM 2005, Santa Barbara, CA, USA, pp. 147–150 (2005)
20. Huang, Y., Zhang, C.: New data types and operations to support geo-streams. In: Cova, T.J., Miller, H.J., Beard, K., Frank, A.U., Goodchild, M.F. (eds.) GIScience 2008. LNCS, vol. 5266, pp. 106–118. Springer, Heidelberg (2008)
21. Kemp, K.: Fields as a framework for integrating GIS and environmental process models. Part 1: representing spatial continuity. Trans. GIS **1**(3), 219–234 (1996)
22. Laurini, R., Paolino, L., Sebillo, M., Tortora, G., Vitiello, G.: A spatial SQL extension for continuous field querying. In: International Computer Software Applications Conference (COMPSAC 2004) vol. 2, pp. 78–81 (2004)
23. Liang, Q.: Towards the continuous spatio-temporal field model for sensor data streams. Dissertation, University of Maine (2015)
24. Nittel, S.: Real-time sensor data streams. SIGSPATIAL Newslett. Spec. Issue "Geosens. Netw." **7**(2), 22–28 (2015)
25. Sanchez, L., Galache, J., Gutierrez, V., Hernandez, J.M., Bernat, J., Gluhak, A., Garcia, T.: SmartSantander: the meeting point between future internet research and experimentation and the smart cities. In: Future Network and Mobile Summit (FutureNetw), pp. 1–8 (2011)
26. Stonebraker, M., Çetintemel, U., Zdonik, S.: The 8 requirements of real-time stream processing. ACM SIGMOD Rec. **34**(4), 42–47 (2005)
27. Whittier, J., Liang, Q., Nittel, S.: Evaluating stream predicates over dynamic fields. In: Proceedings of 5th International ACM SIGSPATIAL Workshop on GeoStreaming, pp. 2–11 (2014)
28. Whittier, J., Nittel, S., Liang, Q., Plummer, M.: Towards window stream queries over continuous phenomena. In: Proceedings of 4th International ACM SIGSPATIAL Workshop on GeoStreaming, Orlando, FL, pp. 1–10 (2013)

Point Partitions: A Qualitative Representation for Region-Based Spatial Scenes in $\mathbb{R}^2$

Joshua A. Lewis[✉] and Max J. Egenhofer

School of Computing and Information Science, University of Maine,
5711 Boardman Hall, Orono, ME 04469-5711, USA
joshua.lewis@umit.maine.edu, max@spatial.maine.edu

Abstract. A complete qualitative scene description should be such that it captures the essential details of a configuration so that a topologically correct depiction can be recreated. This paper models a spatial scene through sequences of point partitions, that is, how embedding space and objects are distributed around the intersections of the boundaries of regions. Twenty-three base patterns are identified, which suffice to capture complex scenes, including configurations with holes. To demonstrate the diagrammatic depiction of a spatial scene from point partition patterns, such a scene is recreated using the developed model. The paper also provides a means of transitioning between these more complex relations and the eight coarse topological relations of the 4-intersection.

Keywords: Spatial scenes · Regions · Complex objects · Partition · Topology

1 Introduction

In order to model a spatial scene—a collection of spatial objects and their qualitative spatial relations [2, 15–17] —there are several considerations to be made, such as which embedding space, objects, and qualitative measures could be of interest, and to what degree existing models are able to provide meaningful solutions. An ideal model should allow a qualitative description of such a scene to produce a topologically correct representation of that scene, in the form of a diagram or a graphic. Conversely, such a depiction should be able to generate the original scene description without ambiguity, so that two topologically equivalent configurations yield the same spatial scene.

The most familiar models for representing the topological relations between spatial regions—the 4-intersecton [8] and RCC-8 [18] —involve the set of eight binary relations in $\mathbb{R}^2$ (Fig. 1).

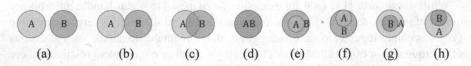

Fig. 1. The eight region-region relations in $\mathbb{R}^2$, described by the 4-intersection: (a) *disjoint*, (b) *meet*, (c) *overlap*, (d) *equal*, (e) *inside*, (f) *coveredBy*, (g) *contains*, and (h) *covers*.

J.A. Miller et al. (Eds.): GIScience 2016, LNCS 9927, pp. 195–209, 2016.
DOI: 10.1007/978-3-319-45738-3_13

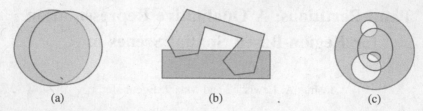

(a)   (b)   (c)

Fig. 2. Three configurations that map onto the same binary relation using 9-intersection or RCC-8: (a) *overlap* between two simple regions, (b) two regions overlapping to form a hole, and (c) two holed regions that overlap.

These coarse qualitative models alone, however, may be insufficient to handle the complexities that may be present within a scene [14] (Fig. 2).

Binary relations with holed and separated regions have been addressed in various ways. RCC-8 [18] accommodates regions with holes and separations, yet it does not differentiate between regions in the outer exterior or enclosed exterior of a holed region. The use of the vanilla 9-intersection [9] to capture relations between complex regions [17] has similar shortcomings. The 9^+-intersection [12, 13] allows the interior, boundary, and exterior components of the 9-intersection [9] to be split, enabling more refined objects to be modeled, such as those with separations of interiors, boundaries, or exteriors (e.g., separated regions, holed regions, or directed lines), thereby capturing more details than the coarse models. Likewise, the compound object model [6] allows for the construction of holed regions or regions with cuts via set difference of basic objects, as well as separations and regions with spikes through the union of basic objects. Other approaches focus on particular domains of relations (e.g., holed regions [10, 20] and separations [5]) or on specific relations, such as types of *overlap* [11], types of *surrounds* [4] and the interplay between complex points, lines and regions [19].

None of these models, however, accounts for the different types of boundary-boundary intersections and the sequences of such intersections along the objects' boundaries, which are germane to capturing essential details of a spatial scene so that a topologically correct depiction can be reconstructed from the symbolic qualitative representation. For binary relations, types and sequences of boundary-boundary intersections have been addressed [7], but these aspects have not been fully explored to capture the potential complexities of scenes with arbitrary numbers of complexly structured spatial objects. These constraints for line-like boundaries have been applied to line-line relations for complex scenes comprised of line segments [3], but that work is not immediately extensible to the boundaries of areal objects in a manner that allows specific region-region relations to be derived.

While *o-notation* [14] (and its extension *i-notation* [15]) can handle an arbitrary number of regions, regions with holes and separations, and situations where an ensemble of regions comes together to surround regions, they are unable to handle such situations as the sequence for objects that all meet at a single point or containment relations where the containing region is divided. Furthermore, *o-notation* and *i-notation* are verbose and redundant, as they repeat the specifications of boundary-boundary intersections for each involved object. This paper builds on the previous models, while overcoming their shortfalls. To ensure a proper grounding, only two objects will be considered at a time, instead of a scene of many objects, but each object will be allowed an arbitrary number of disconnected separations and holes. In the process a set of detailed region-region relations is developed, as well as a bridge to connect them to the more familiar coarse relations (Fig. 2). This process would enable a simple natural-language description of space to be modeled more robustly [21]. The opposite should also be true, representing a detailed scene as something less complex, and easier to understand (Fig. 3).

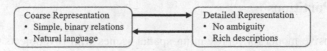

Fig. 3. Dependencies among course representations and detailed representations.

A detailed representation of a spatial scene enables the capture of an arbitrary degree of complexity between regions—potentially much more than is possible with the base relations. To varying degrees of specificity, recent models have attempted to represent the complexity of spatial scenes between two, or sometimes an arbitrary number of, regions. Maptree utilizes combinatorial maps to represent the structure of a scene [22], however a means of teasing coarse relations out of such a structure has not been developed.

As an alternative to just building more complex structures, one might also consider representing more complex relations. For instance, one can model *overlap* through an enumeration of connected components under union and set difference in order to represent the relation along with additional complexities, such as the number of partitions the exterior is divided into [11]. Further refinements to coarse relations might involve recording the sequence of intersections between regions, whether they form a crossing or a touching configuration, their dimension (qualitative length) and the relation to the objects' compliments (indicating whether the exterior is partitioned, for instance) [7]. For example, a scene described by a touch-cross-cross sequence (Fig. 4a) is distinct from a scene described by a cross-touch-cross sequence (Fig. 4b). Without such specification it is sometimes impossible to represent a scene uniquely.

The remainder of this paper develops a new model for capturing the detailed relations between regions, a model that is both relatable to the familiar 9-intersection

Fig. 4. Two simple scenes with different forms of *overlap*: (a) a *touch-cross-cross* sequence, and (b) a *cross-touch-cross* sequence.

and less ambiguous than contemporary theories of spatial scenes, such as *o-notation*. The paper is structured as follows: Section 2 introduces point partitions to capture topological relations. Section 3 demonstrates the construction of spatial scene diagrams from point partitions. Section 4 relates the coarse topological relations with the detailed point partition relations. Section 5 provides the additional structure for modeling regions with holes. Section 6 extends the model to scenes with more than two regions. Section 7 draws conclusions and provides insights into future lines of research.

2   Point Partitions

A comprehensive model for representing spatial scenes with two objects needs to not only capture any number of boundary intersections existing in concert, but also their sequence and dimension [9] — and without the potential ambiguity and redundancy of other approaches. We introduce *point partitions* (P^2) to allow any point within a spatial scene to be characterized in terms of how space around that point is partitioned. A partition of a space X is a collection of mutually disjoint subsets of X whose union is X [1]. When considering an intersection point between the boundaries of two spatial regions, these point partitions provide a language for describing their localized spatial interactions. The space around a point of interest (such as a boundary intersection point) is partitioned into radial slices, called *cells*. Any boundary segments that touch or cross in a point form the boundaries of these cells.

2.1   Representing Spatial Scenes

In a spatial scene (Fig. 5a), the boundary-boundary intersections contain critical information, which can be analyzed in isolation (Fig. 5b–d). The neighborhood of each boundary-boundary intersection also contains metric information (e.g., the shapes of the boundaries). Such information will be discarded and only a diagrammatic representation will be used for each intersection (Fig. 5e–g).

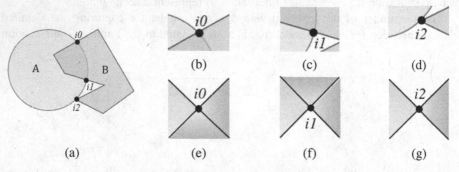

Fig. 5. The partitioning around boundary-boundary intersections: (a) a scene where two regions intersect at three points, (b-d) cells around intersection *i0, i1, i2,* and (e-g) the diagrammatic representations of *i0*, *i1*, and *i2*.

To represent the point partitions that contextualize an intersection it is possible to take a diagrammatic representation and turn it into a symbolic representation: when the space around any point is partitioned the cells that define that space form a distinct sequence that allows that space to be described and identified as participating in a specific spatial relation.

2.1.1   Sequence of Point-Like Boundary-Boundary Intersections

Consider the scene where two regions, A and B, intersect (Fig. 5a). These regions share three distinct points: their boundary intersections at $i0$, $i1$, and $i2$. Walking around the boundary of A clockwise one visits intersection $i0$, followed by intersection $i1$, then $i2$, and back to $i0$ again. This loop can be permuted cyclically, i.e., the sequences ($i0$, $i1$, $i1$), ($i1$, $i2$, $i0$), and ($i2$, $i0$, $i1$) refer to the same configuration). Alternatively, starting with B at $i0$, heading to $i2$, then $i1$, and finally back to $i0$ completes the sequence for that region. These sequences, ($i0$, $i1$, $i2$) for A and ($i0$, $i2$, $i1$) for B, define the boundaries of those objects, allowing us to relate A and B through specific shared points.

2.1.2   Point-Like Boundary-Boundary Intersections

For point-like boundary-boundary intersections, the space immediately surrounding each point in this case is divided into four cells (Fig. 5b-d). For $i0$ this results in a cell corresponding to A, a cell corresponding to the exterior, a cell corresponding to B, and a cell corresponding to the intersection of A and B. Fully written this sequence is {A}, {B}, {A, B},{ }, where each cell is listed as the set of object interiors that it contains. Following in this manner the series of cells around $i1$ is {A}, {A, B}, {B}, { }, and the sequence around $i2$ is {A}, { }, {B}, { }.

The illustrated scene is but one of many possibilities. Between two objects, A and B, their intersection, and the exterior, it is possible to construct sequences by partitioning space into one to four cells for regions that do not self-intersect. In total there are 23 such patterns (Sects. 2.2.1, 2.2.2, 2.2.3 and 2.2.4), each a unique sequence of interior cells surrounding an intersection point, which are called *point partitions* (P^2). For ease of reference, each pattern will be referred to by name instead of sequence whenever possible going forward. These patterns can then be combined to form a bridge between *scene notation* and other approaches. Such a correspondence is important when the vast measure of spatial data is underspecified relative to such a detailed representation.

2.1.3   Linear Boundary Intersections

While sequence has been shown to be central to the construction of P^2 patterns, the dimension of an intersection between regions is also important when generating topologically correct results from a scene description. Intersections that are 0-dimensional are immediately relatable within the context of P^2 relations, whereas 1-dimensional intersections are more involved. Since this model is based on point-intersections, each 1-dimensional intersection (Fig. 6a) is split into a starting endpoint depicted by a distinct sequence of cells that surround it (Fig. 6b), and a finishing endpoint depicted by its own sequence of cells (Fig. 6c).

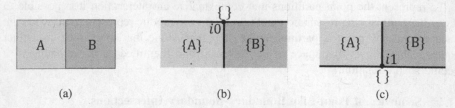

Fig. 6. The representation of a 1-dimensional intersection: (a) the 1-dimensional intersection as two regions *meet*, (b) the start of a *meet* relation, and (c) the end of a *meet* relation.

By splitting line segments into pairs of endpoints the set of P^2 relations will be able to handle 1-dimensional intersections in the exact same manner as 0-dimensional intersections.

2.2   The 23 Base Patterns

The following sections describe relations with one cell, two cells, three cells, and four cells respectively. Since the alphabet of P^2 patterns is comprised of $\{A\}$, $\{B\}$, $\{A, B\}$, and $\{\}$, and no region intersects with itself, $\{A\}$, $\{A\}$, $\{B\}$, $\{B\}$ is not a valid sequence. It is impossible to produce meaningful sequences of more than four cells between pairs of regions.

2.2.1   Single Partitions

Relations with a single cell (Fig. 7) will never appear in the construction of a scene, but may nonetheless describe the context of a specified point. If one starts with a relation where A and B *meet* at a point *i0* and regions A and B are then shrunk, that point of interest, *i0*, at the former intersection then becomes defined by a cell that only contains the exterior (Fig. 7d), for instance. If A grows and B shrinks, that same point may instead be defined as being *inside* A (Fig. 7a) — it is no longer representative of a boundary intersection, but it may be important to the history of the scene to know that that point has been subsumed fully by the interior of A.

If a point does not sit on a boundary these four trivial point partitions may arise: (1) a point is *inside* A's interior only, *inside* A, (2) a point is inside B *'s interior* only, *inside* B, (3) a point is *inside* the intersection of A and B, *inside* AB, and (4) the point is fully in the exterior, and therefore *inside exterior* of A and B.

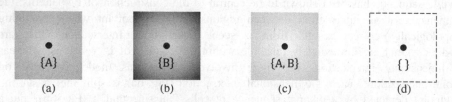

Fig. 7. The four P^2 patterns with a single cell: (a) *inside* A, (b) *inside* B, (c) *inside* AB, and (d) *inside exterior*.

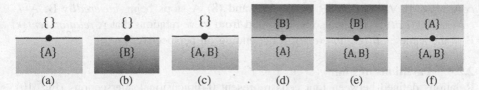

Fig. 8. The six P^2 relations between two cells with respect to regions A and B: (a) A *localDisjoint* B, (b) B *localDisjoint* A, (c) A *localEqual* B, (d) A *localAttach* B, (e) A *localInside* B, and (f) A *localCovers* B.

2.2.2 Partitions of Two
Among the simplest P^2 patterns are those that are defined between two cells. There are six of these relations, labeled B *localDisjoint* A, A *localDisjoint* B, A *localEqual* B, A *localAttach* B, A *localInside* B, and A *localCovers* B (Fig. 8). These relations are prefixed with *local** because within the context of a single point it is impossible to know if the relation holds for the entire object or just at the specified intersection without first considering the other intersections the objects participate in.

2.2.3 Partitions of Three
Relations between three cells are special in that they represent 1-dimensional intersections. Each of the 3-cell relations represent either the start or end of a boundary segment (Fig. 9).

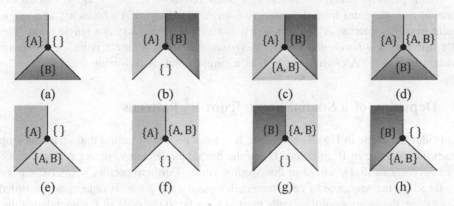

Fig. 9. The eight P^2 relations between three cells with respect to regions A and B: (a) A *1-meet (enter)* B, (b) A *1-meet (exit)* B, (c) A *1overlap (enter)* B, (d) A *1-overlap (exit)* B, (e) A *1-covers (enter)* B, (f) A *1-covers (exit)* B, (g) A *1-coveredBy (enter)* B, and (h) A *1-coveredBy (exit)* B.

Point patterns between three cells may involve: (1) B and A *meet* along a line in a clockwise traversal, A *1-meet (enter)* B, (2) A and B *meet* along a line in a clockwise traversal, A *1-meet (exit)* B, (3) A *crosses* over B's boundary, A *1-overlap (enter)* B, (4) A *crosses* out of B's boundary, A *1-overlap (exit)* B, (5) A starts to *cover* B, A *1-covers (enter)* B, (6) A stops *covering* B, A *1-covers (exit)* B, (7) A starts being *coveredBy* B, A *1-coveredBy* (*enter*) B, and (8) A stops being *coveredBy* B, A *1-coveredBy* (*exit*) B. It can also be reasoned from these relations that A *relation* (*enter*) B is the same as *B relation* (*exit*) *A*, depending on perspective.

2.2.4 Partitions of Four

Relations defined between four cells represent 0-dimensional intersections (Fig. 10). There are five of these relations, which correspond to: *0-meet*, *0-covers*, *0-coveredBy*, and *0-overlap* (enter/exit). The initial example (Fig. 5) depicted A *0-overlap* (*enter*) B, A *0-overlap* (exit) B, and A *0-meet* B. Reading this sequence of names builds up a description of the scene more readily than listing a sequence of 12 sets of elements, illustrating the usefulness of meaningful relation names.

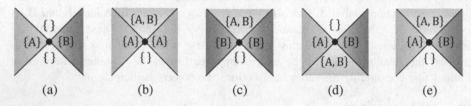

(a)   (b)   (c)   (d)   (e)

Fig. 10. The five P^2 relations between four cells with respect to regions A and B: (a) A *0-meet* B, (b) A *0-covers* B, (c) A *0-coveredBy* B, (d) A *0-overlap* (*enter*) B, and (e) A *0-overlap* (*exit*) B.

Point partition patterns between four cells may involve: (1) A and B's boundary *meet* at a single point from the exterior, *0-meet*, (2) B *meets* A's boundary at a single point from A's interior, A *0-covers* B, (3) A *meets* B's boundary at a single point from B's interior, A *0-coveredBy* B, (4) A *crosses into* B at a single point, A *0-overlap* (*enter*) B, and (5) A *crosses out* of B at a single point, A *0-overlap* (*exit*) B.

3 Depiction of a Spatial Scene from P^2 Patterns

Consider the scene in Fig. 5 once more. It is now possible to build that scene back up from its sequence of P^2 relations. Using the boundary sequence, object A was defined by intersections $i0$, $i1$, and $i2$ in that specific order. Furthermore, $i0$, $i1$, and $i2$ can be replaced by the sequence of cells recorded around each point. If objects are recorded clockwise, the interiors of their cells must always be to the right of the boundaries that define them. Utilizing just the sequence information this means that the boundary segment will start with the first element containing A and end after the final element containing A. Therefore, {A}, {A, B} is on one side of the boundary, while {B}, {} is on the other.

Keeping all cells containing A to the right, the first segment of A's boundary can be drawn. This line segment is continued to intersection $i2$ where it connects between {A, B} and {B} once more, and continues between {A} and {}, and the same process is repeated for intersection $i2$, keeping A always to the right of the line segment that defines it (Fig. 11a). The boundary sequence is cyclic, so $i2$ is connected back to $i1$ and the

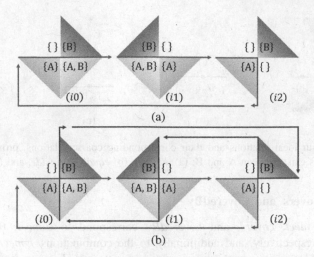

Fig. 11. Building the depiction of a scene from P^2 patterns: (a) drawing A's boundary and (b) drawing B's boundary.

boundary of A is completed. Completing the boundary of B is a matter of repeating the process, with the benefit that the scene is already half drawn (Fig. 11b).

4   Mappings Between Coarse and Detailed Relations

Now that the 23 P^2 relations have been defined the next step is to relate them to the familiar 9-intersection relations (Fig. 1). The similar naming convention from the onset is an additional benefit.

The connection between these two models is threefold: (1) the relations *disjoint*, *inside*, *contains*, and *equal* are each represented by one of the two-cell P^2 patterns; (2) *meet*, *covers*, and *coveredBy* are each be represented by a single 4-cell pattern or a pair of 3-cell patterns (depending on whether the intersection is 0-dimensional or 1-dimensional); and (3) *overlap* can be realized in nine different ways.

4.1   Disjoint, Inside, Contains, and Equal

The relations *localDisjoint*, *localInside*, *localContains*, and *localEqual* correspond to their 9-intersection counterparts for a pair of regions A and B if and only if they hold for all intersections between A and B. This is the simplest correspondence (Fig. 12).

If two regions, A and B, share one of these four *local* relations but additionally share a different relation with each other, the coarse relations *disjoint*, *inside*, *contains*, and *equal* will not hold. This counterexample is particularly relevant if scene changes over time are considered.

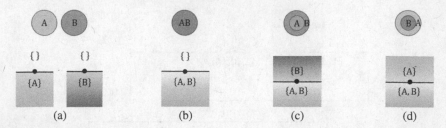

Fig. 12. The four local relations and their corresponding coarse relations, provided no other detailed relations exist between A and B: (a) *disjoint*, (b) *equal*, (c) *inside*, and (d) *contains*.

4.2   Meet, Covers, and CoveredBy

The relations *meet*, *covers*, and *coveredBy* correspond to *0-meet*, *0-covers*, and *0-coveredBy*, respectively, and additionally to the combinations *1-meet* (*enter*) plus *1-meet* (*exit*), *1-covers* (*enter*) plus *1 covers* (*exit*), and *1-coveredBy* (*enter*) plus *1-coveredBy* (*exit*). This distinction is due to the set of P^2 relations being able to distinguish dimension, while the vanilla 9-intersection is dimension agnostic (Fig. 13).

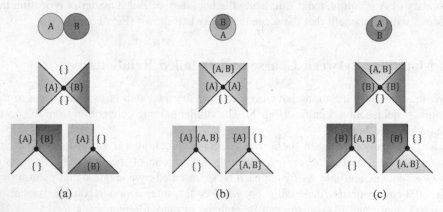

Fig. 13. The correspondence between the P^2 relations and the following 9-intersection relations: (a) *meet*, (b) *covers*, and (c) *coveredBy*.

4.3   Overlap

The most complex correspondence comes in the form of *overlap*. The set of P^2 relations accommodates not only *0-overlap* (*enter*) and (*exit*), but also *1-overlap* (*enter*) and *1-overlap* (*exit*), and varying combinations. Hybrid pairs, such as *1-overlap* plus *1-meet* and *1-covers* plus *1-coveredBy*, may substitute for pure *overlap* configurations when they occur in sequence and represent relations on opposing sides of a boundary (*in* versus *out*). *Overlap* is distinct that in this sense it may be split into more primitive relations. Due to this flexibility, the model is able to produce nine valid, basic overlap configurations via various P^2 relations (Fig. 14). This is distinct from other detailed

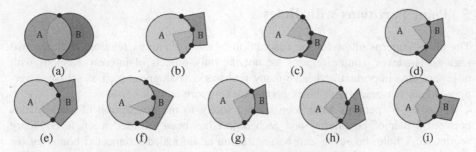

Fig. 14. The 9 basic forms of *overlap* identifiable with point-partitions.

representations of *overlap* in that it only includes cases with a single *crossing* into and out of a boundary, not fewer or more [11].

This example shows that A *0-overlap* (*enter*) B as well as A *1-meet* (*enter*) B plus A *1-overlap* (*enter*) B, and A *1-covers* (*enter*) B plus A *1-coveredBy* (*enter*) B are all valid combinations of P^2 relations that can start an *overlap* relation. Similarly, A *0-overlap* (*exit*) B as well as A *1-covers* (*exit*) B plus A *1-coveredBy* (*exit*) B, and A *1-overlap* (*exit*) B plus A *1-meet* (*exit*) B are all valid combinations of P^2 relations that can end an *overlap* relation.

The various overlap configurations are constructed as follows: (Fig. 14a) A *0-overlap* (*enter*) B ∧ A *0-overlap* (*exit*) B; (Fig. 14b) A *1-meet* (*enter*) B, A *1-overlap* (*enter*) B ∧ A *1-covers* (*exit*) B, A *1-coveredBy* (*exit*) B; (Fig. 14c) A *1-covers* (enter) B, A *1-coveredBy* (*enter*) B ∧ A *1-overlap* (*exit*) B, A *1-meet* (*exit*) B; (Fig. 14d) A *0-overlap* (*enter*) B ∧ A *1-overlap* (*exit*) B, A *1-meet* (*exit*) B; (Fig. 14e) A *1-meet* (*enter*) B, A *1-overlap* (*enter*) B ∧ A *0-overlap* (*exit*) B; (Fig. 14f) A *1-meet* (*enter*) B, A *1-overlap* (*enter*) B ∧ A *1-overlap* (*exit*) B, A *1-meet* (*exit*) B; (Fig. 14g) A *0-overlap* (*enter*) B ∧ A *1-covers* (*exit*) B, A *1-coveredBy* (*exit*) B; (Fig. 14h) A *1-covers* (*enter*) B, A *1-coveredBy* (*enter*) B ∧ A *0-overlap* (*exit*) B; and (Fig. 14i) A *1-covers* (*enter*) B, A *1-coveredBy* (*enter*) B ∧ A *1-covers* (*exit*) B, A *1-coveredBy* (*exit*) B. For further detail, any number of P^2 *coveredBy*, *localInside* and *localEqual* relations can occur between an overlap start and an overlap finish, and any number of P^2 *meet*, *localContains*, *localEqual*, or *covers* relations can occur between an overlap finish and an overlap start.

4.4 From Detailed to Coarse Relations

The conversion from the P^2 relations to 9-intersection relations is simpler. If any set of intersections between two objects forms one of the overlap configurations the resulting 9-intersection relation is only *overlap*. If there are no *overlap* relations, the existence of a *meet, covers*, or *coveredBy* P^2 relation corresponds to the same 9-intersection relation. If none of these relations exist *equal*, *contains*, or *inside* take precedence. Finally, if there are no other relations present the two regions are *disjoint*.

5 Point Partitions with Holes

The 23 P^2 patterns allow for the description of many complex relations between two regions. However, simple regions are not the only objects of interest. Regions with holes are also important and have many real world analogues, such as sensor coverages, where reasoning with holes necessitates a more complex model.

Thus far P^2 patterns have been limited solely to the description of intersections between a pair of boundaries and each object has been afforded a single boundary sequence. A hole, however, can be seen as an additional disconnected boundary (or boundaries) within an object. To accommodate holed regions each object must be allowed any number of boundary sequences, and a notation for describing such scenes must be able to handle this new case.

Two types of holes exist within this context: (1) holes that have boundary intersections with another object (A's hole, Fig. 15a) and (2) holes that do not have any boundary intersections with anything (B's hole, Fig. 15a). Holes with boundary intersections require the addition of a new boundary sequence to the sequence that describes its containing region. The original sequence (for the region) and the sequences for any further boundary is distinguished by adding a semicolon (; $i3$, $i4$, Fig. 15b). The strategy for recording the sequence is the same, save for the distinction that the interior of the containing region will be to the left of the boundary in the clockwise traversal, instead of the right.

The second scenario involves a hole that does not intersect with any other boundaries; it sits freely within its containing region. Such a hole cannot be placed in an exact location as directly, but it is still conceivable to do so. For holes without boundary intersection the sequence for the hole's container is listed *in lieu* of a sequence. In the most trivial circumstance this will be the object's name to which the hole belongs, but more complex cases require listing the sequence describing the simplest and smallest cell the hole resides in. To distinguish a hole's sequence, it is placed in parentheses (; $(i2, i1, i0)$, Fig. 15b).

To demonstrate the use of P^2 relations for describing holed regions, the previous scene (Fig. 15) is constructed next with the associated notation. Taking the P^2 relations in the correct sequence, region A is constructed from its outer boundary (Fig. 16a).

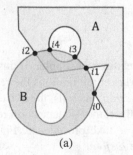

A: $i0,i1,i2$; $i3,i4$

B: $i2,i4,i3,i1,i0$; $(i2,i1,i0)$

$i0$: $\{\},\{A\},\{\},\{B\}$

$i1, i4$: $\{\},\{B\},\{A,B\},\{A\}$

$i2, i3$: $\{\},\{A\},\{A,B\},\{B\}$

(a)   (b)

Fig. 15. A scene with holes where: (a) region A contains holes a hole with a boundary intersection, and (b) region B contains a hole without a boundary intersection.

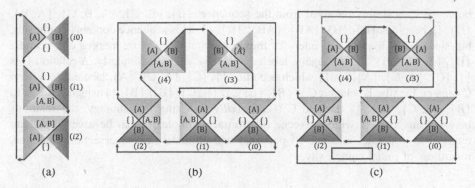

Fig. 16. An incremental construction of a scene starting with: (a) A's boundary, (b) B's boundary, and (c) the holes in each object.

Similarly, the boundary of region B can be constructed (Fig. 16b). Next, the hole in A and the hole in B are added (Fig. 16c).

6 Beyond Scenes with Two Holes

Relations between two spatial regions, including those with holes, are within the descriptive bounds of the 23 P^2 patterns that have been developed, along with the accompanying notation. The framework provided thus far can be extended to model the relations between more than two objects within a single spatial scene. While the P^2 patterns for *localDisjoint*, *localContains*, and *LocalInside* each describe various pure containment relations (instances without a proper boundary sequence), placing an arbitrary number of objects within a scene adds complexity. This problem, however, is the same one that arises when placing holes without boundary intersections within an object—a problem that has been resolved. If an object participating in a scene with more than two regions is wholly contained within some other region for which it has no boundary intersections, its container or the sequence describing its container must be recorded (using the parenthesis notation). As with holed regions, this information allows any region to be exactly placed within a scene.

The next consideration is for objects that do have boundary intersections. The set of P^2 patterns currently is restricted to an alphabet of four cells: {A}, {B}, {A, B}, and { }. To accommodate a third (or n^{th}) region the patterns will necessarily change. A pattern such as {B}, {A, B}, {B}, { } could become {B}, {B, C}, {A, B, C}, {A, B}, {B}, { } if a region C is added to the scene. With a coarse approach (e.g., the compound object model) A and B would be related, B and C would be related, and A and C would be related—each with a separate matrix. This process essentially asks three distinct questions that can generate ambiguity, since the description of B relation C is blind to the existence of A, and this method relies on inferences to draw the complete scene. Using P^2 patterns, however, the scene can be reduced to answer the same three questions, but leading to a consistent result every time.

The scene is still built from the same sequence information and the expanded P^2 patterns that includes as many objects or cells as necessary. To relate just A to B, however, C is simply removed from the sequence: {B}, {B, C}, {A, B, C}, {A, B}, {B}, {} becomes {B}, {B}, {AB}, {AB}, {B}, {}. This sequence contains duplicates, but since no region can self-intersect, these consecutive cells are merged and become {B}, {A, B}, {B}, {} once again, labeled A *0-coveredBy* B. Similarly, A relation C is {}, {C}, {A, C}, {A}, {}, {} which becomes {}, {C}, {A, C}, {A}, labeled A *overlap* C (*enter*). Finally, B relation C is {B}, {B, C}, {B, C}, {B}, {B}, {} which reduces to {B}, {B, C}, {B}, {}, labeled C *0-coveredBy* B. Using this strategy, the relations between any objects within a scene of arbitrary complexity can be ascertained and reduced to a simple sequence P^2 patterns, and ultimately a 9-intersection analogue, regardless of initial complexity.

7   Conclusions and Future Work

Point partitions are a means for capturing spatial phenomena between two (or more) regions regardless of the complexity of their relation or the existence of holes. By recording both the sequence of intersections around an object, as well as the sequence of objects around an intersection, the provided notation is so fine-grained than a topologically correct and unique depiction can be recreated.

These features combine to produce a set of 23 new region-region relations, called P^2 patterns. Any intersection, regardless of how many objects are involved, can be reduced to one of these P^2 patterns by culling unwanted regions from the relation and merging any duplicate cells that form side by side. P^2 relations can also describe relations between regions with holes. Furthermore, the set of detailed P^2 patterns can be reduced to the familiar set of eight region-region relations, and vice versa, allowing for a representation that best suits the provided information and the required task.

As a future consideration, it is also possible to use the sequences produced by the P^2 relations in order to identify the number of separations in the exterior formed by two regions, and the number of distinct cells formed by the intersection of two regions when multiple separations, holes or exterior partitions exist and develop a prototype that automatically draws scene diagrams from P^2 patterns and sequences.

Acknowledgments. Joshua Lewis is supported by a teaching assistantship at the University of Maine. Max Egenhofer's work was partially supported by NSF grants IIS-1016740 and IIS-1527504.

References

1. Adams, C., Franzosa, R.: Introduction to Topology: Pure and Applied. Pearson Prentice Hall, Upper Saddle River (2008)
2. Bruns, T., Egenhofer, M.: Similarity of spatial scenes. In: 7th International Symposium on Spatial Data Handling, Delft, The Netherlands, pp. 31–42 (1996)
3. Clementini, E., Felice, P.D.: Topological invariants for lines. IEEE Trans. Knowl. Data Eng. **10**(1), 38–54 (1998)
4. Dube, M.P., Egenhofer, M.J.: Surrounds in partitions. In: Huang, Y., Schneider, M., Gertz, M., Krumm, J., Sankaranarayanan, J. (eds.) ACM SIGSPATIAL 2014, pp. 233–242. ACM Press, New York (2014)

5. Dube, M.P., Egenhofer, M.J., Lewis, J.A., Stephen, S., Plummer, M.A.: Swiss Canton Regions: a model for complex objects in geographic partitions. In: Fabrikant, S.I., Raubal, M., Bertolotto, M., Davies, C., Freundschuh, S., Bell, S. (eds.) COSIT 2015. LNCS, vol. 9368, pp. 309–330. Springer, Heidelberg (2015). doi:10.1007/978-3-319-23374-1_15

6. Egenhofer, M.J.: A reference system for topological relations between compound spatial objects. In: Heuser, C.A., Pernul, G. (eds.) ER 2009. LNCS, vol. 5833, pp. 307–316. Springer, Heidelberg (2009)

7. Egenhofer, M.J., Franzosa, R.D.: On the equivalence of topological relations. Int. J. Geogr. Inf. Syst. 9(2), 133–152 (1995)

8. Egenhofer, M.J., Franzosa, R.D.: Point-set topological spatial relations. Int. J. Geogr. Inf. Syst. 5(2), 161–174 (1991)

9. Egenhofer, M.J., Herring, J.: Categorizing binary topological relationships between regions, lines and points in geographic database. Technical report, University of Maine (1991)

10. Egenhofer, M., Vasardani, M.: Spatial reasoning with a hole. In: Winter, S., Duckham, M., Kulik, L., Kuipers, B. (eds.) COSIT 2007. LNCS, vol. 4736, pp. 303–320. Springer, Heidelberg (2007)

11. Galton, A.: Modes of overlap. J. Vis. Lang. Comput. 9(1), 61–79 (1998)

12. Kurata, Y.: The 9^+-Intersection: a universal framework for modeling topological relations. In: Cova, T.J., Miller, H.J., Beard, K., Frank, A.U., Goodchild, M.F. (eds.) GIScience 2008. LNCS, vol. 5266, pp. 181–198. Springer, Heidelberg (2008)

13. Kurata, Y., Egenhofer, M.J.: The 9^+-Intersection for topological relations between a directed line segment and a region. In: Gottfried, B. (eds.) Workshop on Behaviour and Monitoring Interpretation, Technical report 42, Technologie-Zentrum Informatik, University of Bremen, Germany, pp. 62–76 (2007)

14. Lewis, J.A., Dube, M.P., Egenhofer, M.J.: The topology of spatial scenes in R^2. In: Tenbrink, T., Stell, J., Galton, A., Wood, Z. (eds.) COSIT 2013. LNCS, vol. 8116, pp. 495–515. Springer, Heidelberg (2013)

15. Lewis, J.A., Egenhofer, M.J.: Oriented regions for linearly conceptualized features. In: Duckham, M., Pebesma, E., Stewart, K., Frank, A.U. (eds.) GIScience 2014. LNCS, vol. 8728, pp. 333–348. Springer, Heidelberg (2014)

16. Nedas, K.A., Egenhofer, M.J.: Spatial-scene similarity queries. Trans. GIS 12(6), 661–681 (2008)

17. Papadimitriou, C.H., Suciu, D., Vianu, V.: Topological queries in spatial databases. In: ACM PODS, pp. 81–92 (1996)

18. Randell, D.A., Cui, Z., Cohn, A.G.: A spatial logic based on regions and connection. In: Nebel, B., Rich, C., Swartout, W.R. (eds.) KR92, pp. 165–176. Morgan Kaufmann, San Francisco (1992)

19. Schneider, M., Behr, T.: Topological relationships between complex spatial objects. ACM Trans. Database Syst. 31(1), 39–81 (2006)

20. Vasardani, M., Egenhofer, M.J.: Comparing relations with a multi-holed region. In: Hornsby, K.S., Claramunt, C., Denis, M., Ligozat, G. (eds.) COSIT 2009. LNCS, vol. 5756, pp. 159–176. Springer, Heidelberg (2009)

21. Vasardani, M., Timpf, S., Winter, S., Tomko, M.: From descriptions to depictions: a conceptual framework. In: Tenbrink, T., Stell, J., Galton, A., Wood, Z. (eds.) COSIT 2013. LNCS, vol. 8116, pp. 299–319. Springer, Heidelberg (2013)

22. Worboys, M.: The maptree: a fine-grained formal representation of space. In: Xiao, N., Kwan, M.-P., Goodchild, M.F., Shekhar, S. (eds.) GIScience 2012. LNCS, vol. 7478, pp. 298–310. Springer, Heidelberg (2012)

Fine Scale Spatio-Temporal Modelling of Urban Air Pollution

Xiaoxiao Liu[⊠] and Stefania Bertazzon

Department of Geography, University of Calgary,
2500 University Dr. NW, Calgary, AB T2N 1N4, Canada
xiaoxili@ucalgary.ca

Abstract. Urban air pollution is a leading environmental health concern. However, the association between urban air pollution and health outcomes is not consistently reported in the literature, likely because of inaccurate exposure assessment induced by spatial error. In this study, a spatio-temporal model is presented, which integrates harmonic regression and land use regression (LUR) to estimate urban air pollution at fine spatio-temporal scale. The space-time field is decomposed into space-time mean and space-time residuals. The mean is estimated by linear combinations of harmonic regression components, and the spatial field is modelled with LUR. The residuals account for spatio-temporal deviation from the mean model. Using data from a regulatory monitor network and geographic covariates from a LUR model, the study yields monthly nitrogen dioxide estimates at the postal code level for Calgary, Canada. The model yields a satisfactory fit ($R^2 = 0.78$). The space-time residuals exhibit non-significant to moderate spatial and temporal autocorrelation.

Keywords: Spatio-temporal model · Harmonic regression · Land use regression · Nitrogen dioxide · Fine scale estimates

1 Introduction

Urban air pollution is a leading problem in environmental health and a potential risk factor for adverse health effects, including cardiovascular disease [1]. Numerous studies provide substantial evidence of association between urban air pollution and cardiovascular diseases [2–6]; however, the link is not consistent across studies [7]. For example, D'Ippoliti et al. [2] found a positive association between nitrogen dioxide (NO_2) air pollution and hospitalizations for myocardial infarction (MI) in Rome (Italy). This positive association is supported by Lanki et al. [4], a European study; however, a study about the short-term exposure effects on MI in France [3], found no association between NO_2 exposure and MI occurrence. Jerrett et al. [7] show that inconsistent association between exposure and health outcome may be caused by inaccurate exposure assessment, due to spatial misalignment, a common issue in environmental studies. Usually, air pollution data are acquired from sparse monitoring stations, whereas disease data are generally available at small area level (e.g., postal code) [8]. The mismatch between data measured at different resolutions results in spatial misalignment [9], which can induce error and biased estimates of risk. The earliest exposure predictions rely on city-wide averages, which fail to consider the spatial

© Springer International Publishing Switzerland 2016
J.A. Miller et al. (Eds.): GIScience 2016, LNCS 9927, pp. 210–224, 2016.
DOI: 10.1007/978-3-319-45738-3_14

variation within a city [10]. More recent studies assign exposures using nearest monitor interpolation [11], land use regression [12], or geostatistical methods, such as universal kriging [13]. These relatively simple spatial statistical techniques rely on data from existing regulatory networks [14].

To accurately assess the association between exposure and health outcomes, it is important to develop more refined models which produce accurate predictions of air pollution at fine spatial scale [15, 16]. Traditionally, researchers have used time series analysis, on the assumption that pollutants are spatially homogeneous. However, it is now widely recognized that different pollutants have different spatial distributions [17]. For example, regional pollutants, such as ozone (O_3) and particulate matter (PM), are often relatively homogeneous over space due to relatively consistent concentration levels and temporal fluctuations. However, other pollutants, such as NO_2 and other traffic related pollutants, are likely to display spatial heterogeneity [17]. Therefore, it is important to estimate spatial variability of pollutants, to reduce exposure measurement error especially for spatially heterogeneous pollutants [18, 19]. Spatial and temporal variability of pollutant concentration are not independent of one another: for example, spatial variability can be greater in the fall, when air masses travel more rapidly, due to changing weather patterns and stronger winds. Spatial-only and temporal-only models fail to capture these spatio-temporal trends, yielding spatially and/or temporally auto-correlated residuals, and hence unreliable and potentially biased risk estimates. Despite their increased complexity, improved spatio-temporal models are a preferable alternative, as they can yield more accurate and reliable exposure estimates at fine spatial scale (e.g., postal code) [15].

In this paper, we develop a spatio-temporal model to estimate NO_2 concentrations in Calgary, Alberta, Canada. By modelling spatial and temporal dependences, the analysis yields reliable exposure predictions at unobserved locations.

2 Methods

This study builds on the methods of Kyriakidis and Journel [20], Lindstrom et al. [15], Sampson et al. [21], and Szpiro et al. [14] to model NO_2 annual patterns at the postal code level in Calgary. A land use regression (LUR) of NO_2 in Calgary, recently published by Bertazzon et al. [24], provides the spatial model embedded in this study. The space–time field is decomposed into two parts: space–time mean and space–time residuals. The space–time mean accounts for spatially varying seasonal and long-term trends, which depend on geographical covariates. The space–time residuals account for spatio-temporal deviation from the mean model. The space–time mean is obtained by linear combinations of harmonic regression components at each monitoring station. Harmonic regressions account for temporal variability; to account for the spatial variability in the temporal structure, a land use regression model is embedded in each harmonic coefficient. As the temporal variability is explained by the harmonic regressions, the literature typically assumes a separate spatial temporal covariance structure for the space–time residuals [14, 15, 21]. This study relies on the same assumption.

2.1 Study Area and Data

Calgary is the 5th largest city in Canada, with a population of 1.2 million and a land area of 725 km^2 [22]. Located just east of the Rocky Mountains, it is exposed predominantly to north winds, carrying cold and dry arctic air, and to wèst winds, carrying warm, moisturized air from the Pacific Ocean [23]. Calgary's metropolitan area hosts a variety of industries which are mainly located in the east side of the city [24]. The largest contributor to NO_2 emissions is transportation, followed by industrial sources (largely oil and gas), power plants, and natural gas combustion [25].

Air Pollution. Monthly average NO_2 concentrations between July 2011 and July 2014 were acquired from Alberta's Airdata Warehouse (AEMERA), a central repository for archived ambient air quality data collected in Alberta. In Calgary, these data are managed by the Calgary Region Airshed Zone (CRAZ), which runs the regulatory monitor network in the city. The data used in this study were recorded at this monitor network, which consists of eight passive stations and two continuous stations,[1] for a total of ten locations (Fig. 1a). Figure 1b shows the monitor network deployed for the LUR study [24] in Summer 2010 and Winter 2011.

Both the continuous stations yielded a complete series of 37 observations, while the eight passive stations had gaps between July 2013 and February 2014, resulting in 20 missing values from the expected 370 monthly observations. The continuous

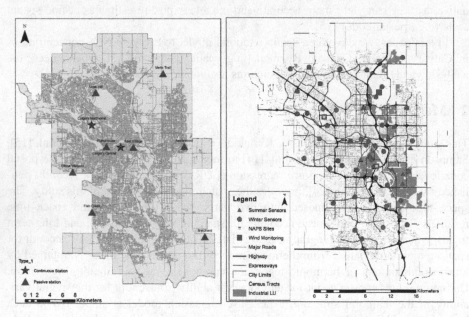

Fig. 1. Calgary regulatory monitoring network (a); 2010–2011 LUR monitoring network (b). (Color figure online)

---

[1] One additional continuous station, Calgary Southeast, was decommissioned in April 2011 and relocated in April 2014. For this reason, data from this station were not used in this analysis.

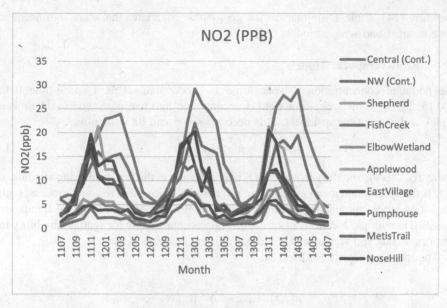

Fig. 2. Time series of NO_2 (ppb) from July 2011 to July 2014. (Color figure online)

stations (red stars in Fig. 1a) provide nearly instantaneous measurements of ambient concentrations, and data are stored in one-hour average time blocks. Passive stations (blue triangles in Fig. 1a) collect samples over a period of about one month. NO_2 time series for the ten stations over the three years are plotted in Fig. 2.

Geographic Variables. Land use regression is a form of multiple linear regression that uses geographic covariates as independent variables to predict air pollution at unobserved locations. Bertazzon et al. [24] calculated summer and a winter LUR models for Calgary. That study deployed a spatial network of 50 monitors within the

Table 1. Significant variables for summer and winter LUR models

Model	WS_N	NO₂_EM_dist	EXPW _dist	MRD _200	EXPHW_400	LU_ind _1000	POP_den _2500
Summer	√	√		√	√	√	
Winter	√		√	√		√	√

Those covariates are: *North wind speed* (WS_N) in winter and summer; *Distance from industrial NO_2 emissions* (NO_2_EM_dist), *Distance from expressways* (EXPW_dist); *Major roads within a 200 m buffer* (MRD_200); *Sum of primary highways and expressways within a 400 m buffer* (EXPHW_400); *Industrial land use within a 1000 m buffer* (LU_ind_1000); and *Population density within a 2500 m buffer* (POP_den_2500). All these variables[2] are considered in this study.

city (Fig. 1b), where 2-week sampling campaigns were run in August 2010 and February 2011. Data acquisition, model selection, and other details are published elsewhere [24]. Table 1 summarizes the geographic covariates that were significant in those summer and winter models [24].

2.2 Spatio-Temporal Models

The pollutant concentration in space-time is conceptualized as a space-time field [14, 15, 20, 21]. This space-time field is decomposed into two parts: space–time mean model, and space–time residuals. This decomposition can be written as:

$$Y(s,t) = \mu(s,t) + \varepsilon(s,t). \tag{1}$$

where $Y(s, t)$ is the monthly average NO_2 concentration at the ten monitoring stations, μ (s, t) is the space–time mean process and $\varepsilon(s, t)$ is the space–time residual process, with s denoting space (monitoring stations) and t time (in months). As temporal trends are modelled at multiple locations in space, this approach accounts for spatial variability in temporal trends and spatial non-stationarity in the residuals [21].

The mean process is modeled as:

$$\mu(s,t) = \beta_{0s} + \sum_{i=1}^{m} \beta_{is} f_i(t). \tag{2}$$

where $\{f_i(t)\}$ ($i = 1$ to m) is a set of harmonic components and β_{1s} is the spatially varying coefficient of the temporal components. Typically the number of harmonic regression components, m, is small [14, 26]. The method developed here differs from others [15, 14, 21], which use empirical orthogonal functions (EOF), a variant of principal component analysis, to model the space–time mean. Here we directly use harmonic components, a key tool in time series analysis, on the assumption of stationarity of the process [26]. Annual periodic cycles of NO_2, $f_{is}(t)$, are modeled by a harmonic regression at each of the ten monitoring stations. For example, if the harmonic regression model has two significant component, the mean process can be rewritten as:

$$\mu(s,t) = \beta_{0s} + \beta_{1s} \cos\left(\frac{2\pi t}{T}\right) + \beta_{2s} \sin\left(\frac{2\pi t}{T}\right) \tag{3}$$

which can also be written as [27]:

$$\mu(s,t) = \beta_{0s} + \sqrt{\beta_{1s}^2 + \beta_{1s}^2} \cos\left(\frac{2\pi t}{T} - \varphi\right). \tag{4}$$

where T denotes the harmonic period, 1/T the frequency, $\sqrt{\beta_{1s}^2 + \beta_{1s}^2}$ the amplitude, and φ the phase angle [27].

---

[2] The variable *Winter north wind speed* was not included, due to its high correlation with the corresponding summer variable.

The spatial fields of the β_{1s} coefficients are amplitudes of temporal patterns [21]. They are modelled with land use regression [24, 28, 29], which predicts spatial variability through a linear regressions on q geographic covariates [21]:

$$\beta_1 = \sum_{j=1}^{q} \alpha_{ij} X_{ij} + e_i \tag{5}$$

where X_{ij} denotes the jth geographical covariate for the ith coefficient β_i. Each β_i may have different LUR coefficients and geographic covariates [14]. The α_{ij} are LUR regression coefficients and e_i denotes LUR residuals for each spatial field β_i.

As the temporal variability is captured by the harmonic regression, the space–time residual process can be assumed to be temporally independent and spatially correlated [15, 21].

3 Results

3.1 Harmonic Regression Model

As shown in Fig. 2, NO_2 exhibits a prominent annual cycle (12-month period), as well as minor, seasonal cycles with shorter periods. The initial model all these cycles into consideration, with independent harmonic components covering periods from 2 to 12 months:

1. $\cos\left(\frac{2\pi t*1}{12}\right) + \sin\left(\frac{2\pi t*1}{12}\right)$; 2. $\cos\left(\frac{2\pi t*2}{12}\right) + \sin\left(\frac{2\pi t*2}{12}\right)$; 3. $\cos\left(\frac{2\pi t*3}{12}\right) + \sin\left(\frac{2\pi t*3}{12}\right)$;
4. $\cos\left(\frac{2\pi t*4}{12}\right) + \sin\left(\frac{2\pi t*4}{12}\right)$; 5. $\cos\left(\frac{2\pi t*5}{12}\right) + \sin\left(\frac{2\pi t*5}{12}\right)$; 6. $\cos\left(\frac{2\pi t*6}{12}\right) + \sin\left(\frac{2\pi t*6}{12}\right)$;

However, only the harmonic components $\cos\left(\frac{2\pi t}{12}\right)$ and $\sin\left(\frac{2\pi t}{12}\right)$ are significant in the model, suggesting that the annual cycle is the most prominent temporal pattern for NO_2. Therefore, the final harmonic regression model at a specific monitoring station s is simply:

$$\mu(t) = \beta_0 + \beta_1 \cos\left(\frac{2\pi t}{T}\right) + \beta_2 \sin\left(\frac{2\pi t}{T}\right) + \epsilon_t \tag{6}$$

Table 2. Harmonic regression estimates at 10 stations in Calgary

	Central	North-west	Shepherd	Fish creek	Elbow wetland	Apple wood	East village	Pump-house	Metis trail	Nose hill
β_0	15.8*	11.2*	3.93	2.2	3.59*	9.14	9.15	7.33	7.18	3.5
β_1	−5.78*	−4.55*	−2.92*	−1.69*	−2.6*	−7.59*	−7.01*	−5.4*	−4.88*	−1.94*
β_2	−9.11*	−4.99*	0.03	−0.22	−0.46*	−0.81	−0.6	−0.5	0.13	−0.08
R^2	0.94	0.92	0.83	0.67	0.84	0.82	0.81	0.75	0.84	0.72
ρ	0.37	0.12	0.28	0.29	0.25	0.13	0.12	0.15	0.25	0.39
D-W	1.15	1.73	1.44	1.43	1.50	1.74	1.74	1.71	1.50	1.20
p	0.01	0.25	0.05	0.04	0.06	0.29	0.26	0.17	0.08	0.00

where β_0 is the constant, β_1 is the parameter of the cosine component, and β_2 is the parameter of the sine component.

Table 2 summarizes the harmonic regressions at the ten monitoring stations. For the cosine component, all the 10 coefficients are negative and statistically significant (95 %, denoted by asterisks). For the sine component, most coefficients are negative, relatively small in absolute value, and only three are significant (denoted by asterisks). Nonetheless, both components are included in the model, because cosine and sine together form a single harmonic motion. The model is satisfactory for all the ten stations, with R^2 ranging from 0.65 (Fish Creek) to 0.93 (Central). As $\sqrt{\beta_1^2 + \beta_2^2}$ denotes the peak value of the NO_2 trend at each station, it is easy to tell Central and Northwest (the two continuous stations) have higher peak NO_2 concentrations, which is consistent with time series of NO_2 in Fig. 2.

Consistently, β_2 is significant only at these two stations and at Elbow Wetland, which exhibits a lower but pronounced peak concentration, most discernible in the 2014 cycle (Fig. 2). Table 2 also shows the harmonic regression residual autocorrelation (ρ), and the Durbin–Watson statistic (D–W) with associated p values, which indicate that the residual temporal autocorrelation is not significant in any of the ten stations.

Table 3. Summary of land use regression model for each spatial field β

	Intercept	NO$_2$_EM_dist	MRD_200	WS_N	EXPW_dist	R^2	Adj. R^2	Moran's I	p(Z (I))
β_0	15.02	−0.002	0.007			0.82	0.77	0.01	0.34
Std. β_0		−1.04	0.64						
Res. β_0								−0.03	0.32
β_1	−0.02			−0.66	−0.001	0.78	0.72	0.13	0.06
Std. β_1				−0.49	−0.76				
Res. β_1								−0.37	0.68
β_2	−6.26	−0.001	0.005			0.54	0.41	−0.29	0.17
Std. β_2		−0.82	0.6						
Res. β_2								−0.18	0.58

3.2   LUR for Spatial Fields β

All the significant geographic covariates from Bertazzon et al. [24] are included in the LUR models for the β coefficients; however, due to their high cross-correlation (0.74), *Distance from industrial NO$_2$ emissions* and *Distance from expressways* are not included simultaneously, to control for multicollinearity in ecological multiple regression [29, 30]. The model results are shown in Table 3.

For the harmonic LUR model of β_0, significant covariates are *Distance from industrial NO_2 emissions* and *Major roads within 200 meters*. For the model of β_1, they are *North wind speed* and *Distance from expressways*. The model of β_2 has the same significant covariates as the intercept model. The model of β_0 has the highest R^2 (0.82), not far from the R^2 of β_1 (0.78), whereas the model of β_2 has the lowest R^2 (0.54). Remarkably, a high R^2 is associated with β_1, which is significant in all the harmonic models, whereas a much lower R^2 is associated with β_2, which is only significant in three of the ten harmonic models (Table 2). The poor spatial fit of the sine coefficient, along with its lack of significance in the temporal model, may be explained by the relatively flat temporal trends and relative spatial homogeneity of the stations located in rural areas and in proximity of major green spaces. This may also be the reason why the same geographic covariates are significant for the intercept and the sine coefficient, versus a different set of significant covariates for the cosine coefficient. Spatial autocorrelation of the β spatial fields was assessed by Moran's I, with a spatial relationship defined by the inverse distance method. Spatial autocorrelation in the model residuals was calculated using Moran's I test, based on a spatial weights matrix defined on row-standardized 2 nearest neighbours [31]. As shown in Table 3, spatial autocorrelation was not significant for any of the spatial fields or residuals.

3.3 Space-Time Residuals

The last part of the model is the estimation of the space–time residuals after fitting both harmonic regression and land use regression models. The estimated $\hat{\mu}_{st}$ is constructed by plugging the estimated $\hat{\beta}_0$, $\hat{\beta}_1$ and $\hat{\beta}_2$, from the land use regression model into Eq. (3) at each location s.

The normality assumption for the residual field is analyzed by a normal QQ plot of the combined distribution of the residuals from the ten stations, and shown in Fig. 3. Although the QQ plot has a heavy right tail, suggesting skewness in the distribution,

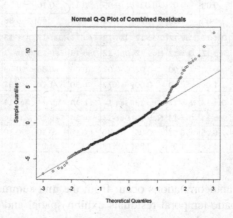

Fig. 3. Combined Normal QQ plot for space-time residuals

overall the normality assumption appears to be a reasonable approximation. The R^2 value is 0.78, indicating a satisfactory goodness of fit.

Once the space–time residuals were obtained, their temporal autocorrelation was tested using the Durbin–Watson statistic. The residuals of seven of the ten stations exhibit moderate, yet not statistically significant, temporal autocorrelation at one-month lag. Therefore, in comparison with the residuals of the harmonic regression model (paragraph 3.1), the increased residual temporal correlation can be ascribed to the error

Table 4. Spatial autocorrelation for space-time residuals at 37 month time points

Month	Jul-11	Aug-11	Sep-11	Oct-11	Nov-11	Dec-11	Jan-12	Feb-12	Mar-12	Apr-12	May-12	Jun-12	
Moran's I	0.10	0.28	-0.02	0.49	0.06	-0.28	-0.07	-0.05	-0.19	-0.26	-0.28	-0.15	
Z(I)	0.90	1.66	0.37	2.55	0.72	-0.83	0.19	0.30	-0.37	-0.62	-0.72	-0.16	
Month	Jul-12	Aug-12	Sep-12	Oct-12	Nov-12	Dec-12	Jan-13	Feb-13	Mar-13	Apr-13	May-13	Jun-13	
Moran's I	0.06	0.40	0.30	-0.18	-0.21	-0.08	-0.13	-0.26	0.00	-0.35	-0.21	-0.07	
Z(I)	0.73	2.16	1.76	-0.30	-0.41	0.15	-0.09	-0.85	0.49	-1.23	-0.46	0.16	
Month	Jul-13	Aug-13	Sep-13	Oct-13	Nov-13	Dec-13	Jan-14	Feb-14	Mar-14	Apr-14	May-14	Jun-14	Jul-14
Moran's I	-0.04	-0.08	-0.13	0.43	0.18	-0.09	-0.46	-0.19	-0.17	-0.24	-0.19	0.31	0.42
Z(I)	0.32	0.14	-0.06	2.59	1.30	0.11	-1.86	-0.41	-0.33	-0.65	-0.37	1.86	2.32

introduced by plugging the LUR estimated spatial fields, $\hat{\beta}_0$, $\hat{\beta}_1$ and $\hat{\beta}_2$,, in the space-time mean model, in place of the original harmonic parameters. Moran's I test with a spatial weights matrix defined on row-standardized 2 nearest neighbours was applied to assess the residual spatial autocorrelation for the 37 months (Table 4). The index is significant (95 %) for only four of the 37 months (four more at 90 %). Most of

Table 5. Bivariate spatial correlation of space-time residuals

$X_{i,t}$	Aug-11	Sep-11	Oct-11	Nov-11	Dec-11	Jan-12	Feb-12	Mar-12	Apr-12	May-12	Jun-12	Jul-12
$X_{j,t-1}$	Jul-11	Aug-11	Sep-11	Oct-11	Nov-11	Dec-11	Jan-12	Feb-12	Mar-12	Apr-12	May-12	Jun-12
Moran's I	-0.10	-0.07	-0.09	-0.07	-0.02	-0.05	-0.11	-0.10	-0.10	-0.10	-0.09	-0.10
Z(I)	-1.19	1.31	0.00	1.09	1.98	0.00	-1.78	1.43	-0.88	1.14	1.11	1.20
$X_{i,t}$	Aug-12	Sep-12	Oct-12	Nov-12	Dec-12	Jan-13	Feb-13	Mar-13	Apr-13	May-13	Jun-13	Jul-13
$X_{j,t-1}$	Jul-12	Aug-12	Sep-12	Oct-12	Nov-12	Dec-12	Jan-13	Feb-13	Mar-13	Apr-13	May-13	Jun-13
Moran's I	-0.09	-0.07	-0.03	0.00	-0.07	-0.07	-0.09	-0.86	-0.09	-0.07	-0.10	-0.08
Z(I)	-0.65	1.11	-1.04	0.13	0.00	-0.77	0.90	-1.10	-0.91	1.08	1.66	0.81
$X_{i,t}$	Aug-13	Sep-13	Oct-13	Nov-13	Dec-13	Jan-14	Feb-14	Mar-14	Apr-14	May-14	Jun-14	Jul-14
$X_{j,t-1}$	Jul-13	Aug-13	Sep-13	Oct-13	Nov-13	Dec-13	Jan-14	Feb-14	Mar-14	Apr-14	May-14	Jun-14
Moran's I	-0.09	-0.10	-0.05	-0.07	-0.06	-0.09	-0.10	-0.10	-0.10	-0.11	-0.80	-0.10
Z(I)	-0.96	0.00	-1.08	1.05	0.95	0.73	-0.87	-1.07	0.89	0.77	-1.01	1.07

the significant spatial autocorrelations occur from the mid-summer to the fall.

As some of the spatio-temporal residuals exhibit spatial and/or temporal autocorrelation, multivariate spatial autocorrelation was also assessed. However, because

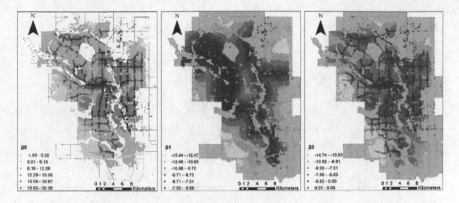

Fig. 4. Predicted spatial coefficients of β_0 (left), β_1 (middle) and β_2 (right)

temporal autocorrelation was observed only for one-month lag, a bivariate measure was deemed appropriate. For each pair, the index compares the residual at one location with its spatial neighbour, the latter lagged by one month. The results are shown in Table 5.

All the bivariate spatial correlations are negative, indicating residual dispersion, and most of them are between 0.00 and 0.11. Only two values are between 0.80 and 0.86 (February–March 2013, and May–June 2014), yet not statistically significant.

3.4 Predictions

Figure 4 shows maps of the spatial fields β_0, β_1, and β_2 calculated by the LUR models with the respective geographic covariates. β_0, the constant of the space-time model, exhibits mostly positive values, whereas β_1, the cosine parameter, and β_2, the sine parameter, exhibit mostly negative values. These negative values are multiplied by alternating positive and negative values of the harmonic components in the various seasons.

The models for β_0, and β_2 have the same set of independent variables (Table 3) and exhibit very similar spatial patterns: the effect of *Major roads within 200 meters* is discernible, as a gridline pattern, whereas the effect of *Distance from industrial NO_2 emissions* is shown by higher pollution levels in the east quadrants. For the β_1 model, the independent variables are *North wind speed*, recognizable as a north-south pattern, and *Distance from expressways*, identifiable in a roughly Y-shaped pattern (also shown in Fig. 1a), formed by Deerfoot Trail, Crowchild Trail, and Highway1.

Using these estimated parameters, prediction maps of NO_2 were created for each of the 37 months and each of the 28,980 Calgary postal codes. Figure 5 shows, as examples, the predictions for November 2011, February 2012, and May 2012.

Predicted concentration levels are highest in February, which is consistent with the peak exhibited by most stations in Fig. 2, and can be ascribed to more intense residential heating and more people using motorized vehicles to commute, in addition to industrial activities. Consistently, the association between NO_2 concentration and traffic is visible in all three months, yet more pronounced in November and May, when

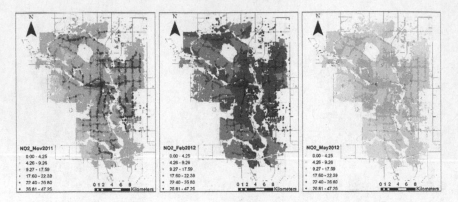

Fig. 5. Predicted NO₂, postal code level: November 2011 (a), February 2012 (b), May 2012 (c)

pollution levels are relatively lower. As in Fig. 4, the association with major roads is shown by a grid-like pattern, along with the pattern of major highways. The association with industrial emissions is discernible through higher concentration levels in the east side, particularly in November and February.

Leave-one-out-cross-validation (LOOCV) for the spatio-temporal model yields a wide range of root mean square error (RMSE) values, with the lowest values (< 1.1) at Elbow Wetlands, Applewood, and East Village. The average RMSE for the 8 passive stations is 1.24, which rises to 1.33 when the continuous stations are included. In addition, we attempted a comparison of the model predictions with independent concentration records. Unfortunately, the only available independent records were from the Calgary East continuous station, which was decommissioned in April 2011, therefore providing data from January to April 2011. The model predictions were extrapolated and compared with the recorded NO₂ concentrations in early 2011. The percentage difference between predicted and recorded concentrations [24] indicates that the model tends to under-predict, yielding and an average error of 0.13.

4 Discussion

The method presented in this paper yields predictions of NO₂ concentration at high spatial and temporal resolution (28,980 locations at monthly intervals) for 3 years over a large urban area (725 km²). These predictions are obtained through a multistep method integrating harmonic regressions at each station of a regulatory network with a land use regression that links geographic covariates with pollutant concentration. A major assumption of the model is the independence of the residuals over time and space.

Indeed, the assumption of temporal independence is met by both the harmonic regression model and the spatio-temporal model. In the former model, temporal dependences are accounted for by the regression itself, whereas in the latter model, the βs estimated within the harmonic regression are replaced with those estimated by the LUR model. Thus, the model embeds an extra layer of error, which cannot easily be accounted for; hence, some residuals exhibit moderate, yet not significant one-month-lag temporal autocorrelation. Likewise, the assumption of spatial independence of the residual is essentially met by both the spatial and the spatio-temporal models: significant spatial autocorrelation is exhibited by none of the LUR residuals, and by only four for the 37 monthly values of the spatio-temporal model. Spatio-temporal correlation across pairs of stations over one-month lag is not significant, though values were high for two of the 37 pairs. The few significant spatial autocorrelations occurred mostly between mid-summer and fall, possibly suggesting that greater spatial variability of pollution is associated with unstable weather patterns at that time of the year. Low spatial and spatio-temporal autocorrelation can be ascribed to the sparseness of the monitoring network, which is also consistent with the large number of negative, yet non-significant, spatial autocorrelation values. Conversely, the denser and more regular temporal sample can be associated with more significant residual temporal autocorrelation. To address the latter problem, the literature suggests a Bayesian hierarchical approach [19].

Regression diagnostics indicate that the models yield satisfactory goodness of fit. Interpretation of spatial field models is less straightforward than standard LUR models, which directly estimate pollution concentrations: here, pollution concentration predictions are obtained as a combination of these sets of estimated coefficients and the harmonic components. The proposed method can effectively capture variations in the pollutant concentration that are intrinsically spatio-temporal. For example: lower NO_2 concentrations were recorded at rural stations and close to major parks (Elbow Wetland, Fish Creek, Shepherd, and Calgary Nose Hill). NO_2 in these locations exhibits similar, flatter temporal trends, in comparison with higher concentrations and larger seasonal fluctuations at inner-city stations. As well, NO_2 is a traffic-related air pollutant, exhibiting higher concentration near major roads: the association between NO_2 and traffic is captured by the model, which estimates higher NO_2 values along the main roads and highways, and a more diffused pollution pattern in the winter, when Calgary residents tend to drive more, even over short distances.

The model was validated by LOOCV cross validation, which yields encouraging results, yet suggests that the model can be improved, and further indicates that the proposed model works better for passive stations. The model predictions were further assessed against independent concentration records. This informal comparison only provides a broad indication, as the benchmark records lie outside the temporal prediction range and were measured at a continuous station. Some limitations of this study are related to data availability. Spatial data for the LUR models [24] were collected in August 2010 and February 2011, whereas temporal data collection by CRAZ did not start until July 2011. Additionally, data from passive and continuous stations were lumped together, using monthly interval data from continuous stations: further analysis should confirm the consistency between the two sets of data. In this study, geographic covariates from two seasonal LUR models (summer and winter) were combined in the estimation of a single spatial field model. In this process, important seasonal differences were obscured, including the effect of the summer vs. winter wind speed, and the effect of population density, ascribed to residential heating in the winter LUR model [24]. Addressing this limitation requires a more complex modelling framework, and that it be complemented by a careful analysis of seasonal and meteorological patterns over the analyzed time period. The harmonic regression models yield detailed estimates of the seasonal cycles of NO_2, but do not accurately estimate the temporal path through the years. Future work shall address this limitation, by exploring additional temporal functions, for example by integrating the harmonic components with a trend function or using Fourier decomposition.

The method presented in this paper offers a powerful tool to estimate pollution levels at fine spatial and temporal scales. Monthly air pollution estimates at the postal code level yield a much greater spatio-temporal detail than data recorded by regulatory networks. For this reason, these estimates can reduce the spatial error currently associated with risk assessment and health models. Once refined, the method will be employed to estimate other pollutants, including particulate matter (PM), volatile organic compounds (VOC), and polycyclic aromatic hydrocarbons (PAH). Like previous LUR models [24], these estimates will be shared with health research partners, and used in the analysis of health outcomes, including cardiovascular disease, pediatric asthma, and gastrointestinal diseases.

5 Conclusion

This paper presented a method to estimate NO_2 concentration at high spatial and temporal resolution. Estimates were computed for the urban area of Calgary (Canada) at the postal code level (28,980 locations) at monthly intervals, based on data obtained from a regulatory network of ten stations over three years. The multistep statistical spatiotemporal model integrates harmonic regression and land use regression to estimate the mean and residuals of a space–time field. By explicitly accounting for spatial and temporal errors, the model yields satisfactory goodness-of-fit and more reliable estimates. These fine-scale estimates can reduce spatial misalignment errors, leading to improved exposure assessment in health models.

Acknowledgements. We would like to acknowledge Health Canada and our colleagues of the Air Health Science Division for their contribution to the spatial monitoring campaigns and the development of the land use regression models used in this paper. We are grateful to students and colleagues in our lab for their contributions to data management, visualization, and collegial discussions. We wish to thank our health research partners for their continuous encouragement, stimulating ideas, and constant support of this project. Finally, we are grateful to Alberta Innovates – Technology Futures (AITF) for supporting Xiaoxiao Liu's doctoral work.

References

1. Brunekreef, B., Holgate, S.T.: Air pollution and health. Lancet **360**(9341), 1233–1242 (2002)
2. D'Ippoliti, D., Forastiere, F., Ancona, C., Agabiti, N., Fusco, D., Michelozzi, P., Perucci, C. A.: Air pollution and myocardial infarction in Rome: a case-crossover analysis. Epidemiology **14**(5), 528–535 (2003)

3. Ruidavets, J.-B., Cournot, M., Cassadou, S., Giroux, M., Meybeck, M., Ferrières, J.: Ozone air pollution is associated with acute myocardial infarction. Circulation **111**(5), 563–569 (2005)
4. Lanki, T., Pekkanen, J., Aalto, P., Elosua, R., Berglind, N., D'Ippoliti, D., Kulmala, M., Nyberg, F., Peters, A., Picciotto, S., Salomaa, V., Sunyer, J., Tiittanen, P., von Klot, S., Forastiere, F.: Associations of traffic related air pollutants with hospitalisation for first acute myocardial infarction: the HEAPSS study. Occup. Environ. Med. **63**(12), 844–851 (2006)
5. Beckerman, B.S., Jerrett, M., Finkelstein, M., Kanaroglou, P., Brook, J.R., Arain, M.A., Sears, M.R., Stieb, D., Balmes, J., Chapman, K.: The association between chronic exposure to traffic- related air pollution and ischemic heart disease. J. Toxicol. Environ. Health. A **75** (7), 402–411 (2012)
6. Tonne, C., Wilkinson, P.: Long-term exposure to air pollution is associated with survival following acute coronary syndrome. Eur. Heart J. **34**(17), 1306–1311 (2013)
7. Jerrett, M., Burnett, R.T., Beckerman, B.S., Turner, M.C., Krewski, D., Thurston, G., Martin, R.V., van Donkelaar, A., Hughes, E., Shi, Y., Gapstur, S.M., Thun, M.J., Pope, C. A.: Spatial analysis of air pollution and mortality in California. Am. J. Respir. Crit. Care Med. **188**(5), 593–599 (2013)
8. Greco, F.P., Lawson, A.B., Cocchi, D., Temples, T.: Some interpolation estimators in environmental risk assessment for spatially misaligned health data. Environ. Ecol. Stat. **12** (4), 379–395 (2005)
9. Banerjee, S., Gelfand, A., Carlin, B.: Hierarchical Modeling and Analysis for Spatial Data, vol. 101. Chapman and Hall/CRC, Boca Raton (2003)
10. Pope III, C.A., Burnett, R.T., Thun, M.J., Calle, E.E., Krewski, D., Thurston, G.D.: Lung cancer, cardiopulmonary mortality and long-term exposure to fine particulate air pollution. J. Am. Med. **287**(9), 1132–1141 (2002)
11. Miller, K.A., Siscovick, D.S., Sheppard, L., Shepherd, K., Sullivan, J.H., Anderson, G.L., Kaufman, J.D.: Long-term exposure to air pollution and incidence of cardiovascular events in women. N. Engl. J. Med. **356**(5), 447–458 (2007)
12. Basagaña, X., Aguilera, I., Rivera, M., Agis, D., Foraster, M., Marrugat, J., Elosua, R., Künzli, N.: Measurement error in epidemiologic studies of air pollution based on land-use regression models. Am. J. Epidemiol. **178**(8), 1342–1436 (2013)
13. Jerrett, M., Burnett, R.T., Ma, R., Pope, C.A., Krewski, D., Newbold, K.B., Thurston, G., Shi, Y., Finkelstein, N., Calle, E.E., Thun, M.J.: Spatial analysis of air pollution and mortality in Los Angeles. Epidemiology **16**(6), 727–736 (2005)
14. Szpiro, A.A., Sampson, P.D., Sheppard, L., Lumley, T., Adar, S.D., Kaufman, J.D.: Predicting intra-urban variation in air pollution concentrations with complex spatio-temporal dependencies. Environmetrics **21**, 606–631 (2010)
15. Lindstrom, J., Szpiro, A.A., Sampson, P.D., Oron, A.P., Richards, M., Larson, T.V., Sheppard, L.: A flexible spatio-temporal model for air pollution with spatial and spatio-temporal covariates. Environ. Ecol. Stat. **21**(3), 411–433 (2014)
16. Hoek, G., Krishnan, R.M., Beelen, R., Peters, A., Ostro, B., Brunekreef, B., Kaufman, J.D.: Long-term air pollution exposure and cardio- respiratory mortality: a review. Environ. Health **12**(43), 1–15 (2013)
17. Gryparis, A., Paciorek, C.J., Zeka, A., Schwartz, J., Coull, B.A.: Measurement error caused by spatial misalignment in environmental epidemiology. Biostatistics **10**(2), 258–274 (2009)
18. Kim, S.-Y., Sheppard, L., Kim, H.: Health effects of long-term air pollution: Influence of exposure prediction methods. Epidemiology **20**(3), 442–450 (2009)
19. Sheppard, L., Burnett, R.T., Szpiro, A.A., Kim, S.-Y., Jerrett, M., Pope, C.A., Brunekreef, B.: Confounding and exposure measurement error in air pollution epidemiology. Air Qual. Atmos. Health **5**(2), 203–216 (2012)

20. Kyriakidis, P.C., Journel, H.G.: Stochastic modeling of atmospheric pollution: a spatial time-series framework. Part I: Methodol. Atmos. Environ. **35**, 2331–2337 (2001)

21. Sampson, P.D., Szpiro, A.A., Sheppard, L., Lindström, J., Kaufman, J.D.: Pragmatic estimation of a spatio-temporal air quality model with irregular monitoring data. Atmos. Environ. **45**(36), 6593–6606 (2011)

22. Statistics Canada: Calgary, Alberta (Code 4806016) and Canada (Code 01) (table). Census profile Statistics Canada catalogue no. 98-316-XWE, Ottawa, 24 October 2012

23. Nkemdirim, L.C., Leggat, K.: The effect of Chinook weather on urban heat islands and air pollution. Water, Air Soil Pollut. **9**, 53–67 (1978)

24. Bertazzon, S., Johnson, M., Eccles, K., Kaplan, G.G.: Accounting for spatial effects in land use regression for urban air pollution modeling. Spat. Spatiotemporal. Epidemiol. **14–15**, 9–21 (2015)

25. Calgary Region Airshed Zone: 2012 Calgary region airshed zone annual report (2012)

26. Fuentes, M., Guttorp, P., Sampson, P.: Using transforms to analyze space-time processes. In: Statistical Method for Spatial temporal systems, pp. 78–147 (2007)

27. Prado, R., West, M.: Time Series: Modeling, Computation, and Interface. CRC Press, Boca Raton (2010)

28. Zhang, J.J.Y., Sun, L., Barrett, O., Bertazzon, S., Underwood, F.E., Johnson, M.: Development of land-use regression models for metals associated with airborne particulate matter in a North American city. Atmos. Environ. **106**, 165–177 (2015)

29. Hoek, G., Beelen, R., de Hoogh, K., Vienneau, D., Gulliver, J., Fischer, P., Briggs, D.: A review of land-use regression models to assess spatial variation of outdoor air pollution. Atmos. Environ. **42**(33), 7561–7578 (2008)

30. Graham, M.H.: Confronting multicollinearity in ecological multiple regression. Ecology **84** (11), 2809–2815 (2003)

31. Bivand, R., Piras, G.: Comparing implementations of estimation methods for spatial econometrics. J. Stat. Softw. **63**(18), 1–36 (2015)

Modeling Checkpoint-Based Movement with the Earth Mover's Distance

Matt Duckham[1], Marc van Kreveld[2], Ross Purves[3], Bettina Speckmann[4], Yaguang Tao[1(✉)], Kevin Verbeek[4], and Jo Wood[5]

[1] School of Science, RMIT University, Melbourne, Australia
matt.duckham@rmit.edu.au, s3553285@student.rmit.edu.au
[2] Department of Computing and Information Sciences, Utrecht University, Utrecht, The Netherlands
m.j.vankreveld@uu.nl
[3] Department of Geography, University of Zurich, Zürich, Switzerland
ross.purves@geo.uzh.ch
[4] Department of Mathematics and Computer Science, TU Eindhoven, Eindhoven, The Netherlands
{b.speckmann,k.a.b.verbeek}@tue.nl
[5] Department of Computer Science, City University London, London, UK
J.D.Wood@city.ac.uk

Abstract. Movement data comes in various forms, including trajectory data and checkpoint data. While trajectories give detailed information about the movement of individual entities, checkpoint data in its simplest form does not give identities, just counts at checkpoints. However, checkpoint data is of increasing interest since it is readily available due to privacy reasons and as a by-product of other data collection. In this paper we propose to use the Earth Mover's Distance as a versatile tool to reconstruct individual movements or flow based on checkpoint counts at different times. We analyze the modeling possibilities and provide experiments that validate model predictions, based on coarse-grained aggregations of data about actual movements of couriers in London, UK. While we cannot expect to reconstruct precise individual movements from highly granular checkpoint data, the evaluation does show that the approach can generate meaningful estimates of object movements.

1 Introduction

Throughout the years, interest in spatial data has shifted from static planar maps, to space-time [19] and 3D GIS [1], and to movement data [21,29]. The study of movement data has grown explosively due to the availability of tracking

B. Speckmann and K. Verbeek are supported by the Netherlands Organisation for Scientific Research (NWO) under project nos. 639.023.208 and 639.021.541, respectively. This paper arose from work initiated at Dagstuhl seminar 12512 "Representation, analysis and visualization of moving objects", December 2012. The authors gratefully acknowledge Schloss Dagstuhl for their support.

© Springer International Publishing Switzerland 2016
J.A. Miller et al. (Eds.): GIScience 2016, LNCS 9927, pp. 225–239, 2016.
DOI: 10.1007/978-3-319-45738-3_15

devices and their increased quality. Movement is essential for modeling many types of spatial interaction, one of the central concepts in spatial analysis.

Movement data is often available in the form of *trajectories:* sequences of time-stamped locations acquired through GPS or other devices that can determine the location of an individual entity. There are a host of computational tools to analyze trajectories, for example, to determine similarity, to cluster, or to find specific patterns in the trajectories (such as flocks or leadership) or the underlying space (like hotspots) [9,16].

Recently, a different type of movement data has become of increasing interest, namely *checkpoint data* [3,4,6,11,25,26,28]. Here the entities themselves need not be equipped with GPS, but rather their presence at a location or neighborhood is recorded by a stationary sensor. Such sensors include street cameras counting passing pedestrians, check-in gates at metro stations, inductive loops counting cars, RFID sensors in mass participation sporting events, and mobile phone cell towers and wifi access points counting the number of connections in their vicinity. The resulting type of movement data is typically either anonymous or anonymized before being made available for analysis. Hence, frequently the only data available is counts of entities at certain times or in certain intervals.

Checkpoint data is usually much less information-rich than trajectory data. This is partly due to the typically coarse spatial granularity of fixed checkpoint locations, but also due to the lack of heading, speed, chosen route, and stops that are not recorded nor so easily derived from aggregate counts. We can identify several types of checkpoint data based on the spatial extent of acquisition of the data (point-based or area-based) and the movement space (network or more general). Examples of the resulting four classes are given in Table 1.

Table 1. Examples of various types of checkpoint data.

	Network movement	Areal movement
Point-based check (cameras, gates, inductive loops)	Road traffic, subway	Indoor movement (airport, hall)
Area-based check (cell towers, satellite)	Pedestrians (street)	Pedestrians (square, park)

The coarse-grained aspect of the data makes it suitable only for coarse-grained pattern analysis. Perhaps the most important one of these patterns is global *flow* of entities. But since no identity, heading, or speed data is available, flow must be reconstructed from the counts. Reconstruction of flow can be based on any of various *spatial interaction models*. Spatial interaction models describe the flow of people, goods, infections, or information between locations in geographic space, and are therefore studied in various fields of geography.

In this paper we assume a tessellated geographic space and a number of time stamps as a model for area-based checkpoint data. At each time stamp or snapshot, we have a count of the number of entities in each region of the tessellation (termed "temporal checkpoints" in [25], akin to a function from time to a spatial field). Such data may arise from mobile phone connection counts in cell tower regions, for example, aggregated over time intervals.

We will study the possibilities of reconstructing flow consistent with this data using the *Earth Mover's Distance* [23], a well-known measure for capturing the distance (or its inverse, similarity) between two images or weighted point sets. It has also been used in GIS for similarity assessment (see, for example, [10,15,17]). Let $R = \{(r_1, w_1), \ldots, (r_n, w_n)\}$ be a set of n tuples consisting of points r_i and corresponding weights w_i. Let $W = \sum w_i$. Similarly, let $B = \{(b_1, v_1), \ldots, (b_m, v_m)\}$ be a set of m tuples, and let $V = \sum v_i$. The Earth Mover's Distance between R and B is defined if $W = V$, and is the minimum total effort to transport all the weight from R to B. The effort to transport weight w from a point r to a point b is defined as $w \cdot \text{dist}(r, b)$, where $\text{dist}(r, b)$ is a distance measure, for example the Euclidean distance. The Earth Mover's Distance is a metric, also known as the Wasserstein metric. Since the total weight in R and B is the same, we must transport all weight from R to give all points of B the correct weight. Any point in R can give its weight to multiple points in B, and any point in B may receive its weight from one or more points in R. Therefore, a minimum effort transportation corresponds to a flow from R to B.

Reconstructing flows allows us to make effective visualizations including OD maps [27] and flow maps [2,5]. Figure 1 shows a typical output of our model estimating flows of people based on granular mobile phone data.

Results and Organization. In Sect. 2 we overview spatial interaction models and argue that the Earth Mover's Distance is suitable for reconstructing flow from checkpoint data. We recap a linear-programming formulation to compute the Earth Mover's Distance. In Sect. 3 we use the Earth Mover's Distance

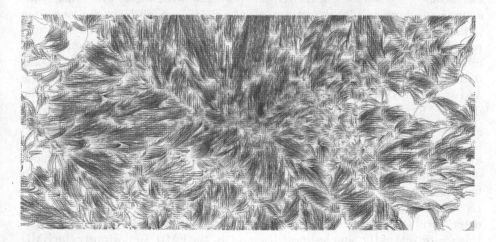

Fig. 1. Estimated flows of people between 9:05 and 9:10 am in central London, 3rd June 2012. Flow estimations based on least cost movement between mobile telephone density surfaces over the 5 min period.

to reconstruct flow in typical scenarios like mobility in a city. We show that environmental situations like obstacles (rivers) and metro stations can easily be incorporated by adapting the objective function and constraints of the linear program. In Sect. 4 we analyze the success of the Earth Mover's Distance to reconstruct flows. To this end we evaluate our approach using data about real trajectories of couriers in London, UK, by converting them to tessellated counts at time stamps and then trying to reconstruct the flows present in the original trajectories. Section 5 summarizes the contribution of this work, as well as indicating further possibilities and improvements for future work.

2   Spatial Interaction and the Earth Mover's Distance

Spatial interaction models of flow are commonly associated with the gravity model [8,20,22], which in its original form relates the trade flow F_{ij} between two countries i, j using their economic masses M_i and M_j and their distance d_{ij}:

$$F_{ij} = c \cdot \frac{M_i^{\beta_1} M_j^{\beta_2}}{d_{ij}^{\beta_3}},$$

where $c, \beta_1, \beta_2, \beta_3$ are constants. The distance may be influenced by the cost of transportation but also by trade barriers. Many extensions of the gravity model have been described, taking into account more factors or compensating for weaknesses. Besides economics, the gravity model is also popular in transportation, migration, and mobility modeling. Other spatial interaction models include the radiation model [24] and Huff's probabilistic model [13,14].

While these models could be used to model movement in checkpoint data, the Earth Mover's Distance [23] (EMD) has potential advantages. The other models aim to represent global patterns of interaction, established over long time periods (over which small variations are smoothed out), and focus on economic principles such as supply and demand. There is little reason to believe that such models would work well for reconstruction of movement based on checkpoint data, which has a much finer time resolution and may vary rapidly in both time and space. Furthermore, models like the gravity model attempt to explain the degree of interaction based on (economic) masses without taking local patterns into account. A gravity function can be fitted to the data, but such a function will be global and apply to the whole grid. Geographically weighted regression approach [18] has been taken recently to support local spatial interaction modeling. While more location specific parameters were introduced to reach a better fit, existing flow data is required in training the models. Our objective is to reconstruct deviations from global movement behavior, or random patterns, and detect local trends of movement that exist in specific areas at specific times. For this we use one of the simplest possible models, the EMD. Importantly the EMD conserves mass in flow, although we purposefully adapt it to account for loss or gain of mass, for example, because of sensor error or movements not detected by sensors.

We consider a specific instance of checkpoint data where we have counts at time steps $t_1, \ldots, t_s$ at all checkpoints. For descriptive purposes we assume that the checkpoints provide counts in regions of a regular grid. When a grid of counts at time t_i and a grid of counts at time t_{i+1} are known, we can infer movement from entities in cells at time t_i to cells at time t_{i+1}, see Fig. 2. In particular, if some cell c contains 10 entities at time t_i and 6 entities at time t_{i+1}, we are certain that at least 4 entities have left the cell. Possibly, all 10 entities have left and 6 other entities appeared. It is also possible that yet other entities passed all the way through cell c between times t_i and t_{i+1} and were never counted.

Checkpoint data does not allow us to completely reconstruct flow, since, for example, it is difficult to identify flow between two cells of the same magnitude, because they cancel out. However, we can still hope to determine flows at a somewhat more global level if there is a trend. To this end, we make an assumption of minimum cost movement. We do not claim that this is realistic, but it does provide a lower bound on the total flow. Minimum cost flow can be derived from the EMD, as described in the introduction. We let the location of an entity be the center of the cell the entity is in. So an entity sits in the same cell at time t_i and t_{i+1} has exactly the same location despite that it might moved slightly. When a minimum cost flow lets entities move to the same cell as the one they started, the cost of the movement is zero because the movement distance is zero. Movement of entities to an adjacent cell has cost equal to the product of the cell size and the number of entities moving. In Fig. 2 there are two minimum flow solutions.

The minimum cost flow problem can be formulated as a linear program. Here the flow from a cell j at t_i to a cell k at time t_{i+1} becomes a variable F_{jk}. The objective function, to be minimized, is the summation of all flows times the distance:

$$\sum_{j,k} F_{jk} \cdot d_{jk} \tag{1}$$

where the distance is assumed to be the distance between the cell centers. To ensure that the flow transports the correct numbers, we use constraints. They come in three types:

– Non-negativity constraint: $F_{jk} \geq 0$ for all j, k
– Origin constraint: $\sum_k F_{jk} =$ count of cell j at time t_i, for all j
– Destination constraint: $\sum_j F_{jk} =$ count of cell k at time t_{i+1}, for all k

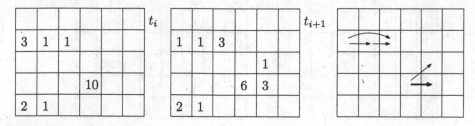

Fig. 2. Grid with counts at times t_i and t_{i+1}, and a possible flow indicated to the right by arrows. In another minimum cost movement, two of the three entities at the top left moved two cells to the right, and the one entity in the middle did not move.

In principle, no flow is negative, the whole count must exit each cell at time t_i, and the resulting count at each cell has arrived at time t_{i+1}. We can replace the two equalities by inequalities and obtain a linear program with the same solution.

3   Modeling and Computation of Flow in Specific Cases

To demonstrate the versatility of linear programming to compute the EMD flow, we show how to incorporate various situations in a natural way. We consider flow in an urban environment based on mobile phone data and a time interval of 5 min. This is a typical situation in practice. It allows identification of the main flows during morning and evening rush hours, flow during big events, and generally flow patterns at different times. We can imagine a grid of, for instance, 20×20 cells, each of 100×100 m.

Urban Movement from Area-Based Counts. The basic computation of flow using the EMD follows the three linear programming constraints given in the previous section. The EMD is in principle mass preserving, but we can expect that in our situation of urban movement there will be different total counts at times t_i and t_{i+1}. There are two main reasons for this:

- People at the edges of the area of interest move to the outside, or people just outside the area of interest move inside. We can assume that this movement influences the counts in the cells close to the boundary.
- People can at any time switch on or off their device, and they may also lose connection or acquire a connection.

To incorporate the former we extend the grid with an extra ring of cells surrounding the original grid, see Fig. 3. The extra ring does not have data, so there are no counts for these cells. In our model we allow these cells to produce extra entities moving into the core grid, or take up entities departing from the core grid. This models the boundary effects in a simple and elegant manner as a (potentially infinite) sink/source.

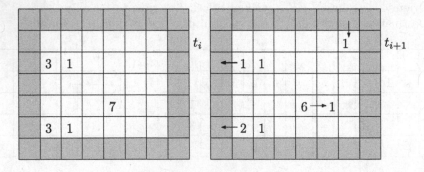

Fig. 3. An extra ring of cells (grey) around the core grid (white) allows us to model movement of entities to and from the outside area.

To incorporate the latter we allow entities in every cell to disappear or appear in a count. Since we prefer to "explain" changing numbers by movement, the cost of appearing or disappearing will be significantly higher than that of movement. Technically, we add one extra "cell" to t_i and to t_{i+1}, which does not have a location. The extra cell in t_i (and t_{i+1}) allows movement of any number to (from) any cell in t_{i+1} (and t_i) at the same, high cost per unit. That is, we set d_{cj} and $d_{kc'}$ to high values in Eq. (1) when c and c' are the extra cells.

In many big cities, a major reason for losing cell tower connection is going underground to take a subway. This can be incorporated easily in our model. Grid cells that contain an entrance to the subway have a lower cost (captured in d_{jk}) of appearing and disappearing. The same applies to cells from which a subway entrance can be reached, naturally incorporating the distance between the cell and the subway entrance.

Another common feature is the presence of obstacles in a city, like a river or a stretch of train tracks that does not have crossings. Such situations can cause two nearby grid cells to be much further apart by travel distance than by Euclidean distance. So again, we need only change the distance function d_{jk} in Eq. (1) to accommodate for the increased distance. It is reasonable to use the *geodesic distance,* the length of the shortest path that does not cross obstacles, as the altered distance.

In the model we can choose to favor many small movements over fewer larger movements or vice versa. With the linear conversion of distance to cost in Eq. (1) we observe that five unit-distance movements cost as much as one movement over five units. By raising d_{jk} to a power γ we can favor smaller movements by setting $\gamma > 1$ or larger movements by setting $\gamma < 1$. The parameter γ is closely related to β_3 in the gravity model and corresponds to the concept of *distance decay.* Also note that the LP remains linear in its variables, so this adaptation does not influence efficiency.

We observe that it is generally not possible in our scenario to get from any cell to any other cell in a given time interval. By assuming a maximum travel speed in the city, we can limit the number of cells that can be reached from any cell. This has a positive effect on both the resulting flow (we forbid long-distance, unrealistic flows) and on the efficiency. Since the LP has a flow variable F_{jk} for every cell pair j at t_i and k at t_{i+1} between which flow is possible, we can reduce the number of variables drastically this way.

Finally, we observe that the assumption of a grid is not necessary for EMD and its LP-based algorithm. For any partition into regions we can use a representative point inside (the cell tower location) instead of the grid cell center.

Other Movement. Movement monitored by gates or cameras leads to point-based counts rather than area-based counts. With toll gates on highways and with check-in gates of subways, we know the direction and precise count of entities accessing a particular area; with cameras this is less precise. Previous research on traffic management in combination with checkpoints concentrated on toll gate placement and pricing [7,11,28], travel time estimation [3,26], or traffic flow modeling in general, see, for example [12].

We briefly discuss movement described by point-based counts, because it is considerably different from movement described by area-based counts. We assume a network is given with certain positions where check-in and check-out is possible. Again our objective is to determine flow, which is closely related to matching up in-flow of the network with out-flow. For example, if there is a large check-in count at checkpoints a and i in Fig. 4, and a large check-out count later at checkpoints f and d, it is interesting to try and determine if entities mostly went from a to d and from i to f, or if they mostly went from a to f and from i to d.

With area-based checkpoints every entity – in theory – is counted once by a checkpoint at any time. With point-based checkpoints, time plays a different role. To be able to compute a matching also for point-based checkpoint data, we can generate check-in counts and check-out counts in 5-minute intervals. This results in two sets of weighted points, where the points are a combination (c, i) of checkpoint c and time interval t_i, and the weights are the corresponding counts. Thus,

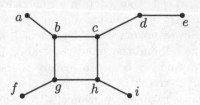

Fig. 4. A simple network with checkpoints.

we can again use the EMD to reconstruct flow. This results in flows of the form $F_{c,i,c',j}$, where $F_{c,i,c',j}$ describes the potential flow from any check-in point c at time interval t_i to any check-out point c' at time interval t_j. In our LP, we need a variable for $F_{c,i,c',j}$ only if $j \geq i$, or more generally, if the trip from c to c' is possible in $t_j - t_i$ time, plus the sampling interval. It is natural to use the typical travel times between checkpoints to obtain the most likely matching. For a particular flow $F_{c,i,c',j}$ we set the cost (d_{jk} in Eq. (1)) to capture the likelihood that an entity that checked in at c in time interval t_i will check out at c' in time interval t_j. This likelihood can be modeled using various factors.

We observe that also with anonymous point-based checkpoint data in a network, we can potentially reconstruct flow using the EMD. However, the number of variables needed may be large, especially if we use fine granularity of time.

4   Evaluation

This section provides an experimental validation of the use of our approach in estimating flows from checkpoint snapshots. The evaluation uses real movement data as its "ground truth", generates granulations of this data at different timestamps based on spatial tessellations as input to the EMD LP, and evaluates the accuracy of the estimates based on comparison with the original movement data. It is important to note that the experiments described in this section are not overly concerned with actually reconstructing precise moving object flows and trajectories from granular snapshots—the snapshots are too information-poor for any method to reliably achieve that. Rather, the evaluation attempts to demonstrate the extent to which our approach can capture the broader flows, directions, and distances, and show that it is flexible enough to accommodate a range of other information about constraints to movement.

Experimental Setup. Our evaluations use a real data set of courier movement trajectories in central London, UK, in 2007 (the Ecourier data set[1]). The location-update frequency of trajectories varies between one coordinate every 10–30 s.

As discussed above, the EMD LP takes as input two snapshots of the granular distribution of spatial objects, generating a matching between the cells in one snapshot to the cells in the next snapshot as output. This matching can be directly interpreted as flow. Each input snapshot summarizes the number of objects in each cell at that time. Thus in our experiments, we spatially granulate the trajectory data by aggregating courier locations at specified times according to a raster grid of user-defined size and location. Based on preliminary studies of the data, a 22 km squared area of central London was chosen for this study, and decomposed into a 40 × 40 raster grid for the purposes of trajectory aggregation (i.e., each cell is square with a 550 m side length). Each trajectory was snapped at the relevant snapshot times to the nearest grid center, yielding a rounded ground truth that can in theory be reconstructed exactly. Our evaluation can then compare these known "ground truth" trajectories with the flows predicted by the EMD LP based only on the counts in cells.

Using the Ecourier data set ensures that our evaluation operates upon realistic movement patterns. However, the limited number of couriers in close proximity at any one time would make the task of unambiguously identifying movements in the raw courier data too simple for the EMD LP (that is, in reality, most grid cells would contain zero or one couriers at any one time). To provide a more challenging simulation of the contemporaneous movements of larger numbers of objects, we densified the trajectory data set by aggregating all courier trajectories over every day over a two month period (May–July 2007) down to a single day (that is, retaining the time-of-day portion of the trajectory time stamps, but discarding the trajectory date). Hence, our evaluation uses approximately 280 courier trajectories in our study area at any one time, ensuring that between zero and 10 couriers may appear in the same cell at a time.

Experiment 1: Flow Accuracy. We begin by comparing the flows estimated by the EMD with the known "ground truth" trajectories of moving objects. Figure 5a shows the changes in accuracy with increasing the time interval between the two input snapshots. Accuracy is measured on a per-object basis as the number of correctly estimated object movements (i.e., correct flow between an origin cell to a destination cell) divided by the total number of objects.

Broadly, Fig. 5 shows EMD estimation accuracy decreasing with increasing time interval between snapshots. This decrease is to be expected, as in longer temporal intervals, objects have a greater range of potential destinations. On average, couriers in our data set travel about 400 m in 1 min, with the fastest objects traveling 1.8 km in that time ($\approx$110 km/h).

---

[1] https://en.wikipedia.org/wiki/Ecourier.

Overall, the model can be said to perform relatively well. At the smallest temporal interval between snapshots (10 s), the model achieves near perfect accuracy of prediction. With snapshots 2 min apart (120 s), the model still achieves 50 % accuracy in predictions.

Experiment 2: Distance and Direction Accuracy. The evaluations in Fig. 5 do not account for "near misses"; only estimated flows that are *exactly* correct contribute to the accuracy or skill scores. In practice, estimations may differ in the *degree* to which they approximate the true flows. Figure 6a shows the accuracy of estimated flow *distance*, in terms of the total number of objects with estimated flows of the correct length, when compared with the total number of moving objects. As might be expected, the accuracy is moderately increased over the accuracy observed in Fig. 5.

Figure 6b shows the accuracy of estimated flow *direction*. Averaging the direction of all flows from each cell provides an overall flow direction for that cell. The accuracy in Fig. 6b is the proportion of cells with an overall estimated flow direction within 30° of the overall true flow direction. Even though the individual estimated flows might not exactly match the main flows, the response curve in Fig. 6b shows that the overall direction of estimated flows closely matches (i.e., is within 30°) of the overall direction of main flows in the majority of cases, even up to and beyond 2 min gaps between snapshots.

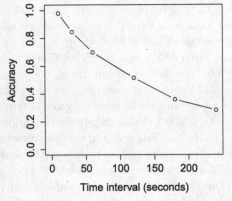

Fig. 5. Estimation accuracy of the flow.

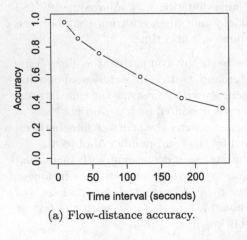

(a) Flow-distance accuracy.

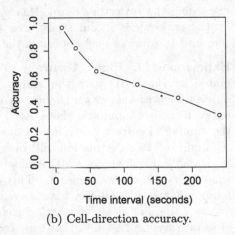

(b) Cell-direction accuracy.

Fig. 6. Flow accuracy for movements with various distances.

Experiment 3: Comparison with Baseline. One further evaluation of the EMD LP flow estimations is to compare with an independent matching baseline. A natural baseline is a randomized, greedy allocation, as summarized in Algorithm 1. In short, based on the two snapshots, the algorithm randomly selects a "provider" cell with a stock of objects that must flow out. It then allocates as much of that stock as possible to the nearest "consumer" cell with a demand for in-flowing objects. The algorithm iterates until all the stocks are exhausted and demands are satisfied.

Algorithm 1. Randomized, greedy allocation Baseline

Data: Set of cells L and numbers of objects in each cell $n_s : L \rightarrow \mathbb{N}$ and $n_e : L \rightarrow \mathbb{N}$ at start and end snapshots respectively

1 Initialize the *stock* of each cell *stock* $: L \rightarrow \mathbb{N}$ as $stock(l) \mapsto n_s(l) - n_e(l)$;
2 Initialize $P = \{l \in L | stock(l) > 0\}$ (providers) ;
3 **while** P *is not empty* **do**
4   Select a random provider cell $p \in P$;
5   Assign as many objects as possible from $stock(p)$ to the nearest consumer cell, $c \in C$ where $stock(p) < 0$;
6   Update remaining *stock* for c and p;
7   If $stock(p) = 0$ remove p from P;

Figure 7a compares the estimation accuracy of the EMD with the estimation accuracy of the Baseline. The response curves of the two estimations, the EMD (also shown in Fig. 5 and the Baseline, show little difference, with perhaps the EMD marginally outperforming the Baseline over shorter time intervals. However, a t-test comparing the per-cell accuracy values at each time interval revealed no significant difference between EMD and Baseline estimations (at the 95 % level).

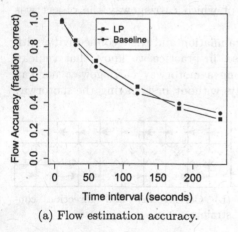

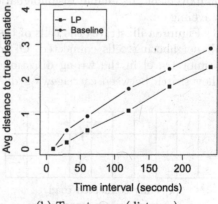

(a) Flow estimation accuracy.   (b) Target error (distance).

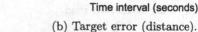

Fig. 7. Distance prediction accuracy.

At first glance, this result is disappointing as it seems to indicate the EMD solution cannot demonstrably outperform the naïve, suboptimal Baseline. However, on closer inspection, both EMD and Baseline are fundamentally matching algorithms, using exactly the same information and constraints. Further, looking more closely at the quality of estimation, in terms of the spatial distance between estimated and true flows, does reveal a performance advantage of using the EMD. Figure 7b shows the average distance (in terms of number of cells) between the estimated target (destination cell) of flows and the true target of flows. The results show that the flows estimated by the EMD have targets that are systematically closer to the targets of the true flows than for the corresponding Baseline estimation. A t-test showed that this difference was statistically significant at the 95 % level for all time intervals, except the shortest (10 s).

Experiment 4: Movement Constraints. One final evaluation examines the addition of movement constraints to the LP model. As discussed in Sect. 3, it is possible to add to the LP known constraints to movement, such as obstacles or barriers to movement. It was not possible to add these constraints to the experimental setup used in the previous experiments, because in central London at a grid size of 550 m, every grid cell is effectively "connected" to every adjacent cell by at least one road. Hence, at this level of granularity, there are no obstacles to movement.

Instead, Experiment 4 "zooms in" on one road, a 16 km section of the M25 London Orbital. This major motorway was frequently used by many couriers, although once again we densified the data, aggregating all the courier trips along that stretch of motorway to a single day, to ensure a sufficiently challenging, large set of contemporaneous movements. The road was then segmented into 20 1.6 km long segments: 10 segments for couriers traveling east to west; 10 segments for couriers traveling west to east. Figure 8a illustrates the cells of the granulation and their connectivity, with all neighboring cells connected. At each timestep, moving objects were assigned to cells in this granulation based on both coordinate location (provides east/west cell location) and on direction of movement, to enable disambiguation of which carriageway the object was traveling on.

Figure 8a illustrates the cells of the granulation and their connectivity, with all neighboring cells connected. Of course, in practice we know that vehicles cannot travel in the wrong direction along a motorway carriageway, nor can they switch between carriageways directly, without first leaving the motorway

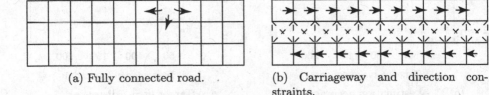

(a) Fully connected road.   (b) Carriageway and direction constraints.

Fig. 8. Cells and connectivity of experiment 4 road granulation.

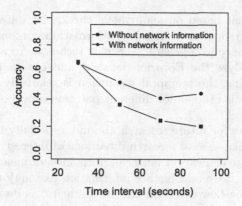

Fig. 9. Effect of movement constraints upon EMD estimation accuracy

and rejoining at an exit. Hence, Fig. 8b illustrates these constraints to movement, encoded through penalizing to the maximum weight disallowed movements between cells (i.e., between carriageways or in the wrong direction along a carriageway).

Figure 9 compares the EMD estimated flows with and without the constraints to disallowed movements along the motorway. The figure shows that the EMD does provide a better estimation of flows when information of underlying movement constraints are provided. As temporal granularity decreases, the difference network information makes tends to be more significant. A t-test suggested that the difference was statistically significant at the 95 % level except for the 30 s time intervals group.

Discussion. The four experiments described above aim to provide a picture of the strengths and weaknesses of our approach, using the EMD to reconstruct flows from granular checkpoint data. In summary, the results of these experiments indicate that the EMD:

1. is capable of regenerating flows from spatially granular checkpoint data with relatively high accuracy, certainly better than chance, especially for shorter temporal intervals where the potential for dispersion are lessened (Experiment 1);
2. is able to provide even greater reliability in generating information about broader distance and directions of flows (Experiment 2);
3. can significantly improve on the quality of estimations when compared with a naïve, suboptimal baseline matching solution, at least in terms of the spatial proximity of estimated flow targets to true flow targets; and
4. is able to incorporate information about constraints to movement, where available, and use that to improve the accuracy of estimates.

5 Conclusions and Future Work

Checkpoint data is becoming increasingly a source of data to be analyzed. This is due to both new data acquisition methods and to privacy considerations. We have shown that movement based on anonymous checkpoint data can be analyzed, and flow reconstructed, despite the low information content. We suggest the Earth Mover's Distance as a general, versatile technique to achieve this. In our experiments we analyze the Ecourier data set and obtain meaningful results on flow, provided that the temporal resolution is relatively small. We cannot reconstruct flow if it is random, or different patterns cancel out the possibilities of detection.

The opportunities for future research abound, especially in experimentation and validation. We list several research directions of interest.

In our data set, we can expect better performance, or meaningful results over longer time periods, if we add further information like major roads. These can be incorporated using flow direction and as obstacles, as described, but also as preferred (faster) routes by lowering the distance costs between certain cells.

Intuitively, network distance is more accurate than other types of distances for network-based movement. The difficulty for applying it lies in choosing representative network nodes for cells based on which network distance can be defined. Such difficulty can be reduced by aggregating movement with a fine-grained space partition schema. Also, with point-based checkpoint data in stead of area-based one as used in this paper, network distance is naturally more suitable than Euclidean distance.

It is also interesting to analyze to what extent we can find flow patterns in other data sets, using similar approaches. These could be data sets based on mobile phone data, as in Sect. 3, or point-based checkpoint data in a network.

We are interested in the spatial and temporal granularities and how they affect the correctness of the flow we find. With a high spatial granularity, we will run into efficiency problems and may need to develop hierarchical methods to approximate the EMD-based flow efficiently.

We can potentially obtain better and more reliable flow when we use more than two snapshots in a single flow reconstruction. This must be modeled first, and then tested against flow reconstruction based on two snapshots only.

References

1. Abdul-Rahman, A., Pilouk, M.: Spatial Data Modelling for 3D GIS. Springer, Heidelberg (2008)
2. Andrienko, N.V., Andrienko, G.L.: Spatial generalization and aggregation of massive movement data. IEEE Trans. Vis. Comput. Graph. **17**(2), 205–219 (2011)
3. Ban, X., Herring, R., Margulici, J.D., Bayen, A.M.: Optimal sensor placement for freeway travel time estimation. In: Lam, W.H.K., Wong, S.C., Lo, H.K. (eds.) (ISTTT18), pp. 697–721. Springer, New York (2009)
4. Both, A., Duckham, M., Laube, P., Wark, T., Yeoman, J.: Decentralized monitoring of moving objects in a transportation network augmented with checkpoints. Comput. J. **56**(12), 1432–1449 (2013)
5. Buchin, K., Speckmann, B., Verbeek, K.: Angle-restricted steiner arborescences for flow map layout. Algorithmica **72**(2), 656–685 (2015)

6. Giudice, N.A., Walton, L.A., Worboys, M.: The informatics of indoor, outdoor space: a research agenda. In: Proceedings of 2nd ACM SIGSPATIAL International Workshop on Indoor Spatial Awareness, pp. 47–53 (2010)
7. Goh, M.: Congestion management and electronic road pricing in Singapore. J. Transp. Geogr. **10**, 29–38 (2002)
8. Greene, R.P., Pick, J.B.: Exploring the Urban Community - A GIS Approach. Prentice Hall, Upper Saddle River (2006)
9. Gudmundsson, J., Laube, P., Wolle, T.: Movement patterns in spatio-temporal data. In: Shekhar, S., Xiong, H. (eds.) Encyclopedia of GIS, pp. 726–732. Springer, Heidelberg (2008)
10. Gunopulos, D., Trajcevski, G.: Similarity in (spatial, temporal and) spatio-temporal datasets. In: Proceedings of 15th International Conference on Extending Database Technology, EDBT, pp. 554–557 (2012)
11. Ho, H.W., Wong, S.C., Yang, H., Loo, B.P.Y.: Cordon-based congestion pricing in a continuum traffic equilibrium system. Transp. Res. Part A: Policy Pract. **39**, 813–834 (2005)
12. Hoogendoorn, S.P., Bovy, P.H.L.: State-of-the-art of vehicular traffic flow modelling. Proc. Inst. Mech. Eng. Part I: J. Syst. Control Eng. **215**(4), 283–303 (2001)
13. Huff, D.: Defining, estimating a trade area. J. Market. **28**, 34–38 (1964)
14. Huff, D., Black, W.: The Huff model in retrospect. Appl. Geogr. Stud. **1**, 83–93 (1997)
15. Jeszenszky, P., Weibel, R.: Measuring boundaries in the dialect continuum. In: Proceedings of AGILE (2015)
16. Laube, P.: Computational Movement Analysis. Springer Briefs in Computer Science. Springer, Heidelberg (2014)
17. Mao, B., Harrie, L., Ban, Y.: Detection and typification of linear structures for dynamic visualization of 3D city models. Comput. Environ. Urban Struct. **36**, 233–244 (2012)
18. Nakaya, T.: Local spatial interaction modelling based on the geographically weighted regression approach. GeoJournal **53**(4), 347–358 (2001)
19. Ott, T., Swiaczny, F.: Time-Integrative Geographic Information Systems. Springer, Heidelberg (2001)
20. Reilly, W.J.: The Law of Retail Gravitation. Knickerbocker Press, New Rochelle (1934)
21. Rense, C., Spaccapietra, S., Zimányi, E. (eds.): Mobility Data - Modelling, Management, and Understanding. Cambridge University Press, Cambridge (2013)
22. Rodrigue, J.-P., Comtois, C., Slack, B.: The Geography of Transport Systems. Routledge, Abingdon (2006)
23. Rubner, Y., Tomasi, C., Guibas, L.J.: The earth mover's distance as a metric for image retrieval. Int. J. Comput. Vis. **40**(2), 99–121 (2000)
24. Simini, F., Gonález, M.C., Maritan, A., Barabázi, A.-L.: A universal model for mobility and migration patterns. Nature **484**, 96–100 (2012)
25. Wang, J., Duckham, M., Worboys, M.: A framework for models of movement in geographic space. Int. J. Geogr. Inf. Sci. **30**, 970–992 (2016)
26. Wood, J.: Visualizing personal progress in participatory sports cycling events. IEEE Comput. Graph. Appl. **35**(4), 73–81 (2015)
27. Wood, J., Dykes, J., Slingsby, A.: Visualisation of origins, destinations and flows with OD maps. Cartographic J. **47**(2), 117–129 (2010)
28. Zhang, X., Yang, H.: The optimal cordon-based network congestion pricing problem. Transp. Res. Part B: Methodological **38**, 517–537 (2004)
29. Zheng, Y., Zhou, X. (eds.): Computing with Spatial Trajectories. Springer, Heidelberg (2011)

User-Generated Data and Linked Data

Exploratory Chronotopic Data Analysis

Benjamin Adams[✉] and Mark Gahegan

Centre for eResearch, The University of Auckland, Auckland, New Zealand
{b.adams,m.gahegan}@auckland.ac.nz

Abstract. The intrinsic connection between place, space, and time in narrative texts is the subject of *chronotopic* literary analysis. We take the notion of the chronotope and apply it to exploratory analysis of unstructured big data. Exploratory chronotopic data analysis provides a data-driven perspective on how place, space, and time are connected in large, crowdsourced text collections. In this study, we processed the English Wikipedia text to find all co-occurrences of named places and dates and discovered that times are linked to places in a large majority of cases. We analyzed these millions of connections between places and dates and discovered a number of interesting trends. Because of the scale of the data involved, we suggest that chronotopic data analysis will lead to the development of new data models and methods for geographic information science and related fields, such as digital humanities.

Keywords: Place · Time · Chronology · Historical geographic information science · Big data · Volunteered geographic information

1 Introduction

Although human history is a continuum of events and processes happening over time and space, when writing about history people structure historical information using discrete times and places as anchors. Wars are fought between countries, cities specialize in industries, historical eras are described at the granularity of centuries, and decades are characterized by particular cultural or social movements. In popular historical writing it is common to talk about places having golden ages like Athens, Greece in the 4th century BCE or important seminal events in the history of place, such as the D-Day invasion in Normandy. How we refer to places and times together helps to create a conceptual framework for history. But how do we refer to places and times? There is scant research on this question from a data-driven perspective, looking at the integrated dynamics of spatial and temporal references in a large corpus of text. The availability of many such corpora, improvements in geographic and temporal parsing of natural language, and the ability to support the associated algorithms and data structures on high-performance computing infrastructure means we have an unprecedented opportunity to explore this topic in new ways.

The deep-rooted connection between representations of time and space in literature has been a focus of literary narrative analysis. The Russian literary

© Springer International Publishing Switzerland 2016
J.A. Miller et al. (Eds.): GIScience 2016, LNCS 9927, pp. 243–258, 2016.
DOI: 10.1007/978-3-319-45738-3_16

theorist, Mikhail Bakhtin, introduced the concept of the *chronotope* to describe how literary genres are characterized by modes of language, which reflect specific spatio-temporal configurations [4]. For example, ancient Greek romances operate on "adventure-time" and are characterized by highly abstract, interchangable representations of times and places in an "alien world" that is not connected with a concrete, familiar landscape and historical timeline. Other works in contrast have more concrete and substantial spatial and temporal structure based on the life course of an individual. In later works there was an effort to merge historical time sequences describing the life of cities, nations, and other social organizations with individual life sequences, though the two sequences are not fused in the sense that they focus on different types of events. The changing ways that people have represented time and space in literature reflect changing conceptualizations of how people live their lives, and shifting cultural attitudes and ideas about the role of the individual and society [5]. Fundamentally, what differentiates chronotopic analysis from other kinds of investigations of place or time in literature is that it is predicated on the idea that spatial and temporal relations and structures in narrative texts are *intrinsically connected*. Thus in chronotopic analysis time and space are not analyzed independently and neither takes precedence over the other. The term chronotope, being an amalgamation of the Greek words for time and space, was inspired by the space-time theories developed in relativity physics in the early 20th century. Although Bakhtin first wrote about chronotopes in 1937, his essay on chronotopes was not published until the 1970s and not translated into English until 1981. But since that time chronotopic analysis has flourished into a broad and heterogeneous field of literary theory.

The development of data models, e.g., space-time prisms, and geographic information systems designed to enable analysis of spatio-temporal phenomena has also been an ongoing research area in GIScience for some years [18,29]. Conventionally, these models extend existing spatial models to include time ('three-plus-one' representations), though there has been some exploration of fully four-dimensional models as well (see [7]). One of the key application areas for such systems is the representation and understanding of human activities and interactions [28]. The application of geographic information science to analyze and represent history has primarily focused on using existing GIS technologies to create historical snapshots of geographic information, e.g., a representation of the boundary of an ancient civilization and the cities within [14]. The use of integrated historical and geographic context can also be used to support geovisual analytics and sensemaking of unstructured information sources [26].

The emergence of new kinds of crowdsourced geographic information (e.g., social media data), which is primarily referenced in terms of named places rather than spatially, has led to research on how to model place-based information [8,25]. In GIScience this recent interest in modeling place (in contrast to space) has included the notion of representing places in terms of their temporal signatures [27]. And there are examples of using machine learning to infer spatio-temporal patterns in the themes that people write about in social media, for example to detect events [16,22]. However, most of the research on place in GIScience has focused on gazetteer development as well as the spatial and thematic (or affordance-based) elements of their representation, not in an integrated way that combines space and time [1,11,13]. An analytic approach that incorporates the intrinsic connectedness between time and place (or space) in collections of unstructured texts remains largely underdeveloped.

Meanwhile, in recent years there has been growing interest in the use of corpus studies and the exploration of big data to understand broad cultural and sociological trends through the written word and other kinds of media. The Google n-grams project which looks at trends in word use in millions of published books has shown that data-driven analysis can uncover shifts in language use over time and examples of social forces acting to change how people write because of policies, such as censorship [17]. Spatial analysis has also grown in prominence in digital humanities [10].

A research program on chronotopic analysis of large text corpora will provide great value, helping us understand the varied ways in which people conceptualize place and time in an integrated way, which in turn can be used to help us organize historical geographic information. In this paper we carve off a preliminary slice of this research. We report on an exploratory analysis of the millions of references to places and times that are found to co-occur in the English Wikipedia corpus. This analysis provides a window into understanding how the semantics of time are structured in the context of crowdsourced, encyclopedic historical content about places. This work can be viewed as a first step toward developing a broader methodology of data-driven chronotopic analysis of unstructured text.

In the following section we describe our data processing workflow to match place and temporal references in Wikipedia. In Sect. 3 we discuss patterns around the use of temporal references alone, and in Sect. 4 look at patterns in how place and time references co-occur. In Sect. 5 we discuss the larger implications of this exploratory study for GIScience research and point to future research directions in exploratory chronotopic data analysis.

2   Data Processing Methods

In this section we describe our methods for identifying place and temporal references and how we matched these references in the text. We leveraged existing open source tools to accomplish this task, but due to the large size of the data, custom analytic scripts were developed to explore the results. For our experiments we used the August 8, 2015 dump of the English Wikipedia, which consists of 7,131,349 articles of which 4,659,056 are actual article pages (i.e., not category, image, or disambiguation pages). The numbers of place and temporal references (detailed below) are both of the same order–in the tens of millions.

2.1   Temporal Tagging

The narrative-style HeidelTime temporal tagger was used to identify temporal tags in the articles [24]. In total **68,657,749 temporal references** were identified within all the main article pages of the English Wikipedia. The existing methods for matching of temporal entities in text are not perfect. There are some false positives that we noticed. For example, references to AM radio station frequencies are often identified as dates. We endeavored to identify and isolate these incorrectly classified entities, but no doubt some noise is still present in the results because of misclassified entities.

2.2   Place Tagging

In order to find place references in Wikipedia we used DBpedia data to find all *place* pages and used the links to those pages to identify georeferences in other articles [3]. DBpedia organizes place references into classes, including Country, City, and Administrative Unit as well as other feature types like Museums and Parks. We identified all these place types in the texts, but for the analysis performed in this study we focused on two main categories of places: (1) **Countries** and (2) **Populated places**, corresponding to City, Town, Village, and Administrative Unit features in DBpedia. Table 1 shows the statistics on number of matched places by type, with **31,922,923 place references** identified in total. Since it is customary to make only one link to a referenced page within an entire Wikipedia article, we matched all additional references to place names that were linked at least once in an article. For example, if a page contains a link to the "Rome" page in the abstract, then we also find all other references to Rome in other paragraphs in the article and match those as well. Once these links were identified we removed all Category pages to focus on references in the narrative text of actual article pages.

Table 1. Summary statistics on the occurrences of named place references in the English Wikipedia. The *Instances* column is the count of distinct named places, and the *References* column list the count of how many times a reference of that type is made in the corpus. *Number of articles* shows the total count of articles that reference at least one instance of the place category in the text, and *Pct. articles* is the percentage of all articles that contain a reference of that type.

Place type	Instances	References	Articles	Pct. articles
Country	255	6,330,851	1,998,273	42.9%
Populated places (cities, towns, etc.)	273,329	12,450,520	2,527,910	54.3%
Other place types (DBpedia)	351,453	13,141,552	1,900,407	40.8%
Any place types	625,037	31,922,923	3,480,667	74.7%

2.3   Matching Places and Times

Although Wikipedia articles are crowdsourced and thus can vary in terms of writing style, in most cases the encyclopaedic format of Wikipedia articles is fairly standardized. As such, paragraphs tend to be self-contained to the degree that we can begin our chronotopic analysis using the simple heuristic of matching places with times if they are found in the same paragraph. In addition to these matches based on co-occurrence in paragraphs, we also matched temporal references to places when found anywhere within an article about that place (e.g., all dates within the main Wikipedia page for New Zealand are matched to New Zealand). While this undoubtedly leads to some false positives in the sense that a place and time might be considered connected even when they are unrelated in the text, it serves as a useful starting point. Using this method, 29,265,607 or **42.6 % of all temporal references in the English Wikipedia are associated with some named place**, and 19,998,504 or **62.6 % of all place references are associated with a temporal reference**. It is clear that place and time are connected concepts across a wide variety of encyclopaedic content. These statistics alone lend credence to the idea that integrated data-driven analysis of time and place references in large text corpora has the potential to lead us to a richer understanding of the semantics of place and time more generally. In addition, it demonstrates that temporality is at least as important, if not more so, for understanding and representing place as place is for understanding and representing time.

3 Dynamics of Date References

In this section we begin the analysis by looking at patterns found in the temporal information on its own. Temporal taggers capture some of the diversity of ways that times are referenced in text. In the TIMEX3 format generated by Heidel-Time, a temporal reference type can be DATE, TIME, DURATION, or SET [21]. A TIME reference refers to a time in a day, e.g., 3:45 pm. A DURATION refers to a length of time, such as "for 2 h". A SET reference is a collection of dates, such as the second Thursday of every month or "annual". A DATE reference is a relative or absolute date based on the Gregorian calendar. The temporal granularity of DATE references ranges from centuries to decades to years through to seasons, months and weeks to individual days and days of the week. In this work our analysis focuses on DATE references, which make up the vast majority of all the temporal references found in Wikipedia. Table 2 shows the summary statistics for these different granularities of date references in the text.

Table 2. Summary statistics on the temporal references in the English Wikipedia.

Temporal type	Count	Number of articles	Pct. articles	Avg. per article
DATE	59, 225, 232	4, 282, 056	91.9 %	12.71
TIME	1, 029, 268	422, 923	9.1 %	0.22
DURATION	6, 867, 967	1, 876, 934	40.3 %	1.47
SET	2, 102, 917	978, 907	21.0 %	0.45
Any type	68, 657, 749	4, 343, 050	93.2 %	14.74

3.1   Decade, Year, Month, and Day Patterns

Figure 1 shows a log scale plot of references to decades from the year 1000 to the 2010s. A remarkable feature of this is the identification that the 10 s decade of every century is referenced on an order of magnitude fewer times than other decades are. The first decade (00s) of the century is referenced more so than others, however that is likely an artifact of the parser not being able to distinguish between century and decade references in those cases. A plausible reason for the reduction in the 10 s is that it reflects the common use of phrases like "the early 1900s" for the first two decades of the century; however, that remains to be evaluated. Ignoring the first two decades of the century, from the early 18th century on there is a steady increase in references to decades, which matches the overall trend for more fine-grained dates as well. In the 20th century a reduction in decade references is found in references to the 1940 s as well, which appears to be a result of the events of World War II dominating the structure of temporal references, so that there are more single year references in that decade than others. This is corroborated by Fig. 2. That Figure illustrates that the U.S. Civil War and the two World Wars are such dominant topics in Wikipedia, that events are described in finer grained (at the level of days and months) detail for those years. Since 2000 the ratio of day references has increased substantially, so that it is on a trend to eclipse year references. It remains to be seen whether this increase is due to the recency of the dates or whether there is a genuine shift in how we are writing about history due to changes in digital technology and our ability to record temporal events at increasing granularity and precision.

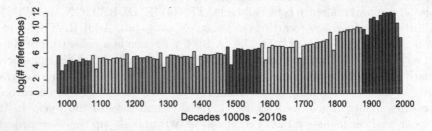

Fig. 1. Log scale plot of the number of references to decades (e.g., "1960s") from 1000 s to 2010s.

3.2   Temporal References and Human Population

The number of temporal references in Wikipedia grows as a function of the date being referenced, which simply means that we've recorded more of our history over time. What is unclear is whether this growth is due to our technical ability to record history with better temporal precision, or if it perhaps reflects other factors as well. To explore this we plotted two ratios in Fig. 3. The red line shows human population relative to the population at 1950, so there are approximately

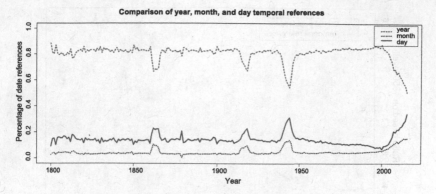

Fig. 2. A comparison of counts of temporal reference types. The U.S. civil war and the two world wars are described in much finer temporal granularity than other years from 1800–2000.

3 times as many people living today as in 1950. The blue line shows the number of temporal references for each year in ratio to the 1950 count. Interestingly, both values grow at the same rate until around 1990, with exponential growth in the number of temporal references until recent years, which presumably is due to a lag in recording contemporary events in Wikipedia (and the data not including the full year of 2015). While this is merely correlation it suggests a hypothesis that as population grows the number of interesting events to record grows in the same way, barring any major technological change.[1] The explosion in temporal references is perhaps due to the advent of the Internet, which revolutionized our ability to record history digitally. Wikipedia was not founded until 2001 around a decade after this increase began.

4 Place and Time Together

Chronotopic analysis is based on the premise that there are characteristic space-time configurations that help us understand categories of written texts and their social context. The first step in approaching this process from a data-driven perspective is to investigate how places and times are expressed together. That is, what are the configurations that exist? In this section we present an exploratory analysis of the connectedness of places and times in the English Wikipedia.

4.1 Historical Trends for Places

Some places have long recorded histories whereas others are more circumscribed due to a combination of factors, including not only the eurocentrism of Wikipedia but also the variations in quality of written historical records from around the

---

[1] Or alternatively, we have increasing time and energy to devote to minutiae!.

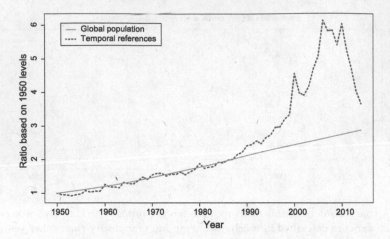

Fig. 3. Based on 1950 levels the number of temporal references for a year grows with global population until the late 1980s where it begins growing much faster.

world that have survived into the modern era [9]. We can use the data we have collected to understand these differences in the historical record of places.

Looking at the changing number of temporal references for a place over time can show trends in how the history of that place has been recorded. We looked at these trends at the granularity of centuries, by aggregating all references to dates at finer granularities (year, day, etc.) into century bins. Then we looked at the average number of references for the countries per century and compared individual countries to that average. Figure 4 shows the results for four countries (Iraq, Greece, France, and China) from 3000 BCE to present day. This chart shows that the region of Iraq is of outsized importance in the 3rd and 2nd millennium BCE as it was the home of many of the earliest civilizations in the fertile crescent. China has a long recorded history, and in Greece there is a clear spike during the 4th century BCE. France in contrast has relatively low numbers of temporal references until after 1000 CE.

Although plotting the timelines of individual places helps us understand the temporality of those places, similarity and clustering techniques for time series data can help to uncover larger trends across a set of places. Figure 5 shows that among countries that have 500 or more century co-references, there is a stand-out group of nine countries that are distinctly different from the others: Egypt, Syria, Greece, Iraq, Iran, Italy, France, China, and India. The plot is a multidimensional scaling (MDS) of the century time series data based on Euclidean distance [6]. Note, that although these countries did not exist as such for much of this time, they are still used in reference to dates long before their founding. This demonstrates that present-day place names (such as Iraq) can operate as metonyms for historical places (e.g., Mesopotamia) in many cases. This has implications for spatio-temporal representation of place in a historical GIS, since we cannot assume that a place name should semantically be restricted to a founding (or ending) timestamp.

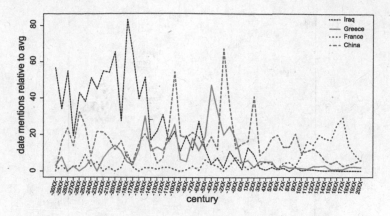

Fig. 4. Changes in the number of temporal references in proportion to the average shows historically important eras for countries.

For different types of DATE references we can also construct histograms for each place, which indicate the distribution of dates for the place. We constructed two histograms of this type based on counts of individual century references from 3000 BCE to the 21st century. The first of these two histograms was built based on counts of pure century references, e.g. "the 14th century." The second was based on counts of references to all century, year, and day binned by century. Therefore, a date like 1941 will be binned into the 20th century as will the day February 3, 1996. Based on these histograms we can calculate the entropy of temporal references for a place, which serves as a measure of how diverse the dates are over time vs. being focused on a few centuries. The entropy measure is shown in Eq. 1, where $H(X)$ is the entropy value ($[0.. \log_2(n)]$, n equals number of classes) and $P(x)$ is probability of date x in the histogram [23].

$$H(X) = -\sum_{i=1}^{n} P(x_i) \log_2 P(x_i) \tag{1}$$

Figures 6 and 7 are quantile choropleth maps showing the century reference entropy results for countries. There is a very strong spatial autocorrelation for the measure when it comes to specific century references (Fig. 6). The highest entropy values run in an east-west band from China through the Middle East to North Africa and southeastern Europe, indicating that references to many centuries at a coarse granularity are made in the context of these countries. This matches the spread of complex state societies out of the fertile crescent [19]. There is less historical record of the pre-Columbian states in the Americas, which is reflected here as well. When more fine-grained dates are included in the century counts (Fig. 7), Western Europe as well as Egypt and parts of the Middle East show the most diversity of centuries represented. This would reflect more historical record across many centuries after around 1000 rather than before, when the recording temporal references became more precise.

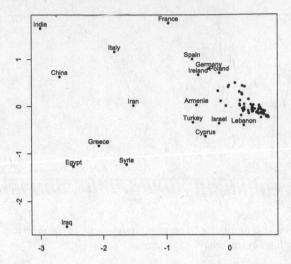

Fig. 5. MDS of countries based on Euclidean similarity of century time series.

In contrast to looking at how centuries are referenced, we can also examine the distribution of different individual years that are referenced in the context of a country. For this measure we look at all the years from 1000 to 2015 and make a similar choropleth map for the countries, shown in Fig. 8. In this case European countries have the most spread of years referenced and in strong contrast to the centuries mapped in Fig. 6 the Middle East is referenced in terms of a relatively small number of individual years.

4.2 Times in Terms of Places, Places in Terms of Times

Not surprisingly, countries on average have more associated temporal references than do populated places such as cities and towns. However, countries and other populated places are similar in that on average there are about equal numbers of century and decade references, on the order of ten times more day references, and about four times again more references to individual years (with no specific day reference). Table 3 breaks down how place types and date types are related in the texts. 61.8 % of all Wikipedia pages have a place and temporal reference that co-occur in a paragraph. Further, this means that out of all pages that reference a place (N=3,480,667), **82.7 %** of those articles have a place and date reference co-occurring in a paragraph. This result points to the potential benefit of using place-time information as fundamental dimensions by which to organize information retrieval systems for large-scale text data, with the further implication that place-based GISs that intrinsically include a temporal dimension will open up significant opportunities for analysis that a spatial (only) GIS cannot [2, 12].

Fig. 6. Information entropy of dates per country by century reference only.

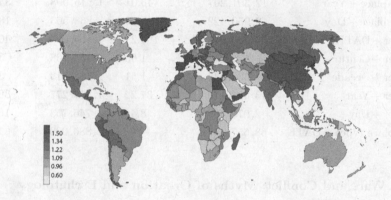

Fig. 7. Information entropy of dates per country by references to all dates aggregated by century.

Fig. 8. Information entropy of dates per country by individual year from 1000 to 2015.

Table 3. Summary statistics on the co-occurrence of named place and date references in paragraphs of the English Wikipedia. The *Avg. per-type* column shows the ratio of count to the number of instances of the place type (i.e., country, populated place, or other).

Place type + Temporal ref	Count	Avg. per type	Articles	Pct. articles
Country + DATE	9,343,550	36,641.37	1,413,690	30.3
Country + Century	177,377	695.60	72,259	1.6
Country + Decade	204,481	801.89	94,109	2.0
Country + Year	4,951,018	19,415.76	1,073,744	23.0
Country + Day	1,418,277	5,561.87	588,293	12.6
Pop. place + DATE	22,687,527	83.00	2,029,940	43.6
Pop. place + Century	508,843	1.86	153,415	3.3
Pop. place + Decade	475,722	1.74	179,734	3.9
Pop. place + Year	12,301,207	45.01	1,616,998	34.7
Pop. place + Day	2,950,422	10.79	858,563	18.4
Other + DATE	36,626,672	104.21	1,865,095	40.0
Other + Century	571,218	1.63	154,708	3.3
Other + Decade	530,834	1.51	183,913	3.9
Other + Year	13,081,170	37.22	1,517,277	32.6
Other + Day	2,972,499	8.46	746,753	16.0
All place types + DATE	68,657,749	--	2,880,090	61.8

4.3 Wars and Conflict: Myths of Creation and Eschatology

In his essay on chronotopic analysis, Bakhtin wrote, "For a long time the central and almost sole theme of purely historical narrative was the theme of war" [4]. We examined the top-3 referenced single day pre-2000 dates for each of the 255 countries and found that 65 % of the dates are related to a battle, declaration of war, or peace treaty. It is similar for large cities. This shows that, in the English Wikipedia at least, the theme of war still dominates how we talk about places. The other major category of event is the creation of a new geopolitical entity (often after a period of war). Table 4 shows a sample of the most cited days.

Table 4 also demonstrates that the recording of historical events in the English Wikipedia, no matter where the events have occurred, is heavily skewed to a United States, United Kingdom and commonwealth perspective. For example, the most highly referenced day for Egypt (29 references) is the date of the ANZAC landing during the Gallipoli campaign, which happened in Turkey, though the troops disembarked for the campaign from a station in Egypt. In contrast the beginning of the Yom Kippur War, a date presumably of more interest to the population living in Egypt, is referenced 23 times. This is further evidence of the eurocentric bias in Wikipedia content, which has been well-documented across all language editions [9].

Table 4. Top-2 referenced days from 2001 and earlier for selected places.

Country	Count	Date	Historical event
Argentina	32	1816-07-09	Argentine declaration of independence
Argentina	23	1982-04-02	Falklands War begins
China	101	1949-10-01	Mao speech creating People's Rep. of China
China	56	1997-07-01	Transfer of sovereignty of Hong Kong
Egypt	29	1915-04-25	Landing at Anzac cove (Gallipoli)
Egypt	23	1973-10-06	Yom Kippur War
France	73	1918-11-11	Armistice of 11 November 1918
France	52	1944-06-06	D-Day Normandy landings
Germany	109	1990-10-03	Reunification of Germany
Germany	70	1939-09-01	Invasion of Poland
Greece	34	1940-10-28	Ohi Day (Greco-Italian War)
Greece	31	1941-04-06	Germany invades Greece
India	156	1947-08-15	Independence day (India)
India	81	1950-01-26	Republic day (India)
Indonesia	19	1941-12-07	Dutch East Indies Campaign
Indonesia	16	1949-12-27	Proclamation of Indonesian Independence
Iran	30	1979-11-04	Iran hostage crisis
Iran	22	1988-07-03	Shooting of Iran Air Flight 655
Japan	145	1941-12-07	Pearl Harbor bombing
Japan	88	1945-08-15	Surrender of Japan (V-J Day)
Mexico	27	1848-02-02	Treaty of Guadalupe Hidalgo
Mexico	24	1994-01-01	NAFTA operational, Zapatista uprising
Russia	30	1991-12-25	Dissolution of the Soviet Union
Russia	26	1998-02-02	Russian financial crisis
South Africa	134	1910-05-31	South African independence
South Africa	102	1994-04-27	First democratic elections (Freedom day)
United Kingdom	45	1939-09-03	Britain declares war on Germany
United Kingdom	43	1910-05-31	South African independence
United States	461	2001-09-11	September 11 terrorist attacks
United States	131	1941-12-07	Pearl Harbor bombing
Paris	24	1792-08-10	Insurrection of 10 August 1792
Paris	19	1860-01-01	Annexation of 1860
Rome	19	1944-06-04	Liberation of Rome
Rome	17	1870-09-20	Capture of Rome (Risorgimento)

5 Implications of Chronotopic Analysis for GIScience

The chronotopic analysis we performed in this study point to many interesting relationships between place and time in very large unstructured data collections. Going forward, there are a number of underlying representational and algorithmic challenges that need to be addressed for GIScience to leverage the opportunities provided by big data.

Better discovery of spatial and temporal references in text. Currently, methods to discover spatial and temporal entities in text leave a lot of room for improvement. For example, place name disambiguation still relies on heuristics that could potentially be improved with machine learning classifiers.

Scaling of discovery methods. The document scraping or feature extraction stage of such work can require massive amounts of time and consume large amounts of storage. In the work we describe above, the temporal tagging and creation of the database of temporal references required approximately 50,000 core hours of processing in a single pass (equivalent to approximately 5 years on a single core computer). Fortunately, the tasks are embarrassingly parallel (the task can easily be decomposed into many smaller but separate tasks), so in our case we could make use of a local HPC service, utilizing 3000 compute cores and a GPFS parallel file system, bringing the elapsed time down to a couple of days. In our experience this stage often needs to be repeated many times to train and refine the extraction methods used, so such savings are critical.

Data structures and algorithms. The figures quoted in Table 3 suggest that both spatial and temporal dimensions are useful ways to organize this corpus. In fact a strong case could be made for a combined spatio-temporal index, given that this would cover over 60 % of the documents. Within GIScience there has been some useful work on adding in the temporal dimension [15,20], but less on the data structures and related algorithms that could scale to many millions of objects that have complex, multi-valued relationships to both place and time.

Formalizing complex spatial and temporal references. Given that there may be multiple spatial and temporal references in a document, each taking different forms, more nuanced analyses will require us to describe the 'spatiality' or 'temporality' of a document more formally. How do these map onto human understandings of space and time? What kinds of query operators and interfaces are needed? How do we extend the current formal models of topology and spatial relations to address these more complex, multi-space, multi-time objects?

6 Conclusion

In this paper we introduced the notion of chronotopic data analysis as a methodology to study spatio-temporal structure in a large text corpora. As an example of this kind of analysis we examined the set of all place and date co-references in the English Wikipedia and found that millions of place references have a temporal association. We demonstrated that by exploring places and dates together we can uncover a number of unexpected patterns that shed light on the importance of the temporal dimension in understanding place.

We have just scratched the surface of chronotopic analysis of big data. Our investigation into place and time in Wikipedia was done by looking at statistics for the entire corpus. Chronotopic analysis in literature also looks at how the spatio-temporal configuration relates to other aspects of the narrative. Toward that end, there is much that can be done to extend the methodology, for example looking at how different types of articles within Wikipedia reference place-time differently. In addition, this type of exploratory data analysis can discover regularities or unique characteristics in the spatio-temporal patterns that manifest in different kinds of historical textual collections, such as novels, newspaper collections, and the literature of private life, e.g., diaries and letters.

References

1. Adams, B., Janowicz, K.: Thematic signatures for cleansing and enriching place-related linked data. Int. J. Geogr. Inf. Sci. **29**(4), 556–579 (2015)
2. Adams, B., McKenzie, G., Gahegan, M.: Frankenplace: interactive thematic mapping for ad hoc exploratory search. In: Proceedings of the 24th International Conference on World Wide Web, pp. 12–22. IW3C2 (2015)
3. Auer, S., Bizer, C., Kobilarov, G., Lehmann, J., Cyganiak, R., Ives, Z.G.: DBpedia: a nucleus for a web of open data. In: Aberer, K., et al. (eds.) ASWC 2007 and ISWC 2007. LNCS, vol. 4825, pp. 722–735. Springer, Heidelberg (2007)
4. Bakhtin, M.: Forms of time and of the chronotope in the novel. In: The Dialogic Imagination, pp. 84–258. University of Texas Press, Austin (1981)
5. Bemong, N., Borghart, P., De Dobbeleer, M., Demoen, K., De Temmerman, K., Keunen, B.: Bakhtin's Theory of the Literary Chronotope: Reflections, Applications, Perspectives. Academia Press, Gent (2010)
6. Borg, I., Groenen, P.J.: Modern Multidimensional Scaling: Theory and Applications. Springer Science & Business Media, New York (2005)
7. Galton, A.: Fields and objects in space, time, and space-time. Spat. Cogn. Comput. **4**(1), 39–68 (2004)
8. Goodchild, M.F.: Formalizing place in geographic information systems. In: Burton, L.M., Matthews, S.A., Leung, M., Kemp, S.P., Takeuchi, D.T. (eds.) Communities, Neighborhoods, and Health, pp. 21–33. Springer, Berlin (2011)
9. Graham, M., Hogan, B., Straumann, R.K., Medhat, A.: Uneven geographies of user-generated information: patterns of increasing informational poverty. Ann. Assoc. Am. Geogr. **104**(4), 746–764 (2014)
10. Gregory, I., Donaldson, C., Murrieta-Flores, P., Rayson, P.: Geoparsing, gis, and textual analysis: current developments in spatial humanities research. Int. J. Humanit. Comput. **9**(1), 1–14 (2015)
11. Hill, L.L.: Core elements of digital gazetteers: placenames, categories, and footprints. In: Borbinha, J.L., Baker, T. (eds.) ECDL 2000. LNCS, vol. 1923, pp. 280–290. Springer, Heidelberg (2000)
12. Janowicz, K.: The role of space and time for knowledge organization on the semantic web. Semant. Web **1**(1, 2), 25–32 (2010)
13. Jordan, T., Raubal, M., Gartrell, B., Egenhofer, M.: An affordance-based model of place in GIS. In: 8th International Symposium on Spatial Data Handling, SDH, vol. 98, pp. 98–109 (1998)

14. Knowles, A.K., Hillier, A.: Placing History: How Maps, Spatial Data, and GIS are Changing Historical Scholarship. ESRI Inc., Redlands (2008)

15. Langran, G.: Issues of implementing a spatiotemporal system. Int. J. Geogr. Inf. Sci. **7**(4), 305–314 (1993)

16. Mei, Q., Liu, C., Su, H., Zhai, C.: A probabilistic approach to spatiotemporal theme pattern mining on weblogs. In: Proceedings of the 15th International Conference on World Wide Web, pp. 533–542. ACM (2006)

17. Michel, J.B., Shen, Y.K., Aiden, A.P., Veres, A., Gray, M.K., Pickett, J.P., Hoiberg, D., Clancy, D., Norvig, P., Orwant, J., Pinker, S., Nowak, M.A., Aiden, E.L.: Quantitative analysis of culture using millions of digitized books. Science **331**(6014), 176–182 (2011)

18. Miller, H.J.: Modelling accessibility using space-time prism concepts within geographical information systems. Int. J. Geogr. Inf. Syst. **5**(3), 287–301 (1991)

19. Peebles, C.S., Kus, S.M.: Some archaeological correlates of ranked societies. Am. Antiq. **42**(3), 421–448 (1977)

20. Peuquet, D.J.: Making space for time: Issues in space-time data representation. GeoInformatica **5**(1), 11–32 (2001)

21. Pustejovsky, J., Castano, J.M., Ingria, R., Sauri, R., Gaizauskas, R.J., Setzer, A., Katz, G., Radev, D.R.: TimeML: Robust specification of event and temporal expressions in text. New Dir. Question Answering **3**, 28–34 (2003)

22. Sakaki, T., Okazaki, M., Matsuo, Y.: Earthquake shakes twitter users: real-time event detection by social sensors. In: Proceedings of the 19th International Conference on World Wide Web, pp. 851–860. ACM (2010)

23. Shannon, C.E.: The mathematical theory of communication. Bell Syst. Tech. J. **27**(379–423), 623–656 (1948)

24. Strötgen, J., Gertz, M.: Multilingual and cross-domain temporal tagging. Lang. Resour. Eval. **47**(2), 269–298 (2013)

25. Sui, D., Goodchild, M.: The convergence of GIS and social media: challenges for GIScience. Int. J. Geogr. Inf. Sci. **25**(11), 1737–1748 (2011)

26. Tomaszewski, B., MacEachren, A.M.: Geo-historical context support for information foraging and sensemaking: conceptual model, implementation, and assessment. In: 2010 IEEE Symposium on Visual Analytics Science and Technology (VAST), pp. 139–146. IEEE (2010)

27. Ye, M., Janowicz, K., Mülligann, C., Lee, W.C.: What you are is when you are: the temporal dimension of feature types in location-based social networks. In: Proceedings of the 19th ACM SIGSPATIAL International Conference on Advances in Geographic Information Systems, pp. 102–111. ACM (2011)

28. Yu, H.: Spatio-temporal GIS design for exploring interactions of human activities. Cartography Geogr. Inf. Sci. **33**(1), 3–19 (2006)

29. Yuan, M.: Temporal GIS and applications. In: Shekhar, S., Xiong, H. (eds.) Encyclopedia of GIS, pp. 1147–1150. Springer, New York (2008)

Exploring the Notion of Spatial Lenses

Christopher Allen[2], Thomas Hervey[1], Sara Lafia[1], Daniel W. Phillips[1],
Behzad Vahedi[1], and Werner Kuhn[1(✉)]

[1] Department of Geography, University of California Santa Barbara (UCSB),
Santa Barbara, CA, USA
werner@ucsb.edu
[2] Department of Geography, San Diego State University (SDSU),
San Diego, CA, USA
http://geog.ucsb.edu

Abstract. We explore the idea of spatial lenses as pieces of software interpreting data sets in a particular spatial view of an environment. The lenses serve to prepare the data sets for subsequent analysis in that view. Examples include a network lens to view places in a literary text, or a field lens to interpret pharmacy sales in terms of seasonal allergy risks. The theory underlying these lenses is that of core concepts of spatial information, but here we exploit how these concepts enhance the usability of data rather than that of systems. Spatial lenses also supply transformations between multiple views of an environment, for example, between field and object views. They lift these transformations from the level of data format conversions to that of understanding an environment in multiple ways. In software engineering terms, spatial lenses are defined by constructors, generating instances of core concept representations from spatial data sets. Deployed as web services or libraries, spatial lenses would make larger varieties of data sets amenable to mapping and spatial analysis, compared to today's situation, where file formats determine and limit what one can do. To illustrate and evaluate the idea of spatial lenses, we present a set of experimental lenses, implemented in a variety of languages, and test them with a variety of data sets, some of them non-spatial.

Keywords: Conceptual lenses · Core concepts of spatial information · Spatial analysis · Data usability · Format conversions

1 Introduction

There is an implicit assumption underlying most work with GIS, namely that a data set encodes a certain view of an environment and should therefore be analyzed with tools conforming to that view *only*. For example, a raster data set is normally seen as encoding a field view of space, consequently admitting map algebra operations, while a set of polygons would be seen as encoding an object view of space, and a literary text would in itself not be considered spatial data. While this assumption can guide the choice of analysis tools, it can also stand in

© Springer International Publishing Switzerland 2016
J.A. Miller et al. (Eds.): GIScience 2016, LNCS 9927, pp. 259–274, 2016.
DOI: 10.1007/978-3-319-45738-3_17

the way of more flexible uses and richer interpretations of data. It limits the use of mapping and spatial analysis tools to certain data formats that these tools can handle and it prevents the exploitation of spatial references in data that are not in a GIS format. Furthermore, the assumption can lead to unnecessary data format conversions and information loss.

In this paper, we challenge this assumption by decoupling data from world views. We introduce the notion of *spatial lenses* for data, which we define as pieces of software that interpret data sets in terms of a chosen world view. For example, a field lens interprets a data set as a representation of an environment viewed as a field. The data set to be interpreted in this way can be an image, a set of point measurements, a live sensor network, or anything else that may be interpreted as representing a continuous function from positions in space-time to values (in other words, a field [1]). Spatial lenses defined in this way would normally be built by application programmers or software developers, according to specifications produced by geographic information scientists, ideally in consultation with domain scientists (such as climatologists, archaeologists, or historians).

The decoupling of data from views of environments also allows for (and often requires) introducing some auxiliary information. To interpret a set of point measurements as a field, for example, one obviously needs to supply an interpolation function. If a literary text is to be seen as representing a social network, the nodes (for example, literary characters) and links (for example, their kinship) need to be defined in computable form.

Spatial lenses also reduce the need for explicit data format conversions by users. Spatial analyses often involve multiple switches between different views of space, based on the formats of data sources. With a dedicated choice of how to conceptualize an environment, there is no need to change that lens for computational reasons only, no matter what format the data sources are in. For example, an analysis of night-time lights in certain areas and at multiple levels of granularity does not require alternation between field and object views, though this is often done in practice [2]. Aggregation is handled as a granularity lens operation, which can be layered on top of other content lenses to answer questions about data quality.

Similar to the idea of constructor and observer operators, which are well-known from abstract data types in programming and software engineering [17], using the spatial lens idea allows for separation of analysis from pre-processing. Spatial lenses are constructors that generate instances of core concept representations to which observers can subsequently be applied in order to answer spatio-temporal questions.

Our main goal is to allow for a more flexible view of what is considered "spatial data". A large proportion of data has implicit or explicit spatial references and is therefore in principle amenable to mapping and spatial analysis. However, the gap between the data and the tools is often too large to bridge for those without solid technical GIS expertise. Spatial lenses lift mapping and spatial analysis from the implementation level of data formats and GIS commands to the level of questions about spatial phenomena [5]. Each of the lenses comes with a set of questions they answer. For example, a network lens answers questions about connectivity, centrality, and paths, while an event lens answers questions about temporal sequencing and possible causation.

The paper first surveys previous work in several information sciences, then presents five case studies, each of them proposing and implementing a set of lenses, and ends with conclusions.

2 Previous Work

We take the seminal work of Codd on databases in the early 1970s [3] as inspiration for the (admittedly harder) task of creating a higher-level understanding of spatial information. Codd's relational algebra essentially defined a lens on the world in terms of tables. Once one understands the world as consisting of phenomena that one can represent in rows of tables with columns for their attributes, the power of relational algebra unfolds without a need to understand how the tables are stored and manipulated in a database. The first sentence in Codd's CACM article says it all: "Future users of large data banks must be protected from having to know how the data is organized in the machine (the internal representation)" [3].

Replace "large data banks" with "GIS" (or any other type of spatial computing platform) and ask yourself what you can do without knowing how a system organizes the data internally. Here we do not mean the physical level of data organization, but the logical one, i.e. the data structures. Whole curricula are in fact built on the assumption that you cannot and should not do much without that knowledge. While this assumption creates a cast of GIS experts (generally recognized to be too small, and likely to stay so), it misses out on the vast potential that GIS has for users without the technical skills, time, or financial resources to acquire a thorough understanding of GIS internals before asking a spatial question. In this paper, we do not address the usability of GIS and other tools, but the **usability of data**, improving the means to interpret any data spatially.

A case similar to that about databases could be made for the power (and limitations) of seeing everything stored in a computer as a document of some type (text, table, graphic etc.) or for the idea of Linked Data, built on the simplest possible data model of subject-predicate-object triples [4]. Both paradigms have lifted data manipulation from the level of dealing with data structures to that of dealing with real-world concepts (documents and statements). GIScience has not yet reached similar levels of simplicity and clarity in describing what it is (and GIS are) about.

The core concepts of spatial information have been defined previously to bridge the gap between spatial thinking and spatial computing [5,16]. The following six of them are now explored as concepts to interpret spatial data:

1. **Object** – An individual that has properties and relations with other objects.
2. **Field** – A property with a value for each position in space and time.

3. **Network** – A set of objects (nodes) linked by a binary relation (edges).
4. **Event** – Something that happens and involves fields, objects, and/or networks as participants.
5. **Granularity** – The level of detail in objects, fields, networks, and events.
6. **Accuracy** – The correspondence of spatial information with what is considered a true state of affairs.

Note that the core concept of **Location** underlies the four "content concepts" (object, field, network, event), in the sense that all of these serve to answer 'where' questions. The "quality concepts" of granularity and accuracy, in turn, can be applied to all content concepts. As location is always observed in space and time, all concepts are spatio-temporal (not just the event concept).

In the case of spatial lenses as well as in the cases of databases, documents, and linked data, the idea of a conceptual lens should be understood as a means to **view reality** using some data and software, rather than viewing the data. For example, if one decides to view temperature measurements from a network of weather stations as a representation of a temperature field, this conceptual lens allows for having a temperature value at each position and time of interest (some of which are measured, others interpolated). The field lens is, thus, not applied to the point data set per se (as this would leave out the interpolation function to be supplied), but to the environment, seeing it through the data.

Our notion of Spatial Lenses is not directly related to the Urban Lens[1] from the Senseable City Lab at MIT, which allows users to extract trends from large datasets through the application of a figurative lens. Unlike this data visualization work, our spatial lens notion applies a computational approach that carries a set of associated questions to help users *spatialize* data in a particular way.

In GIScience, attempts to provide clearer conceptual structure to geographic information have so far mainly focused on organizing GIS commands around the sorts of items manipulated by a GIS [6,7], or on finding a single general model to deal with the largest possible range of geographic data [1,8]. Our approach, by contrast, identifies sorts of items or phenomena in the world, together with questions to be asked about them [5]. It then fits the data and operations to these conceptualizations, rather than coercing the understanding of the world to data models.

3   Case Studies

To articulate the flexibility in choosing spatial lenses, we investigate non-trivial domain-specific questions through a set of case studies. These studies highlight lens views of environments based on existing data, spatial or other, taken from diverse sources and mostly available online. The data range from historical newspaper texts through typical GIS data in raster and vector form, to research objects (publications and research data of any kind) (Fig. 1).

---

[1] http://senseable.mit.edu/urban-lens/.

Fig. 1. Spatial lenses applied to a campus environment, illustrating symbolically how a campus can be seen as a set of objects, as observed through various fields, as forming networks, as participating in events, and how all these views come with a certain level of granularity (detail) and accuracy

3.1 Baltimore City Vacant Buildings

Data Source and Uses. Our first data set is an online repository of vacant building features across Baltimore City[2]. Updated monthly, the data set keeps track of parcel properties that are not currently on the market, and are condemned or no longer occupied. Each feature contains attributes for several parent administrative jurisdictions including neighborhood and police districts' as well as a pair of coordinates and a notice date generated upon initial inspection. Available on the web portal OpenBaltimore, the data can be displayed through online mapping tools, or downloaded in tabular form. We know of no spatial analyses that use this data so far. However, for economists and social scientists, the data could be useful when studying spatial urban dynamics or deciding on investments.

Questions and Lens Support. Addressing such possible user perspectives, we can ask the following questions of the data:

---

[2] https://data.baltimorecity.gov.

– Where can one find vacancy clusters?
– Which neighborhoods contain most clusters?

To study clustering, we propose an **object lens**, viewing clustered vacancies as objects generated from vacant parcel points. A user could then, based on the neighborhood attribute, observe how many clusters are within each neighborhood. This information could be valuable for influencing policy. For example, if a user discovers a large number of vacancy clusters within the Sandtown-Winchester neighborhood, they could suggest to the respective council members where to focus and how to allocate rehabilitation funds. Note that this is an example that illustrates how objects are not defined by boundaries (there are none, in this case), but by their identity.

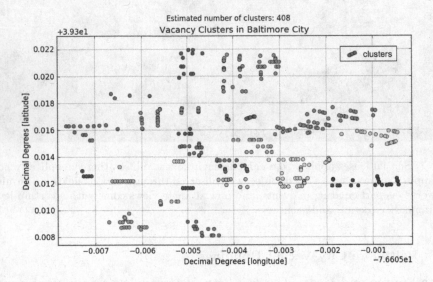

Fig. 2. West Baltimore vacancies plotted as cluster objects (Color figure online)

Constructing the Lens. Since the vacant lots in this data set have a coordinate pair attribute, generating cluster objects is straightforward (for this data, DBSCAN is appropriate [9]), and a user would only need to supply the data in tabular form along with clustering specifications. Upon applying the object lens twice, to individual parcels as well as to the clusters, the following processing steps constitute the object lens constructor:

– Determine and plot point locations based on provided coordinate pairs.
– With user defined settings, run the clustering algorithm.
– Count clusters within each neighborhood.

In our Python implementation, the tabular data is loaded into a matrix, and DBSCAN is run. Several clustering characteristics, including minimum number of samples (in this case five other vacant parcels) and maximum sampling distance (in this case eight houses away), are supplied by the user. The results in Fig. 2 show point objects for each vacant building plotted by location and with clusters differentiated by color. An automated count reveals that the Central Park Heights and Sandtown-Winchester neighborhoods contain the highest number of clusters, with 30 and 22 lots respectively.

Secondary Lens. Viewing vacancies through a **field lens** rather than as a set of objects can provide additional insights. By constructing a field, values could instead represent vacancy density. This "heat map" view could be used to study the spatial distribution of decay rather than just answering the question, where do discrete vacancy clusters occur? An economist or city planner could then suggest to developers where best to launch demolition and development projects. A similar view of vacancies by density is currently being used by The Johns Hopkins Medical Center, which is looking for opportunities to expand further into east Baltimore[3].

3.2   Santa Barbara Communities of Interest

Data Source and Uses. The data set used here was gathered by one of the authors, Daniel W. Phillips, while conducting research about redistricting in Santa Barbara, California. The data were collected to determine how well residents thought the boundaries of the city council district in which they lived reflect what they believed to be their community of interest. They consist of 114 responses to a survey given to residents during the summer of 2015, collected in three of the six city council districts in the city. One of the items on the survey involved each participant taking a base street map of the city and drawing a line around the area that they believed to be their community of interest, defined as a contiguous group of people with shared values, concerns, and cultural traits.

Questions and Lens Support. Analysis of the polygons drawn by residents involved the following questions:

– Do residents of a given district roughly agree about the location and extent of their community of interest?
– Can one identify an area of highest agreement, which might be considered the core of the community of interest within each district?

Applying a **field lens** supports answering these questions. By overlaying the individual polygons for a given district, one can determine the degree of overlap as a field. While the street map survey instrument might have led respondents into more network-based thinking about their communities than field-based, it identified familiar locations better than an aerial image would have and thus enabled people to make more informed decisions; even still, most drew simple

---

[3] https://hub.jhu.edu/gazette/2013/january/east-baltimore-changes-development.

oval shapes that did not conform to the underlying street network. For any given point within this field there is a certain percentage of respondents who included that point in the polygon that they drew. The percentages range from zero to more than 60 %.

Constructing the Lens. After digitizing all drawings, the resulting polygons were merged into a shapefile that served as the input for the constructor operations[4]:

– Compute a count of the overlapping polygons at each point in space.
– Use that count to create an output raster (with 25 m cells).

Constructing classes for degree of overlap allowed for a simpler representation, with a light yellow to dark red color scheme applied to differentiate four classes of agreement (Fig. 3).

Fig. 3. Agreement level of polygons drawn by District 1 residents, with classes from light to dark of 0–39, 40–49, 50–59, and 60+ percent agreement (Color figure online)

Secondary Lens. One could apply an **accuracy lens**, comparing the results of applying different survey instruments. If the survey collectors used a more detailed base map or showed it digitally, allowing for zooming and panning, they might increase the accuracy of the responses obtained.

---

[4] Using a Python script written by Adam Davis, UCSB Department of Geography.

3.3   Yucatan Peninsula Research Data Footprints

Data Source and Uses. Universities increasingly curate data repositories to promote the discovery and reuse of research data. The data considered here expose the spatial extents of research projects from different domains. The first data set examined is UCSB archaeologist Dr. Anabel Ford's archaeological sites and protected areas layer, which contains a collection of point locations and names for 530 archaeological sites on the Yucatan peninsula[5]. The second data set examined is Stanford political scientist Tom Patterson's global disputed border layer, which contains polyline features for disputed areas and breakaway regions derived from the CIA's World Factbook boundary database[6].

Questions and Lens Support. Leveraging the location of researcher data through the creation of footprints for data sets promotes data discovery and integration across disciplines. Researchers working across various domains may have overlapping study areas and would benefit from spatial data discovery. For example, a political scientist interested in contested regions should be able to discover and utilize a relevant protected heritage sites layer contributed by an archaeologist. Applying an **object lens** to study areas makes it possible to ask questions about their spatial properties and relationships, in particular:

– Which data sets overlap with the spatial extent of the area of interest?

A generic method is needed to generate a footprint of any research data type. Spatial metadata often includes an extent attribute, which delineates a minimum rectangular bounding box for the object. However, this extent alone is not a desirable data envelope, as the inclusion of an outlying feature can greatly exaggerate the geometry of the object [10]. Some library resources may also include place names, for example, Library of Congress subject headings. These place names need to be turned into footprints using a gazetteer.

Constructing the Lens. Convex hulls are constructed from an input researcher data set, such as Ford's geocoded archaeological site points and Patterson's disaggregated contested border polylines:

– Determine the spatial extent of each research object.
– Determine overlap between research objects.

Convex hulls are convex sets that contain all points [11] and are constructed for each data object using geoprocessing tools. The intersection of the convex hull shapefiles representing data set footprints reveals overlapping extents.

Figure 4 demonstrates the spatial relationships among the constructed convex hulls. A partial intersection reveals a correspondence between protected areas of archaeological interest and a contested border region in Belize, which may

---

[5] http://discovery.ucsb.opendata.arcgis.com/.
[6] https://earthworks.stanford.edu/.

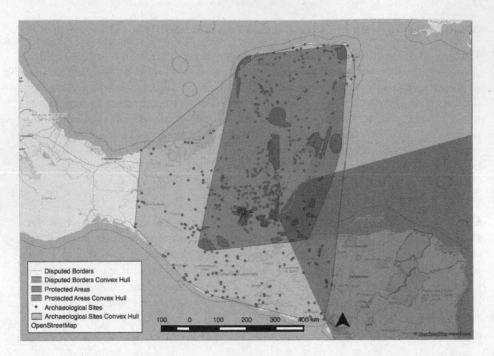

Fig. 4. Convex hulls constructed for research objects reveal intersections in extent.

be of interest to a domain scientist working in the region. Using location as an integrator of contents has potential to enhance the data discovery process, enhancing information retrieval across domains.

Secondary Lens. Another lens view on research data sets that enables data discovery is that of a non-spatial network of linked data triples. The connections between nodes of researchers and the data that they produce are represented as edges for the predicate *isReferencedBy*:

– Which research data sets reference publications in the area of interest?
– Which publications are authored by researchers in the area of interest?

Applying spatial lenses to researcher-generated data sets and to publications authored by researchers exposes undiscovered relationships, providing a basis for data discovery and leading to opportunities for trans-disciplinary research collaboration.

3.4   United States Historical News Archive

Data Source and Uses. The Chronicling America data set[7] is an archive that contains newspaper issues published between 1836 and 1922. Although the database includes newspapers from across the United States, some regions appear

---

[7] http://chroniclingamerica.loc.gov/about/api.

better represented than others. Large-scale efforts to digitize historical archives are a relatively recent phenomenon and thus researchers are just beginning to explore the possibilities that this new data presents. Scholars, for example, have investigated the temporal dynamics of themes found in the Richmond Daily Dispatch during the Civil War [12]. With respect to the Chronicling America data set in particular, there do not appear to be any studies that approach the data set from a geographic point of view so far.

Questions and Lens Support. A **network lens** can be used to investigate questions about relationships between place names in newspaper text:

– Which pairs of place names frequently co-occur in newspaper texts and why?
– What place names are most centrally mentioned (i.e., most connected to others)?

Such a network can be inspected using traditional network analysis tools. For instance, important nodes (i.e., place names) can be identified using centrality measures, and clusters can be found with community detection algorithms. Moreover, it is possible to visualize this network using programming libraries or software tools such as Gephi (see Fig. 5).

Constructing the Lens. Pre-processing is a crucial step in applying spatial lenses to natural language data, as these data are unstructured and often noisy. Natural language processing tools are available to normalize and parse text, as well as to identify potential place names in text. Additionally, such data are fraught with misspellings and other formatting issues resulting from scanning. The Natural Language Toolkit (NLTK) for Python has been used to clean and parse text data. We also take advantage of a tool called CLIFF[8] to recognize place names in newspaper articles. CLIFF accepts unstructured text as input and returns a list of standardized place names that were found. The key steps for constructing a network in this case are:

– Determine proximity parameters for defining co-occurrences.
– Identify place names using the CLIFF tool.
– Iterate over entire text and maintain a list of place names that occur within the specific proximity.
– Create an undirected network using the list of co-occurring place names and weigh each edge by the frequency with which the places co-occur.

Secondary Lens. The Chronicling America data set includes many historical issues of individual newspapers and it is appealing to apply a temporal **granularity lens** to it. Once the mechanisms for constructing a co-occurence network are built, it becomes easy to investigate the data at different temporal scales by aggregating newspaper issues into multiple time windows. Observing changes in

---

[8] http://cliff.mediameter.org/.

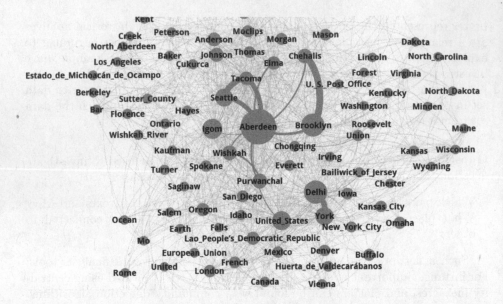

Fig. 5. Cooccurrence network of all 1914 *Aberdeen Herald* issues

the network structure over time can help scholars understand urban processes such as shifting neighborhood demographics or transformations in spatial structure.

3.5 California Wildfires and Land Cover Change

Data Source and Uses. Two sets of data are used in this section. The first is regional land cover data of California in four different years (1996, 2001, 2006, and 2010), produced by the Coastal Change Analysis Program (C-CAP) of the National Oceanic and Atmospheric Administration (NOAA) and updated every five years[9]. Landsat TM 5 satellite imagery is used to produce these data, with 25 different land cover classes. The second data set is a list of the 20 largest California wildfires from 1932 to 2015, produced by California Department of Forestry and Fire Protection[10] (see Fig. 6).

Monitoring land cover change is important for policy decisions, regulatory actions and subsequent land-use activities. These data are frequently used to generate landscape-based metrics and to assess landscape condition and monitor status and trends over a specified time interval [13]. Land cover change sometimes is set in motion by individual landowners and sometimes is driven by environmental forces. Over the past few decades, the most prominent land changes within the U.S. have been changes in the amount and kind of forest

---

[9] https://coast.noaa.gov/ccapftp/.
[10] http://www.fire.ca.gov/communications/downloads/fact_sheets/20LACRES.pdf.

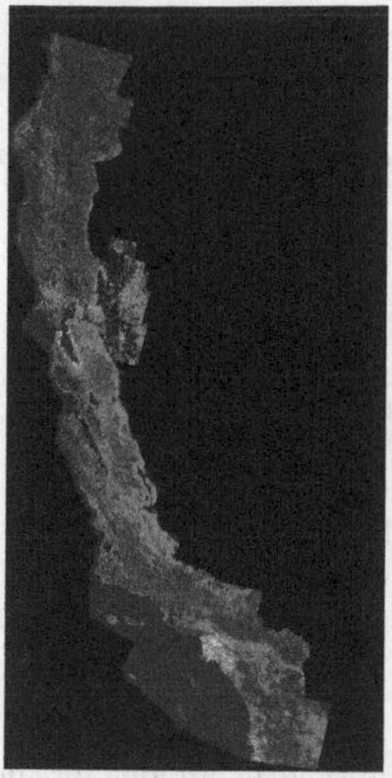

Top 20 Largest California Wildfires

FIRE NAME (CAUSE)	DATE	COUNTY	ACRES	STRUCTURES	DEATHS
1 CEDAR (Human Related)	October 2003	San Diego	273,246	2,820	15
2 RUSH (Lightning)	August 2012	Lassen	271,911 CA / 43,666 NV	0	0
3 RIM (Human Related)	August 2013	Tuolumne	257,314	112	0
4 ZACA (Human Related)	July 2007	Santa Barbara	240,207	1	0
5 MATILIJA (Undetermined)	September 1932	Ventura	220,000	0	0
6 WITCH (Powerlines)	October 2007	San Diego	197,990	1,650	2
7 KLAMATH THEATER COMPLEX (Lightning)	June 2008	Siskiyou	192,038	0	2
8 MARBLE CONE (Lightning)	July 1977	Monterey	177,866	0	0
9 LAGUNA (POWERLINES)	September 1970	San Diego	175,425	382	5
10 BASIN COMPLEX (Lightning)	June 2008	Monterey	162,818	58	0
11 DAY FIRE (Human Related)	September 2006	Ventura	162,702	11	0
12 STATION FIRE (Human Related)	August 2009	Los Angeles	160,557	209	2
13 McNALLY (Human Related)	July 2002	Tulare	150,696	17	0
14 STANISLAUS COMPLEX (Lightning)	August 1987	Tuolumne	145,980	28	1
15 BIG BAR COMPLEX (Lightning)	August 1999	Trinity	140,948	0	0
16 HAPPY CAMP COMPLEX (Lightning)	August 2014	Siskiyou	134,056	6	0
17 CAMPBELL COMPLEX (Powerlines)	August 1990	Tehama	125,892	27	0
18 ROUGH (Lightning)	July 2015	Fresno	119,069	4	0
19 WHEELER (Arson)	July 1985	Ventura	118,000	26	0
20 SIMI (Under Investigation)	October 2003	Ventura	108,204	300	0

*Rough Fire information will change until the fire is contained.

*There is no doubt that there were fires with significant acreage burned in years prior to 1932, but those records are less reliable, and this list is meant to give an overview of the large fires in more recent times.

**This list does not include fire jurisdiction. These are the Top 20 regardless of whether they were state, federal, or local responsibility.

Fig. 6. Data sources: land cover in California in 1996 (left); 20 largest California wildfires (right).

cover. Logging practices, development, urban expansion, and wildfires play the most important roles in this trend [14].

Questions and Lens Support. To analyze the effect of wildfires on land cover change in California between 1996 and 2010, in five-year time periods, we propose to use an **event lens** to look at the data. Both land cover change and wildfires can be seen as events, allowing users to relate and compare land cover change to wildfires, asking questions such as:

– Is there a relationship between the number and the magnitude of wildfires and the amount of land cover change in a given time period?
– What time period has experienced the greatest number of wildfires?
– In what time period has the area of land affected by wildfires been greatest?

Defining environmental phenomena as events enables the users to evaluate the interaction of these phenomena with land cover change and to reason about them. This is not easily done, if at all, with conventional methods.

Constructing the Lens. An event lens enables the user to choose the unit on which they want to study change (the whole field, sub-fields, or even a set of objects). Each of these units could be defined in various sizes and forms, applying a granularity lens. To define land cover change events, we consider each satellite image as representing a field. By comparing two consecutive fields in time, an event is constructed based on whether or not there is a difference between these two fields. The total number of events in which a certain class is changed can then be calculated and this value could give an estimation of the land cover change. Next, each of the 20 wildfires is defined as a separate event that has location (county), name, date, and affected area as properties. To construct events:

- Compare land cover fields for all two consecutive pairs.
- Create an event (and populate its properties) if there is a difference between two fields at a certain location.
- Create wildfire events with corresponding properties.

Having these two sets of events, we can observe their overlap, containment, and time of occurrence. The results of applying the overlap observer, for example, shows that the 2006–2010 period, by having six major wildfires and almost 1116 acres of land affected, has by far the greatest number of wildfires and the largest total area affected among the three periods.

Secondary Lens. Since the wildfire data are at the county level, an idea for the secondary lens would be to use an **object lens**. One can then apply the event lens on these objects to assess the effect of each wildfire on its corresponding county and to study their relationship. Objects can be constructed using Geospatial Object Based Image Analysis (GEOBIA) methods such as supervised maximum likelihood classifier [15].

4   Conclusions

The idea of spatial lenses to view environments through data sets is presented here as a counterpoint to considering data sets as implying singular world views. Data are just data, and while they always result from a certain conceptualization, one can often beneficially interpret them in other ways, including ways not intended by their authors. Today's GIS practice tends to lock users into conceptualizations based on data formats, discouraging the exploration of alternative views of data and, more importantly, of the phenomena under study. Balancing a possible need to restrict what can be done with data against the opportunity to exploit spatiality currently errs on the timid side, i.e. being overly restrictive (while still not really preventing inappropriate uses).

Core concepts are a way of thinking about, encoding, and computing with phenomena in a few intuitive spatial ways. Users of spatial data will ideally approach these lenses knowing what questions they want to answer, and these questions will inform the choice of operations and consequently the lens. Therefore, in order to get meaningful results, data sets should be fed to lenses applied to study areas, rather than applying lenses to data sets. The question what data set can inform what lenses remains to be studied. Implementing more generic constructor operations for each lens will make it more specific by admitting certain data types but not others. Switching between different lenses, on the other hand, appears to be possible even for one and the same data set (for example, seeing point measurements as defining objects or fields).

The five case studies presented in this paper explore, illustrate, and test the idea of spatial lenses. They were chosen to represent a broad spectrum of data sets and applications, as well as to cover all four content and two quality concepts of spatial information defined in our previous work [16]. We have arranged them to start with straightforward object and field lens applications and progress to more elaborate or unusual applications. Each case study furthermore illustrates some less obvious aspects of the chosen data sets. Jointly, they demonstrate the versatility of the spatial lens idea, resulting from the decoupling of spatio-temporal question answering from data and file formats. For example, examining land cover time series from an event perspective is conceptually intuitive and powerful, but not possible using traditional techniques.

Our lenses were straightforward for GIS savvy users to build. Our ambition in ongoing work is to build six generic lenses, usable by domain experts with little or no GIS expertise. Consequently, the next step in this research is to abstract each lens constructor from peculiarities in the data sets and build pieces of software that can be deployed as libraries or web services. These lenses will be parameterized to become applicable to all data sets that allow for an interpretation through them. For example, a field lens, when fed with a point data set with values from a continuously varying phenomenon, would determine the spatial and temporal domain of the field, ask the user for an interpolation method (or apply a default method), and produce an enriched data set that can then be queried by map algebra tools, used to produce isoline maps, or applied to any other field-based analysis.

Acknowledgments. The work presented in this paper (and the writing of the paper) was part of a graduate research seminar at the Geography Department of UCSB. All authors have contributed equally to the paper and are therefore listed in alphabetical order, with the seminar teacher going last. Additional contributions by Carlos Baez, Andrea Ballatore, Chandra Krintz, George Technitis, and Rich Wolski are gratefully acknowledged. The work was supported by the Center for Spatial Studies at UCSB.

References

1. Camara, G., Egenhofer, M.J., Ferreira, K., Andrade, P., Queiroz, G., Sanchez, A., Jones, J., Vinhas, L.: Fields as a generic data type for big spatial data. In: Duckham, M., Pebesma, E., Stewart, K., Frank, A.U. (eds.) GIScience 2014. LNCS, vol. 8728, pp. 159–172. Springer, Heidelberg (2014)
2. Lowe, M.: Night Lights and ArcGIS: A Brief Guide (2014). http://economics.mit.edu/files/8945. Accessed Nov 2015
3. Codd, E.F.: A relational model of data for large shared data banks. Commun. ACM **13**(6), 377–387 (1970)
4. Kuhn, W., Kauppinen, T., Janowicz, K.: Linked data - a paradigm shift for geographic information science. In: Duckham, M., Pebesma, E., Stewart, K., Frank, A.U. (eds.) GIScience 2014. LNCS, vol. 8728, pp. 173–186. Springer, Heidelberg (2014)

5. Vahedi, B., Kuhn, W., Ballatore, A.: Question based spatial computing - a case study. In: Sarjakoski, T., Santos, M.Y., Sarjakoski, L.T. (eds.) AGILE 2016. LNCS. Springer International Publishing, Heidelberg (2016)
6. Albrecht, J.: Universal analytical GIS operations: a task-oriented systematization of data structure-independent GIS functionality. In: Geographic Information Research: Transatlantic Perspectives, pp. 577–591 (1998)
7. Tomlin, C.D.: Geographic Information Systems and Cartographic Modeling. Prentice Hall, Upper Saddle River (1990). (No. 526.0285 T659)
8. Goodchild, M.F., Yuan, M., Cova, T.J.: Towards a general theory of geographic representation in GIS. Int. J. Geogr. Inf. Sci. **21**(3), 239–260 (2007)
9. Ester, M., Kriegel, H.P., Sander, J., Xu, X.: A density-based algorithm for discovering clusters in large spatial databases with noise. In: KDD, vol. 96, no. 34, pp. 226–231 (1996)
10. Hill, L.L.: Georeferencing: The Geographic Associations of Information. MIT Press, Cambridge (2009)
11. De Berg, M., Van Kreveld, M., Overmars, M., Schwarzkopf, O.C.: Convex hulls, mixing things. In: De Berg, M., Van Kreveld, M., Overmars, M., Schwarzkopf, O.C. (eds.) Computational Geometry, pp. 243–258. Springer, Heidelberg (2000)
12. Templeton, C., Brown, T., Battacharyya, S., Boyd-Graber, J.: Mining the dispatch under supervision: using casualty counts to guide topics from the richmond daily dispatch cor. In: Chicago Colloquium on Digital Humanities and Computer Science (2011)
13. Jones, K. Bruce Riitters, K.H., Wickham, J.D., Roger Jr., D., O'Neill, R.V., Chaloud, D.J., Smith, E.R., Neale, A.C.: An ecological assessment of the United States mid-Atlantic region: a landscape atlas (1997)
14. Brown, D.G., Polsky, C., Bolstad, P., Brody, S.D., Hulse, D., Kroh, R., Loveland, T.R., Thomson, A.: Land use and land cover change. In: Melillo, J.M., Richmond, T.C., Yohe, G.W. (eds.) Climate Change Impacts in the United States. The Third National Climate Assessment, pp. 318–332. U.S. Global Change Research Program (2014)
15. Walter, V.: Object-based classification of remote sensing data for change detection. ISPRS J. Photogramm. Remote Sens. **58**(3), 225–238 (2004)
16. Kuhn, W., Ballatore, A.: Designing a language for spatial computing. In: Bacao, F., Santos, M.Y., Painho, M. (eds.) AGILE 2015. LNCS, pp. 309–326. Springer International Publishing, Heidelberg (2015)
17. Liskov, B., Guttag, J.: Abstraction and Specification in Program Development. MIT press, Cambridge (1986)

Moon Landing or Safari? A Study of Systematic Errors and Their Causes in Geographic Linked Data

Krzysztof Janowicz[1(✉)], Yingjie Hu[1], Grant McKenzie[2], Song Gao[1],
Blake Regalia[1], Gengchen Mai[1], Rui Zhu[1], Benjamin Adams[3],
and Kerry Taylor[4]

[1] STKO Lab, University of California, Santa Barbara, USA
janowicz@ucsb.edu
[2] Department of Geographical Sciences, University of Maryland, College Park, USA
[3] Centre for eResearch, The University of Auckland, Auckland, New Zealand
[4] Australian National University, Canberra, Australia

Abstract. While the adoption of Linked Data technologies has grown dramatically over the past few years, it has not come without its own set of growing challenges. The triplification of domain data into Linked Data has not only given rise to a leading role of places and positioning information for the dense interlinkage of data about actors, objects, and events, but also led to massive errors in the generation, transformation, and semantic annotation of data. In a global and densely interlinked graph of data, even seemingly minor error can have far reaching consequences as different datasets make statements about the same resources. In this work we present the first comprehensive study of systematic errors and their potential causes. We also discuss lessons learned and means to avoid some of the introduced pitfalls in the future.

1 Introduction and Motivation

Over the last few years, the Linked Data cloud has grown to a size of more than 85 billion statements, called triples, contributed by more than 9,900 data sources. A cleaned and quality controlled version made available via the LOD Laundromat [2] contains nearly 40 billion triples.[1] The Linked Data cloud (and proprietary versions derived from it and other sources) have brought dramatic changes to industry, governments, and research. For instance, they have enabled question answering systems such as IBM's Watson [3] and Google's new knowledge graph. Linked Data has also increased the pressure on governments to publish open data in machine readable and understandable formats, e.g., via data.gov. Finally, it has enabled the research community to more efficiently publish, retrieve, reuse, and integrate, scientific data, e.g., in the domain of pharmacological drug discovery [16]. The value proposition of Linked Data as a new paradigm for data publishing and integration in GIScience has been recently discussed by Kuhn et al. [10].

---

[1] http://lodlaundromat.org/.

© Springer International Publishing Switzerland 2016
J.A. Miller et al. (Eds.): GIScience 2016, LNCS 9927, pp. 275–290, 2016.
DOI: 10.1007/978-3-319-45738-3_18

Places and positioning information more broadly play a prominent role for Linked Data by serving as nexuses that interconnect different statements and contribute to forming a densely connected global knowledge graph. GeoNames, for example, is the second most interlinked hub on the Linked Data Web, while DBpedia contains more than 924,000 entities with direct spatial footprints and millions of entities with references to places. Examples of these include birth and death locations of historic figures and places where notable events occurred. Many other datasets also contain spatial references such as sightings of certain species on Taxonconcept,[2] references to places in news articles published by the New York Times Linked Data hub,[3] and affiliations of authors accessible via the RKB Explorer,[4] to name but a few. In fact, most Linked Data are either directly or indirectly linked through various spatial and non-spatial relations to some type of geographic identifier.

Nonetheless, current statistics show that about 66 % of published Linked Datasets have some kind of problems including limited availability of SPARQL query endpoints and non-dereferenceable IRIs.[5] A recent study of Linked Datasets published through the Semantic Web journal shows that about 37 % of these datasets are no longer Web-available [6]. In other words, even the core Linked Data community struggles to keep their datasets error-free and available over longer periods. This problem, however, is not new. It has been widely acknowledged that proper publishing and maintenance of data are among the most difficult challenges facing data-intensive science. A variety of approaches have been proposed to address this problem, e.g., providing a sustainable data publication process [15]. Simplifying the infrastructure and publishing process, however, is just one of many means to improve and further grow the Web of Linked Data. Another strategy is to focus on controlling and improving the quality of published data, e.g., through unit testing [9], quality assessment methods such as measuring query latency, endpoint availability, and update frequency [17], as well as by identifying common technical mistakes [5].

Given the importance of places and positioning information on the Linked Data cloud, this paper provides the first comprehensive study of systematic errors, tries to identify likely causes, and discusses lessons learned. However, instead of focusing on *technical* issues such as non-dereferenceable IRIs, unavailable SPARQL endpoints, and so forth, we focus on Linked Data that is technically correct, available, and in (heavy) use. We believe that understanding quality issues in the *contents* published by leading data hubs will allow us to better understand the difficulties faced by most other providers. We argue that the lead issue is the lack of best practices for publishing (geo)-data on the Web of Linked Data. For instance, geo-data is often converted to RDF-based Linked Data without a clear understanding of reference systems or geographic feature types. Our view is not unique and has recently led to the first joint collaboration

---

[2] http://www.taxonconcept.org/.

[3] http://data.nytimes.com/.

[4] http://www.rkbexplorer.com/.

[5] http://stats.lod2.eu/.

of the Open Geospatial Consortium (OGC) and World Wide Web Consortium (W3C) by establishing the *Spatial Data on the Web Working Group*.

In the next sections, we will categorize systematic errors into several types and discuss their impact and likely causes. We will differentiate between (I) errors caused by the triplification and extraction of data, (II) errors that result from an improper use of existing ontologies or a limited understanding of the underlying domain, (III) errors in the design of new ontologies and oversimplifications in conceptual modeling, and (IV) errors related to data accuracy and the lack of an uncertainty framework for Linked Data. Some errors are caused by a combination of these categories. We realize that studies of data quality are often not met with excitement and thus have selected interesting and *humorous* examples that illustrate, with *serious* implications, the far-reaching consequences of seemingly small errors. Finally, we would like to clarify that our work is motivated by improving the quality of the Linked Data to which we contributed datasets ourselves, not in merely blaming errors made by others. We notified the providers of the discussed datasets and some of the issues presented here have been resolved. We hope that our work will help to prevent similar errors in the future.

2 Triplification and Extraction Errors

There are three common ways in which Linked Data is created today. The most common approach is to generate Linked Data from other structured data such as relational databases, comma separated value (CSV) files, or ESRI shapefiles. This approach is often called triplification, i.e., turning data into (RDF) triples. As a second approach, Linked Data is increasingly extracted using natural language processing and machine learning techniques from semi-structured or unstructured data. The most common example is DBpedia [11] which converts (parts of) Wikipedia into Linked Data. Another example is the ontology design patterns-based machine reader FRED that parses any natural language text into Linked Data [14]. Finally, in a small but growing number of cases, Linked Data is the native format in which data are created. This is typically the case for derived data products, such as events mined from sensor observations, metadata records from publishers and libraries, and so on.

The first two approaches share a common workflow. First, the relevant content has to be extracted, e.g., from a tabular representation in hypertext markup language (HTML). Next, the resulting *raw* data have to be analyzed and processed. In a final step, the processed data must be converted into Linked Data by using an ontology. While errors can be introduced during each of these steps, this section focuses on errors introduced during the extraction of data and the conversion into Linked Data, i.e., triplification errors.

One way of studying whether systematic errors have been introduced during the triplification process is to visually map geographic features present in the Linked Data cloud. Figure 1 shows the result for about 15 million features extracted from multiple popular Linked Data sources such as DBpedia, Geonames, Freebase, TaxonConcept, New York Times, and the CIA World Factbook.

Fig. 1. A representative fraction of places in Linked Data (EPSG:4326, Plate Carree).

These features have been selected though SPARQL queries for all subjects that have a W3C Basic Geo predicate, i.e., geo:lat or geo:long. For DBpedia, we included multiple language versions. What is noteworthy about Fig. 1 is the lack of a base map, i.e., the figure is entirely based on point data.[6] In other words, the Linked Data cloud has high spatial coverage. One can easily identify the outlines of continents and most of the land surface is covered by features with varying density. This is also true for regions in the far North and South of the planet. Nonetheless, one can immediately identify significant errors – the most obvious being perpendicular lines crossing in the middle of the map. In this work, we do not focus on random errors (which are expected in a sample of this size and arise from largely unpredictable and thus not easily correctable reasons), but instead on *systematic* errors inherent to the data. These errors are further examined through a set of cases as follows.

Case 1 shows a massive ×-like structure which represents numerous problems with geographic coordinates such as latitudes and longitudes sharing the same single value. This indicates that latitude values were mistaken for longitude values and vice versa. We also found cases were only latitude values or longitude values were given or where multiple appeared such as entities having two latitude values without any longitudes. The quantity of these errors suggests that they are systematic. Most likely, they stem from problems with scraping or parsing scripts. Cases where features were mapped to (0,0) will be discussed below.

Case 2 depicts one of many examples of grid-like structures. From our observations, these are caused by two separate issues. First, features are often merely represented by coarse location information, e.g., by only using degrees and dropping decimals. Second, the vast majority of geo-data on the (Linked Data) Web

---

[6] A high resolution version that gives a better impression of the coverage as well as various errors is available at http://stko.geog.ucsb.edu/pictures/lstd_map.png.

today relies on point geometries. This also includes centroids for regions such as counties, countries, mountain ranges, rivers, and even entire oceans. To give a concrete examples, Geonames places the Atlantic Ocean at (10N, 25W), while DBpedia places it about 1200 Km away at (0N, 30W). Note, however, that many of the features visible in the oceans are not necessarily errors. They include submarine volcanos, mid-ocean ridges, or reports about events such as oil spills. Whether centroids are errors or are simply an inaccurate and largely meaningless way of representing large regions depends on the context in which the data are to be used. However, it is difficult to imagine use cases for centroids of oceans particularly as the two examples above show the arbitrariness of these locations. The same argument can be made for coarse location data, and in fact, we will discuss one example in greater detail below.

Cases 3 and 4 can be seen through block-like structures in China and a second New Zealand in the Northern Hemisphere. The vast majority of these errors are systematic and appear in the DBpedia dataset. We were able to track down a potential reason for them by exploring the different language versions of DBpedia. It appears as though the scripts used by DBpedia curators to extract content from Wikipedia either expected signs, e.g., (34.413,−119.848), or a hemisphere designator, e.g., (34.413 N,119.848 W). Some language versions of Wikipedia, e.g., the Spanish version, use other character designators such as (34.413 N,119.848 O) where O stands for *oeste*. It is likely that the script dropped the O instead of replacing it with a W. Consequently, geographic features in the United States for which a Spanish language version was available in Wikipedia ended up in China. This also explains the lower density of those misplaced features, i.e., the Spanish Wikipedia lists fewer places in the US than the English version. Other, likely non-systematic, errors include the flattening factor for the Earth being reported as 1 which could be caused by a parsing error (or ceiling function) as the data type reported by DBpedia is an xsd:integer.[7]

Case 5 in Fig. 1 is not an error but rather a reminder that despite the overall coverage, certain regions are underrepresented. Interestingly, the Linked Data map bears a remarkable similarity to maps created for different (social media) datasets such as Flickr, Twitter, Wikipedia, and so forth. This highlights two issues. First, and as outlined previously, most Linked Data are created from existing data sources, and secondly the same underlying biases appear to apply for most of these data sources. In other words, most data used in the Linked Data cloud share the same *blind spots*.

Lessons Learned: Two major sources of errors can be differentiated, those introduced during triplification and knowledge extraction as well as those that were part of the original source data. In the first case, errors are typically introduced by software that does not take the full range of possible syntactic variations into account (e.g., *west* versus *oeste*) or fails to accurately distinguish between point-features and bounding boxes. Furthermore

---

[7] SPARQL: ASK WHERE <http://dbpedia.org/resource/Earth> <http://dbpedia.org/property/flattening> 1. *[using DBpedia 2015-04.]*.

the software may confuse latitudes with longitudes for other reasons (causing the ×-like feature in Fig. 1) or parse and cast the data into inappropriate formats (e.g., the flattening factor). For the second type of errors, one could argue that they are not specific to Linked Data but simply a result of errors in the source data. Such argument, however, misses the substantial difference between information embedded in the context of a Web page published for human use with the decontextualized raw data statements that form an interlinked and the machine-available knowledge graph. While the Atlantic Ocean was represented by a point-like feature at (0N, 30 W) in Wikipedia, it is the DBpedia version that allows for inferences such as plotting the place of death of people who are known to have died somewhere in the Atlantic Ocean (e.g., Benjamin Guggenheim) at 0N, 30 W. Summing up, triplification and Linked Data extraction require substantial domain expertise. Approaches such as unit testing and simple integrity constraints could be used to detect many of the errors described above. For instance, most of the places in the US that were duplicated in China also contain topological information such as being part of a county or a state. Thus, checking whether the space-based and place-based information match could be a powerful method to avoid such errors in the future.

3   Ontology Usage and Domain Errors

To improve retrieval and reuse, Linked Data is typically created by using shared ontologies and vocabularies. Most of these, however, are underspecified to a degree where the intended interpretation is largely conveyed by the labels and simple hierarchies rather than a deeper axiomatization. The need for and value of a more expressive formalization is still controversially debated with recent work highlighting the need for stronger ontologies. The following example illustrates the problems that can arise from a lack of deeper axiomatization or the improper use of ontologies outside of their intended interpretation.

Figure 2 shows DBpedia data concerning a lunar crater named after Copernicus. As one can see at the bottom, geo:lat and geo:long are used to represent the centroid of the crater. However, W3C Basic Geo uses WGS84 as a reference datum. Thus, and in contrast to the original Wikipedia data, the information that the crater is not on Earth and that the coordinates use a different, selenographic reference system were lost in the triplification process. Consequently, and as depicted in Fig. 3, systems such as the Fluidops Information Workbench render the crater on the Earth's surface near the city of Sarh, Chad. The same is true for the landing site of Apollo 11 – Tranquility Base – located in the Mare Tranquillitatis. In fact, the same problem occurs for all other locations on distant planets and their moons. Showcasing one consequence of such errors, the current DBpedia version (2015-04) indeed shows that the *moon landing* happened here on Earth, as is evident by the following SPARQL query which returns geographic coordinates in the southern part of Algeria.

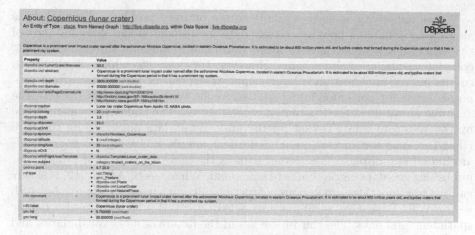

Fig. 2. DBpedia data about the Copernicus crater.

```
SELECT ?lat ?long
WHERE {dbp:Tranquility_Base geo:lat ?lat; geo:long ?long.}
```

lat	long
0.6875	23.4333
0.713889	23.4333
0.6875	23.7078
0.713889	23.7078

Listing 3.1. Query and results showing the location of the moon landing is in Algeria.

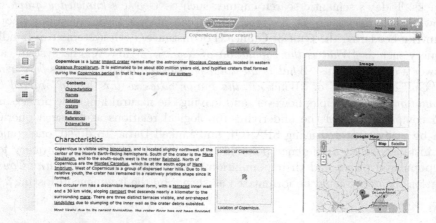

Fig. 3. Fluidops displays Linked data about the Copernicus crater taken from DBpedia.

Three underlying issues contribute to the outlined problems. First, there is an ongoing debate on how to simplify data publishing on the Web and part of this discussion is about how to avoid burdening publishers through enforcing complex vocabularies and schema. However, the degree to which simplification results in oversimplification is largely context-dependent and while current proposals argue for not enforcing spatial reference system identifiers (SRID), the example above illustrated potential consequences. The counterargument made by the Web community is that for the majority of data published on the Web (that has some sort of geographic identifier), simple WGS84 point coordinates are indeed appropriate. The second issue is the lack of a clear *best practice* for publishing geo-data on the Linked Data cloud. While GeoSPARQL [12] is slowly gaining traction, there are various competing or complementary approaches such as the W3C Basic Geo vocabulary or SPARQL-ST [13] which can also handle spatiotemporal data. The third issue lies in the nature of most vocabularies and ontologies themselves as well as a lack of domain expertise. Ontologies cannot fix meaning but only restrict the interpretation of domain terminology towards their intended meaning [10]. Consequently, while the W3C Basic Geo specs identify WGS84 as the reference coordinate system, this is not enforced through the axiomatization, and, thus, there is no way of preventing *geo:lat* and *geo:long* from being used to represent locations on celestial bodies other than the Earth. Finally, as discussed previously, most Linked Data today are created by data enthusiasts from existing data. This typically leads to lost expertise. We expect this problem to disappear with time as more domain experts adopt a Linked Data driven approach to publishing their (scientific) data.

The moon landing error mentioned above arose from using the wrong ontology to annotate data. There are also more subtle cases, however, with more dramatic consequences that arise from a lack of domain knowledge or an unclear scope. Consider, for example, the Gulf of Guinea which is one of the world's key oil exploration regions, recently gaining notoriety through frequent pirate attacks. Today's semantic search engines such as *Google's knowledge graph* or knowledge engines such as *Wolfram Alpha* can answer basic questions about countries bordering the Gulf of Guinea. For instance, both systems can handle a query such as *'What is the population of Nigeria?'*. However, no system can answer a query such as *'What is the total population of all countries bordering the Gulf of Guinea?'* or *'What are the major cities in this region ordered by population?'*. In principle, however, and leaving the natural language processing and comprehension of the underlying topological relations aside, such queries can be easily answered using SPARQL and Linked Data. To do so, one could, for instance, select a reference point in the gulf and use a buffer to query for all populated places and their population. Using PROTON's *populationCount* relation the query could be formulated as shown by the fragment in Listing 3.2.

```
SELECT  (sum(?populationCount)  as  ?totalPopulation)
WHERE  {
  [...]  geo:lat  ?lat  ;  geo:long  ?long  .
  ?place  omgeo:nearby(?lat  ?long"500mi");
  ptop:populationCount  ?populationCount.}
  [...]
```

Listing 3.2. Fragment of a query for the total population of places within a radius of 500 miles around a location in the Gulf of Guinea.

This query, however, will return the population of cities, towns, countries, and so forth, and, thus, will not give a truthful estimate of the population (as citizens of a country and its cities will be counted multiple times). We will revisit the case of towns and cities later and for now will consider all types of geographic features that have a population value, e.g., to rank places by population. The Gulf of Guinea is also home to the intersection of the Equator with the Prime Meridian. Interestingly, and as shown by the results of Listing 3.3, this has surprising implications for the query discussed before. In GeoNames, the Earth, as such, is located in its own reference system at (0,0) together with the statement that its population is 6,814,400,000 and its feature type is *L parks, area*; see Fig. 4. Hence, it is the most populated geographic feature in the Gulf of Guinea and thus causes the gulf to have the world's highest population density. Moreover, these kinds or errors will propagate, e.g., via GeoNames' RDF *nearby* functionally. For instance, we can learn that the United States are nearby the Odessa Church.[8]

One could now argue that placing the Earth at (0,0) is an isolated case, and, thus, not a systematic error. However, this is not the case. Many existing mapping services return (0,0) to indicate geocoding failures. In fact, this is so common that the Natural Earth dataset has created a virtual island at the location called *Null Island* to better flag geocoding failures. Consequently, it is not surprising to find many features on the Linked Data cloud located to (0,0). The second problem, namely the population count, is also systematic. The Linked Data cloud is envisioned as a distributed global graph but it is not yet clear which data should be provided by linking to more authoritative sources and which data should be kept locally. Therefore, for instance, The New York Times Linked Data portal returns a population of 86,681 for Santa Barbara without providing detailed metadata, while GeoNames reports 88,410 (together with a change history). In contrast, DBpedia reports a population of 90,385 as well as corrected data for the latest update, namely 2014-01-01.

```
SELECT distinct ?lat ?long ?populationCount
WHERE {
<http://sws.geonames.org/6295630/> geo:lat ?lat ; geo:long ?long ;
 ptop:populationCount ?populationCount.}
```

lat long populationCount
0   0   6814400000

Listing 3.3. A query for the geographic coordinates of the Earth and its population.

---

[8] E.g. via, wget http://sws.geonames.org/6252001/nearby.rdf.

Fig. 4. The point-feature representation of the Earth.

Lessons Learned: Selecting or creating an appropriate ontology to semantically *lift* data is not trivial and the moon landing example shows some of the potential consequences. As most ontologies are lightweight and thus underspecified, it is important to check the documentation and intended use manually. One proposal is to enforce the explicit specification of coordinate reference systems for all spatial data on the Linked Data cloud. This, however, has been controversially discussed in recent years as it would introduce another hurdle-to-entry for publishers and Web developers. Thus, it has been argued that a layered approach is needed. The second case, namely the population count for the Gulf of Guinea, highlights the need for tighter integration of different data sources based on their scope and authority. Today, a lot of data are published by providers that have limited expertise, cannot provide provenance records, or have no clear maintenance strategy. It is worth noting that the Web (and thus also Linked Data) follows the AAA slogan that Anyone can say Anything about Any topic. While this strategy has enabled the Web we know today, it is a blessing and curse at the same time when it comes to scientific data and reliability. Future work will need to go beyond entity resolution (e.g., via *owl:SameAs*) by providing data conflation services (e.g., to merge/correct population data from different sources).

4   Modeling Errors

Another source of error is introduced by various modeling errors such as ontologies being overly simplistic or overly specific as well as errors that result from how data are semantically lifted using these ontologies. Many of these examples are related to how we assign locations to entities. Clearly, entities typed as *place* (and its subtypes) have a direct spatial footprint such as dbr:Montreal geo:geometry POINT(−73.56 45.5) even though this footprint may be contested, missing, or unknown, such as for the ancient city of Troy. A similar argument can be made for types that describe spatially fixed entities, e.g., statues. In some rare cases this is also true for otherwise mobile entities such as vessels. A common example for this is the HMS Victory that is located on a dry dock in Portsmouth, England. Wikipedia and thus DBpedia assign geographic coordinates to most places, many statues, and some other entities such as the HMS Victory.[9] For many other types of entities, however, this is not an appropriate method for assigning locations. For instance, any living human has a (changing) position at any time. This position is not stable and thus not reported in a resource such as Wikipedia (although it may be stored in a trajectory database). In fact, one would be very surprised to find the up-to-date geographic coordinates for a specific person, car, ongoing event, and so forth in the Wikipedia.

From an ontological modeling perspective, one would expect entities of types such as *event* to be related to a place which in turn is related to a spatial footprint. In fact, the notion that events are located spatially via their physical participants and these participants are temporally located via events, is at the core of the DOLCE foundational ontology. One way of thinking about this is to consider the length of the *property path* that is expected between an entity of a given type and geographic coordinates. For example, Rene Descartes is related to Stockholm which has a spatial footprint: dbr:Rene_Descartes dbp:deathPlace dbr:Stockholm. dbr:Stockholm geo:geometry POINT(18.07 59.33). From this perspective, places are expected to be 0-degree spatial. Persons, events, and so forth, are expected to be 1-degree spatial, and information resources such as academic papers are expected to be 2-degree spatial (via the affiliations of their authors).

Interestingly, performing this experiment on DBpedia yields 1,893 0-degree persons, 371,655 1-degree persons, and 31,182 2-degree persons. Higher degree persons can easily be explained either by a lack of knowledge about their places of birth and death or by the many fictitious persons classified as *Person* in DBpedia. Zero degree persons, however, can be considered modeling errors and will appear in Fig. 1. The same argument can be made for the 5,086 0-degree events, 1,507 0-degree sports teams, 448 0-degree biological species, and so forth.

Let us now illustrate the resulting problems using a concrete example. Figure 5 shows a query for *Terry Fox*. As can be seen on the right side of the figure, there are latitude/longitude coordinates assigned to him directly. The image on the left implies that the information about the *person* Terry Fox may have been accidentally conflated with the *statue* of Terry Fox which indeed may have a fixed location. Checking the geographic coordinates, however, reveals that they point to the Mt. Terry Fox Provincial Park (in the middle of Fig. 5), thereby clearly revealing the modeling error and its consequences.

A second common example related to modeling is the mis-categorization of geographic features. These errors are difficult to quantify as there is no gold standard that would allow us to measure the *semantic accuracy* of type

---

[9] http://dbpedia.org/resource/HMS_Victory.

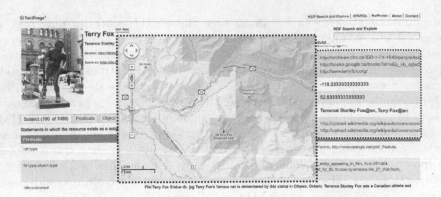

Fig. 5. The spatial footprint of the famous Canadian athlete Terry Fox.

assignment. Nonetheless, some of the clear, e.g., legally defined, cases are worth discussing. For instance, there are 554 places in DBpedia that are classified as being a town while having a population over 100,000, e.g., Stuttgart, Germany, with a population over 500,000, and 3,694 cities with a population below 1,000 such as Eureka, Utah with a current estimated population of 667. The issue here is that the meanings of *city* and *town* varies greatly across countries and even between US states [7]. In Utah, for instance, every settlement with a population below 1,000 is legally considered a town. Hence, Eureka is a town and not a city. In contrast, the class *town* in Pennsylvania is a singleton class that contains Bloomsburg as its sole member. Nonetheless we can find triples such as `dbr:Bloomsburg_University_of_Pennsylvania dbp:city dbr:Bloomsburg,_Pennsylvania` in DBpedia. In both cases, the underlying problem is that the ontologies (which are often semi-automatically learned from data) are overly specific and introduce fine grained distinctions that are not supported through the data; see [1] for more details on feature types in DBpedia.

> **Lessons Learned:** While there is sufficient theoretical work on how entities are located in space and time – namely by modeling location as a *relation* between objects and by spatially anchoring events via their physical participants – there seems to be a gap on how to apply these theoretical results to the practice of data publishing. The case of wrong or overly-specific type assignment is even more difficult to tackle as geographic feature types have spatial, temporal, and culturally indexed definitions as shown by the town and city example. Ongoing work investigates the role of spatial statistics for mining type characteristics *bottom-up* and may help to minimize categorization errors in the future [18].

5   Accuracy and Uncertainty Related Errors

DBpedia also stores 133,941 cardinal direction triples such as the statement, Ventura, CA is to the north of Oxnard, CA: `dbr:Ventura,_California` `dbp:south dbr:Oxnard,_California.`[10] This leads to the interesting question of how accurate these triples are. Testing 100,000 of these triples reveals that 26 % (26,420) of them are inaccurate when using the geometries provided by DBpedia. Our sample only includes triples where subject and object are both of type `dbo:Place` and have valid `geo:geometry` predicates. By considering all 133,941 cardinal triples in DBpedia, we find that 55,928 of them have a subject or object lacking `geo:geometry`, or are not of type `dbo:Place`. Of these, 17,957 triples list a cardinal direction relation to a RDF literal such as an xsd:integer, e.g., `dbr:Harrisburg,_Pennsylvania dbp:north 20 (xsd:integer)`.

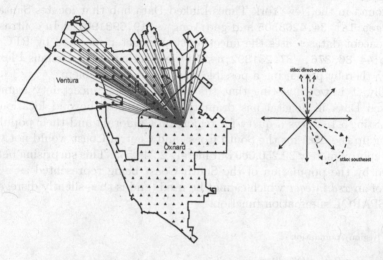

Fig. 6. A 1:n step in the direction computation for 1×1 km grids of Ventura (88 circles) and Oxnard (133 triangles). Grid points in the ocean were removed.

More interesting, however, than discovering these (significant) data errors alone, is the question of how much uncertainty is introduced by using point-features to represent places and how this uncertainty is communicated [4]. Returning to the Ventura and Oxnard example, one can overlay the known administrative areas for both cities with a 1×1 kilometer grid and then pair-wise compare all possible grid points. Figure 6 shows the spatial distribution of those grid points and an 1:n step out of this direction comparison. The direction-ality is determined by testing if the azimuth between two point geometries falls within ω (which is set to $\pi/8$) from the primary angle of the cardinal (N,S,E,W) or the intercardinal direction (NE,SE,SW,NW). For example, SE (*stko:southeast* here) covers the range $5\pi/8$ to $7\pi/8$ which is measured from the positive y-axis. Our results show that the cardinal direction S holds for 34.8 % of the cases in which Ventura is located to the north of Oxnard, while the intercardinal direction SE holds for 50.5 % cases in which Ventura is located to the northwest of

---

[10] The way in which DBpedia uses cardinal directions can be easily misunderstood. The triple states that the entity south of Ventura is the city of Oxnard.

Oxnard. In 0.5 % of the cases the correct direction is NW. This uncertainty (and the fact that SE seems to be the better choice), however, is not communicated by DBpedia.

A similar case, this time based on reporting coordinates and areas beyond a meaningful accuracy, can be found in many other examples. For instance, DBpedia states that the value for dbo:PopulatedPlace/areaTotal for Santa Barbara is 108.69662101458125 km^2. The location for Santa Barbara is given by the centroid POINT(−119.71416473389 34.425834655762) − thus indicating that the exact centroid of Santa Barbara is known at the sub-micron scale. This is not DBpedia specific and thus a systematic error. Similar cases can be found in the New York Times Linked Data hub that locates Santa Barbara at geo:lat 34.4208305 and geo:long −119.6981901.[11] In contrast, the TaxonConcept dataset uses the uncertainty parameter specified by RFC 5870, e.g., geo:44.863876,−87.231892;u=10 for a sighting of the Danaus Plexippus butterfly, thereby presenting a possible solution to the problem.

Finally, it is worth noting that the lack of a clear uncertainty framework for Linked Data in general has dramatic consequences beyond location data alone. Listing 5.1, shows a query for regions in California and their population. Summing up the data for the South Coast and Central Coast would not yield a value of approximately 22,250,000 but merely 2,249,558. This surprising behavior is caused by the population of the South Coast being represented as a *string* instead of an *xsd:integer* which cannot be used (and is thus silently disregarded) by the SPARQL summation function.

```
SELECT ?region ?population
WHERE {
    ?region a yago:RegionsOfCalifornia;
        dbp:population ?population .}
```

region   *population*
[*shortened results*]   ...
South_Coast_(California)   ' ∼ 20*million*'@en //Not recognized as a (approximate) number
Central_Coast_(California) 2249558 //recognized as an xsd:integer

Listing 5.1. Population of (overlapping) regions in California.

Lessons Learned: The cardinal directions example shows the many and massive errors that exist in spatial information on the Linked Data cloud today. Blaming the datasets and their providers, however, is missing the more relevant and underlying problem − namely the effects of decontextualization on data [8] and their transformation into statements in triple form. Consider the following example: The sentence 'Isla Vista, CA is the most populated municipality to the west of the Mississippi.' is meaningful and partially correct. During natural language processing and triplification this sentence would be transfered to a triple such as ex:Isla_Vista dbr:west

---

[11] http://data.nytimes.com/N2261955445337191084.

`ex:Mississippi`. This triple, however, is not only questionable but also leads to exactly those cardinal direction accuracy issues discussed before as the direction will depend on the point coordinates used to represent the Mississippi river. Finally, and as illustrated above, the lack of a general uncertainty framework for Linked Data requires urgent attention in future research.

6 Conclusions

Places and positioning information more broadly play a key role in interlinking data on the Web. Consequently, it is important to study the quality of these (geo-)data. Our work reveals that about 10 % of all spatial data on the Linked Data cloud is erroneous to some degree. We identified major types of systematic errors, discussed their likely causes (some of which have been confirmed by the data providers), and pointed out lessons learned and directions for future research. Some of the identified problems can be easily addressed and prevented in the future, e.g., by unit testing against possible representational choices for geographic coordinates. Other cases remain more challenging such as proper ontological modeling or the representation of uncertainty. Those issues for which a clear best practice can be identified and agreed upon are currently being collected by the joint OGC/W3C *Spatial Data on the Web Working Group*.[12] Finding the right balance between simple models and data publishing processes on the one hand and preventing potentially harmful oversimplifications on the other hand remains the major challenge to be addressed in the future.

Acknowledgements. The authors would like to acknowledge partial support by the National Science Foundation (NSF) under award 1440202 *EarthCube Building Blocks: Collaborative Proposal: GeoLink Leveraging Semantics and Linked Data for Data Sharing and Discovery in the Geosciences*, NSF award 1540849 *EarthCube IA: Collaborative Proposal: Cross-Domain Observational Metadata Environmental Sensing Network (X-DOMES)*, and the USGS award on *Linked Data for the National Map*.

References

1. Adams, B., Janowicz, K.: Thematic signatures for cleansing and enriching place-related linked data. Int. J. Geogr. Inf. Sci. **29**(4), 556–579 (2015)
2. Beek, W., Rietveld, L., Bazoobandi, H.R., Wielemaker, J., Schlobach, S.: LOD Laundromat: a uniform way of publishing other people's dirty data. In: Mika, P., et al. (eds.) ISWC 2014, Part I. LNCS, vol. 8796, pp. 213–228. Springer, Heidelberg (2014)
3. Ferrucci, D.A., Brown, E.W., Chu-Carroll, J., Fan, J., Gondek, D., Kalyanpur, A., Lally, A., Murdock, J.W., Nyberg, E., Prager, J.M., Welty, C.A.: Building Watson: an overview of the DeepQA project. AI Mag. **31**(3), 59–79 (2010)

---

[12] The views presented in this paper belong to the authors and do not necessarily represent the views or positions of the entire working group. A current draft of the best practice report is available at: https://www.w3.org/TR/sdw-bp/.

4. Fisher, P.F.: Models of uncertainty in spatial data. Geograph. Inf. Syst. **1**, 191–205 (1999)

5. Hogan, A., Harth, A., Passant, A., Decker, S., Polleres, A.: Weaving the pedantic web. In: Proceedings of the WWW 2010 Workshop on Linked Data on the Web, LDOW 2010, Raleigh, USA, 27 April 2010 (2010)

6. Hogan, A., Hitzler, P., Janowicz, K.: Linked dataset description papers at the semantic web journal: a critical assessment. Semant. Web **7**(2), 105–116 (2016)

7. Janowicz, K.: Observation-driven geo-ontology engineering. Trans. GIS **16**(3), 351–374 (2012)

8. Janowicz, K., Hitzler, P.: The digital earth as knowledge engine. Semant. Web **3**(3), 213–221 (2012)

9. Kontokostas, D., Westphal, P., Auer, S., Hellmann, S., Lehmann, J., Cornelissen, R., Zaveri, A.: Test-driven evaluation of linked data quality. In: Proceedings of the 23rd International Conference on World Wide Web, pp. 747–758. International World Wide Web Conferences Steering (2014)

10. Kuhn, W., Kauppinen, T., Janowicz, K.: Linked data - a paradigm shift for geographic information science. In: Duckham, M., Pebesma, E., Stewart, K., Frank, A.U. (eds.) GIScience 2014. LNCS, vol. 8728, pp. 173–186. Springer, Heidelberg (2014)

11. Lehmann, J., Isele, R., Jakob, M., Jentzsch, A., Kontokostas, D., Mendes, P.N., Hellmann, S., Morsey, M., van Kleef, P., Auer, S., Bizer, C.: DBpedia - a large-scale, multilingual knowledge base extracted from Wikipedia. Semant. Web **6**(2), 167–195 (2015)

12. Perry, M., Herring, J.: OGC geosparql-a geographic query language for RDF data. Open Geospatial Consortium (2012)

13. Perry, M., Jain, P., Sheth, A.P.: SPARQL-ST: extending SPARQL to support spatiotemporal queries. In: Ashish, N., Sheth, A.P. (eds.) Geospatial Semantics and the Semantic Web - Foundations, Algorithms, and Applications. Semantic Web and Beyond: Computing for Human Experience, vol. 12, pp. 61–86. Springer, Heidelberg (2011)

14. Presutti, V., Draicchio, F., Gangemi, A.: Knowledge extraction based on discourse representation theory and linguistic frames. In: ten Teije, A., Völker, J., Handschuh, S., Stuckenschmidt, H., d'Acquin, M., Nikolov, A., Aussenac-Gilles, N., Hernandez, N. (eds.) EKAW 2012. LNCS, vol. 7603, pp. 114–129. Springer,Y Heidelberg (2012)

15. Rietveld, L., Verborgh, R., Beek, W., Vander Sande, M., Schlobach, S.: Linked data-as-a-service: the semantic web redeployed. In: Gandon, F., Sabou, M., Sack, H., d'Amato, C., Cudré-Mauroux, P., Zimmermann, A. (eds.) ESWC 2015. LNCS, vol. 9088, pp. 471–487. Springer, Heidelberg (2015)

16. Williams, A.J., Harland, L., Groth, P., Pettifer, S., Chichester, C., Willighagen, E.L., Evelo, C.T., Blomberg, N., Ecker, G., Goble, C., Mons, B.: Open phacts: semantic interoperability for drug discovery. Drug Discov. Today **17**(21), 1188–1198 (2012)

17. Zaveri, A., Rula, A., Maurino, A., Pietrobon, R., Lehmann, J., Auer, S.: Quality assessment for linked data: a survey. Semant. Web **7**(1), 63–93 (2015)

18. Zhu, R., Hu, Y., Janowicz, K., McKenzie, G.: Spatial signatures for geographic feature types: examining gazetteer ontologies using spatial statistics. Trans. GIS **20**(3), 333–355 (2016). doi:10.1111/tgis.12232

Automated Cartography and Geovisualization

Circles in the Water:
Towards Island Group Labeling

Arthur van Goethem[1]([✉]), Marc van Kreveld[2], and Bettina Speckmann[1]

[1] TU Eindhoven, Eindhoven, The Netherlands
{a.i.v.goethem,b.speckmann}@tue.nl
[2] Utrecht University, Utrecht, The Netherlands
m.j.vankreveld@uu.nl

Abstract. Many algorithmic results are known for automated label placement on maps. However, algorithms to compute labels for groups of features, such as island groups, are largely missing. In this paper we address this issue by presenting new, efficient algorithms for island label placement in various settings. We consider straight-line and circular-arc labels that may or may not overlap a given set of islands. We concentrate on computing the line or circle that minimizes the maximum distance to the islands, measured by the closest distance. We experimentally test whether the generated labels are reasonable for various real-world island groups, and compare different options. The results are positive and validate our geometric formalizations.

1 Introduction

Map labeling is a fundamental problem in automated cartography which has received a significant amount of attention both in the GIScience and in the algorithms communities [13]. There are a variety of geometric objects to be labeled, ranging from points (representing locations such as cities), over polylines (representing linear cartographic features such as rivers) and polygons (representing areal features such as lakes), to groups of polygons (representing groups of features such as islands). The basis of all algorithmic work in this area is formed by an extensive set of cartographic guidelines which detail the properties of a high quality labeling (see [5,8,14]). These guidelines often lead to optimization problems which can be approached with algorithmic methods.

To generate such guidelines for groups of features, Reimer et al. [12] propose a framework of possible geometric quality measures for the labeling of feature groups. The framework includes the shape of the label, whether it may intersect the features or not, and how the distance between a label and the features is measured. Assuming that the placement of a label is optimal according to some yet unknown measure generates a number of algorithmic optimization problems.

A. van Goethem and B. Speckmann are supported by the Netherlands Organisation for Scientific Research (NWO) under project nos. 612.001.102 and 639.023.208, respectively.

© Springer International Publishing Switzerland 2016
J.A. Miller et al. (Eds.): GIScience 2016, LNCS 9927, pp. 293–307, 2016.
DOI: 10.1007/978-3-319-45738-3_19

Only few of them, namely the simpler ones, can be solved directly with known methods. Reimer et al. [12] conclude that new algorithms are needed to advance the state-of-the-art in automated label placement for feature groups.

In this paper we expand on the research in Reimer et al. [12] by solving several of the algorithmic problems in label placement for groups of features. In particular, we focus on min-max distance measures that minimize the maximum distance to the closest point of each feature. This is arguably more natural than the versions that could be solved with known methods. In this setting, we consider straight-line and circular-arc labels, which may or may not intersect the features. We design new algorithms for these optimization problems, analyze their efficiency, and test implementations on island label groups to know how much the automatically generated labels resemble ones we can find in atlases, originally placed by a cartographer.

Contribution. Our input is a set S of k simple polygons $P_1, \ldots, P_k$, with n vertices in total (the *islands*). We refer to labels that are allowed to overlap islands as *general labels* and to labels that are not allowed any overlap as *water labels*. We focus on placing a single label for an island group, in isolation of other features and labels on the map.

We assume that labels are long enough that we can consider complete lines and circles instead of line segments and circular arcs (see Fig. 1). In practice this assumption may not always be true; we will see some examples in the experiments section. Finding optimal labels that are shorter is a considerably more complex placement problem, resulting in higher running times of solutions. Therefore, in this algorithmic study, we limit ourselves to the simpler case where labels span the island group. We will see that the algorithms we obtain are sufficiently complex already.

We give $O(n \log k)$ and $O(nk + k^2 \log k)$ time algorithms for general and water straight-line labels, respectively, in Sect. 2. For circular-arc labels we give $O(n^2)$ and $O(n^3 \operatorname{polylog} n)$ time algorithms for general and water labels, respectively in Sect. 3. Our solutions are inspired by free placements in motion planning, facility location, minimum-width annulus computation, and dimensional metrology. In all cases we need to carefully capture the geometry of an optimal placement to arrive at an efficient solution.

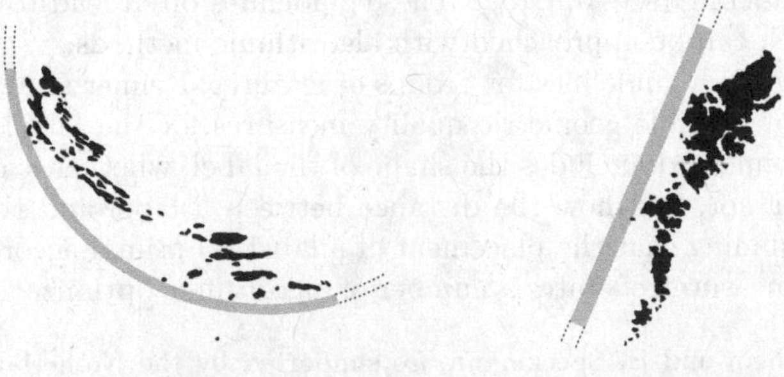

Fig. 1. We assume that labels are long enough to warrant consideration of complete lines and circles. A subsection of the line or circle functions as the final label.

In Sect. 4 we compare the circular-arc labels generated by our algorithms to manually placed labels for a representative set of island groups. The results are positive and often appear already as good as the manually placed labels. We are hence confident that geometric formalization can capture the implicit connection necessary to associate an island label well with its group.

In Sect. 5 we discuss further possible extensions of our algorithms, using the min-sum distance measure and allowing labels with a non-zero height.

Related Work. Many algorithms have been developed for labeling maps, far too many to list here (for an overview see [13]). The labeling of island groups has received little algorithmic treatment so far. An exception is the work by van Kreveld and Schlechter [9] who give an algorithm that places an intersection-free label in the position minimizing the maximum distance from the label to each island of the group using horizontal straight-line labels. Since only straight labels are handled, the solution is not so general. The other exception is the work of Reimer et al. [12], who review what existing geometric algorithms can be adapted directly to island label placement.

2   Straight-Line Labels

In this section we show how to compute a straight-line label – either general or water – that minimizes the maximum distance to the islands. For each island, we consider the distance to its closest point. As we assume labels are long enough to be considered as complete lines, we need not compare different length labels reducing the complexity of the problem. This also implies that the distance to a label can be measured perpendicularly to the label. While this assumption restricts the type of labels we can generate, the resulting reduction of complexity of the problem allows us to formulate good polynomial time algorithms. We discuss how to find the actual label placement on this line in Sect. 4.1.

2.1   General Labels

As we assume that labels are complete lines, the minimum distance between an island and a label can always be measured to a point on the convex hull of

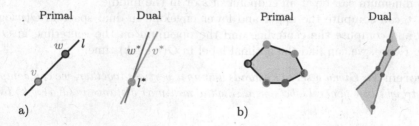

Fig. 2. (a) Two points on a line form two lines intersecting in a point in dual space. (b) The bottom and top of a convex polygon in primal space correspond to the top and bottom of a funnel in dual space.

the island. Hence, to find the optimal general label, we first replace every island by its convex hull, taking $O(n)$ time in total [10]. Then we study the problem in dual space, where lines are represented as points. Each line $l = (y = mx + b)$ becomes a point $l^* = (m, -b)$ in dual space. Similarly, each point $p = (p_x, p_y)$ becomes a line $p^* = (y = p_x x - p_y)$ [4]. The k convex polygons in primal space form k funnels in dual space (see Fig. 2). The *top and bottom boundaries* of each funnel are x-monotone polylines. Funnel boundaries belonging to two different islands only intersect when a line in primal space is tangent to both islands. Between each pair of islands there exist at most four tangents, and at most two of these are tangent to any combination of "upper" and "lower" boundaries of the two islands. Thus, the funnel boundaries form a set of $2k$ x-monotone, pairwise 2-intersecting polylines with $O(n)$ vertices total.

For any fixed rotation of the label to be placed, the furthest closest vertex of an island that is below the label in primal space, is on the upper envelope of the lower boundaries in dual space. As all the funnel-boundaries are 2-intersecting, this upper envelope has complexity $O(n)$ and we can compute it in $O(n \log k)$ time by pairwise merging in $\log k$ phases. A similar argument holds for the furthest closest vertex above and the lower envelope.

If we look at a fixed rotation of the label to be placed, this corresponds to finding the optimal position in dual space on a vertical line. As dualization is distance-preserving on any vertical line, the optimal placement for any fixed rotation is exactly centered between the upper and lower envelope. Thus, the optimal solution for any rotation is located on a *centerline*, which also has $O(n)$ complexity.

For each segment s of the centerline we can compute the optimal position in $O(1)$ time. Let x_{start} be the x-coordinate of the start of s. Let d_{start} be the vertical distance to the upper- (or lower-) envelope at x_{start}. While we move along s the distance to the upper envelope changes by a linear factor c_1 in x. For a shift of δ along the x-axis, the vertical distance in dual space is $f_s(\delta) = d_{start} + c_1 \cdot \delta$. The distance to the closest point in primal space is

$$g_s(\delta) = \frac{(d_{start} + c_1 \cdot \delta)^2}{(x_{start} + \delta)^2 + 1}$$

The optimal position is at the minimum over the domain given by segment s. This minimum can be at an endpoint of s or in the middle.

We can compute the upper and lower envelope in dual space in $O(n \log k)$ time and compute the centerline and the optimum on the centerline in $O(n)$ time. Hence, we can find the optimal label in $O(n \log k)$ time.

Theorem 1. *Given a set of k islands having n vertices together, we can compute the straight-line general label optimizing the min-max distance in $O(n \log k)$ time.*

2.2   Water Labels

An optimal water label need not have an equal distance to the furthest closest point on either side. Positions that have equal distance to both points may result in labels that intersect one or more islands. As a consequence, the dual of the label need not lie on the centerline in between the upper and lower envelope. In this section we describe an algorithm to compute the optimal solution for straight-line water labels in $O(nk + k^2 \log k)$ time.

Compute the Arrangement. We first compute the complete arrangement of all $2k$ funnel boundaries in $O((n + k^2) \log k)$ or $O(nk)$ time.

For the first bound, make an x-sorted list V of funnel boundary vertices in $O(n \log k)$ time. Then do a standard sweep from left to right. The sweep-line intersects $O(k)$ edges simultaneously, leading to the bound.

For the second bound, take the sorted list V and split it into n/k parts such that each part has $O(k)$ vertices. These parts give rise to n/k vertical lines, such that between two vertical lines we need the arrangement of $O(k)$ line segments. We compute the full arrangement of the supporting lines in $O(k^2)$ time and then remove all parts that are on lines but not on the line segments. Then we couple consecutive arrangement parts at the vertical lines into a single arrangement. We spend $n/k \cdot O(k^2) = O(nk)$ time.

Insert the Centerline. The centerline has complexity $O(n)$. As all boundaries and the centerline are x-monotone, and all boundaries are either convex or concave, each edge of the centerline can intersect at most $O(k)$ edges of the arrangement. Consequently, the centerline can cross the arrangement $O(nk)$ times. This bound is also realizable in theory.

To insert the centerline, we sort the boundaries and the centerline by increasing slope of the first segment in $O(k \log k)$ time. This uniquely defines the leftmost face the centerline is in and we can access it in $O(1)$ time. For each face the centerline crosses do the following. Starting at the left-most vertex of the face-boundary, or the last traversed edges of this face, traverse the upper- and lower-boundary of the face, as well as the centerline simultaneously. To get this working, we maintain the last traversed upper and lower edge of every face, and store that with the face in the arrangement. As soon as we enter a face again, we can proceed where we left off. We traverse each edge at most twice and we make at most an additional $O(nk)$ edges by the intersections between the centerline and the arrangement. Hence, the total time complexity is $O(nk)$.

Illegal Placements. A line that intersects an island in primal space dualizes to a point inside the funnel of the island in dual space. Hence, all faces of the arrangement covered by a funnel result in labels that intersect one or more islands. Faces covered by a funnel are defined to be *illegal*. An edge of the arrangement separating two illegal faces is also *illegal*, all other edges are *legal*. When we consider the problem for a fixed rotation of the final label, the legal point vertically closest to the centerline in dual space is optimal. The closest legal points form intervals on the

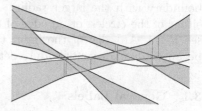

Fig. 3. Three funnels, the illegal cells (grey), the centerline (blue), and the induced closest legal edges (red). (Color figure online)

edges of the arrangement (see Fig. 3). These *subedges* can either be above, below, or on the centerline.

Find the Closest Legal Subedges. We first remove all illegal edges in $O(n + k^2)$ time. Let a *chain* be a maximal sequence of edges connected by vertices of degree two. The resulting arrangement, excluding the centerline, consists of $O(k^2)$ non-intersecting chains with $O(n + k^2)$ vertices in total. We sweep this arrangement with a vertical line, maintaining the intersected edges and the centerline in a balanced binary search tree to determine the closest legal subedges. There are $O(nk)$ possible intersections with the centerline which we can resolve in $O(1)$ time and $O(k^2)$ vertices at which chains start and end requiring $O(\log k)$ time each to update. Hence, we can find all closest legal subedges above and below the centerline in $O(nk + k^2 \log k)$ time.

The closest point to the centerline may be on the centerline or on the closest legal subedge above or below it. In total we get $O(nk)$ legal subedges and legal edges of the centerline, which we can evaluate in $O(1)$ time each.

Theorem 2. *Given a set of k islands having n vertices together, we can compute the straight-line water label optimizing the min-max distance in $O(nk + k^2 \log k)$ time.*

3 Circular-Arc Labels

In this section we consider the problem of computing a circle that minimizes the maximum distance to the islands, measured by the distance to the closest point of each island. We approach the problem by computing an annulus of minimum width that touches all islands. The circle in the middle of this annulus is the required solution. The optimal annulus problem has four degrees of freedom as we can express a solution by the coordinates of the center and two radii.

The minimum-width annulus problem has been studied before for point sets [2,4], and can be solved in $O(n^{3/2+\epsilon})$ time (where $\epsilon > 0$ is a constant that can be chosen). Instead of a point set, we have a set of simple polygons, and the annulus must intersect at least one point of each. We define the *inner circle* as the boundary of the annulus with the smaller radius and the *outer circle* as the boundary with the larger radius. A *contact* is a vertex of the input touching either of the circles, or an edge of the input that is tangent to the outer circle, see Fig. 4. An edge of the input tangent to the inner circle is not a contact, because it cannot contribute to defining the optimal annulus.

3.1 General Labels

We begin with the case of general, unrestricted circular-arc labels, which is the minimum-width annulus problem where the annulus must touch every island. Similar to the minimum-width annulus problem for points, any solution having less than four contacts cannot be optimal (proof omitted).

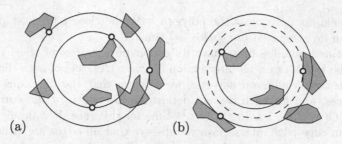

Fig. 4. (a) Annulus touching all islands determined by four contacts (marked), two on the inner circle and two on the outer circle. (b) Annulus partially determined by three contacts with a middle circle that does not intersect any island.

Lemma 3. *The annulus corresponding to the optimal min-max, circular-arc label is determined by at least four contact points.*

The solution to the minimum-width annulus problem for points described in [4] uses the Voronoi Diagram and the Farthest-point Voronoi Diagram to determine the annuli with four contacts that may be the optimal one. The center of each such annulus must lie on a vertex of one of these diagrams, or on an edge-edge intersection of the diagrams. It yields an $O(n^2)$ time algorithm.

We use a similar approach, but need different diagrams. As the annulus is required to overlap or touch each island, the outer radius of the optimal annulus is defined by the closest point of the furthest polygon. Similarly, the inner radius is defined by the polygon with the closest furthest point. We make use of two matching Voronoi Diagrams.

The Farthest-Polygon Voronoi Diagram (FPVD) [6] subdivides the plane in $O(n)$ regions with total complexity $O(n)$. In each region, the same polygon is furthest *and* the same feature (vertex or edge) of that polygon is closest (see Fig. 5(a)). Cheong et al. [6] show that it can be computed in $O(n \log^3 n)$ time.

The Hausdorff Voronoi Diagram (HVD) [7,11] subdivides the plane in a set of regions, with total complexity $O(n)$. In each region, the same polygon is closest *and* the same feature of that polygon is furthest (see Fig. 5(b)). Here the distance between a point x and a polygon P is defined as $\max_{p \in P} d(x, p)$.

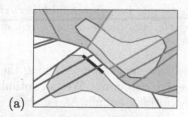
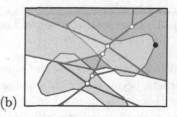

Fig. 5. (a) In each cell of the FPVD the same polygon is furthest and the same feature closest. (b) In each cell of the HVD the same polygon is closest (measured to the furthest point) and the same feature is furthest.

Hence, the closest polygon is the polygon with the closest furthest point. This diagram can be computed in $O(n \log^4 n)$ time [7].

The optimal annulus must have its center on a vertex of the FPVD, on a vertex of the HVD, or on an intersection of two edges, one of each diagram. For annuli with their center on a vertex we observe that the diagrams have only $O(n)$ vertices, and for each we can determine the width of the corresponding annulus in $O(n)$ time, leading to $O(n^2)$ time for this case. For annuli with their center on an edge-edge intersection we observe that all edges are either straight or parabolic and each diagram has $O(n)$ edges. Thus, there can be at most $O(n^2)$ intersections. For each intersection we know the distance to the inner and outer contacts, and we can determine the width of each annulus in $O(1)$ time.

Theorem 4. *Given a set of islands with n vertices together, we can find a circle minimizing the maximum distance to the closest point of any island in $O(n^2)$ time.*

3.2 Water Labels

We next consider the problem of computing a circle that misses all islands and minimizes the maximum distance to them, measured by the distance to the closest point of each island. Phrased in terms of computing an annulus, we want to compute an annulus of minimum width touching every island. For this annulus the additional requirement holds that the *middle* circle, located halfway between the inner- and outer-circle, should not intersect any island.

As before we begin by characterizing properties of an optimal solution by contacts of the three co-centric circles involved and the islands. We can no longer show that there are always four contacts, or two contacts with the inner or outer circle, and hence we cannot use the edges of the FPVD and HVD any longer. Instead we will cover all cases where the annulus is restricted by at least three contacts.

Lemma 5. *An optimal annulus A is of one (or more) of the following types:*

(i)   *A has the contacts as if it were an unrestricted annulus (four contacts on outer and inner circle together);*
(ii) *A has at least two contacts on the outer circle and one on the middle circle;*
(iii) *A has at least two contacts on the inner circle and one on the middle circle;*
(iv) *A has at least one contact on the outer circle, one on the inner circle and one on the middle circle.*

With three contacts, the annulus can move its center while retaining these contacts, because there is still one degree of freedom remaining. We can optimize over this degree of freedom and find all locally optimal annuli. As we check all local optimal annuli we will also check the global optimum solution.

This characterization of optimal annuli gives rise to an algorithmic solution. As a tool we will use a data structure that allows us to test whether a query circle intersects any island, which we describe first. We treat the islands as a set of n line segments and observe that a circle does not intersect any island if and only if it does not intersect any line segment (ignoring the case where the circle lies fully inside an island, which can easily be handled).

Lemma 6 (from [3]**).** *A line segment s intersects a circle C if and only if:*

(i) *exactly one of the endpoints of s lies inside C, or*
(ii) *both endpoints of s lie outside C, the center of C lies in the perpendicular strip of s, and the supporting line of s intersects C.*

The perpendicular strip of s is the strip bounded by the two lines perpendicular to s and each through one endpoint of s (see Fig. 6). Using the lemma we can design a data structure that stores n line segments in a multi-level tree so that for any query circle, we can efficiently decide if it intersects a line segment. Combining the theory of [1] with Lemma 6, and using cutting trees instead of partition trees, we get the following:

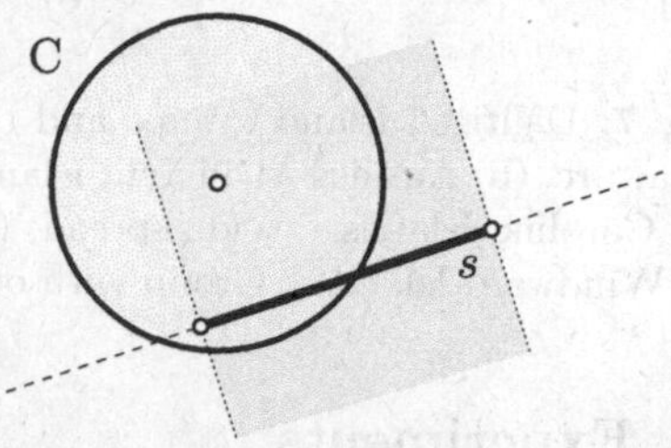

Fig. 6. A line segment s, its supporting line (dashed), its perpendicular strip (yellow), and an intersecting circle C. (Color figure online)

Lemma 7 *A set of n line segments can be stored in a data structure of size and preprocessing time $O(n^3 \operatorname{polylog} n)$, such that for any query circle, we can decide in $O(\operatorname{polylog} n)$ time whether it intersects any line segment of the set, where $\operatorname{polylog} n$ stands for $\log^k(n)$ for some constant k.*

We compute an optimal annulus as follows. First, build the data structure T on the n line segments bounding the islands. Second, use the results of the general case to compute $O(n^2)$ annuli. For each such annulus, determine the middle circle and query with it in T to decide if it intersects any island. If not, it is a candidate solution. Third, take any triple of features of the islands and assume them to be contacts. Choose each of the cases (ii), (iii) and (iv) from Lemma 5 and each assignment of contacts for that case.

To treat any of the $O(n^3)$ choices, compute the loci of centers of the corresponding annulus, and the corresponding function describing the widths of these annuli. This is a one-parameter function because we have fixed three out of four degrees of freedom, and we compute it in $O(1)$ time because only three contacts (features of the islands) are involved. We optimize the width function, finding all $O(1)$ local optima. Each gives rise to an annulus, which we test with our data structure T to see if the middle circle intersects any island.

Since we will be testing $O(n^3)$ circles by querying T, and the construction of T takes $O(n^3 \operatorname{polylog} n)$ time, we obtain:

Theorem 8 *Given a set of islands with n vertices together, we can find a circle that does not intersect any island and minimizes the maximum distance to the closest point of any island in $O(n^3 \operatorname{polylog} n)$ time.*

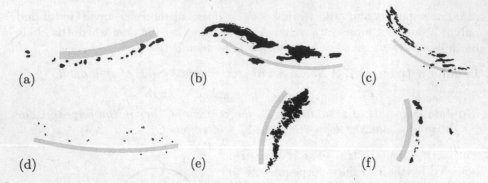

Fig. 7. Digitized island groups and their manual label (grey). (a) Aleutian Islands -
Clear arc. (b) Antilles - Different island sizes. (c) Dalmatian Islands - Arc with spread.
(d) Caroline Islands - Wide spread. (e) Outer Hebrides - Closely interwoven islands.
(f) Windward Islands - Group with outlier.

4   Experiments

We evaluate the quality of the labels generated by our algorithms by performing
both a visual analysis and a comparison in numbers with manually placed labels.[1]

4.1   Setup

To determine the quality of the computed labels we compare them to manually
generated labels. We use digitized island groups from several atlases together
with their respective labels. We selected six candidates to represent a wide range
of possible island configurations (see Fig. 7).

In these experiments we focus on circular arc labels as they are more commonly present in maps. Nevertheless, the straight-line labels generated by our
algorithms give good results (see Fig. 8) that may directly be used as label positions. The rarity of straight-line labels for island groups makes them less suitable
for a direct evaluation though.

Fig. 8. Island groups with their original, manual label (grey) and the computed label
(green). Straight-line water labels under different points of measurement. (a) All points.
(b) Centroid. (c) Closest point. (Color figure online)

[1] Further results are available online: http://www.win.tue.nl/~agoethem/labeling,
May, 2016.

Besides the straight-line and circular-arc labels discussed in this paper, we also tested other possible optimization measures. Specifically, we tested optimal straight-line and circular-arc labels, using the min-max distance measure, when distance is measured to the centroid or to the complete area of each island. The latter is in fact equivalent to measuring distance to the furthest points. The optimal label placements for these settings can be computed using variations on the techniques described in this paper. To ensure we obtained realistic solutions we computed label positions having the same height as the original manual label. The discussed algorithms can easily be extended to take this into account (see Sect. 5).

Finally, our algorithms compute only complete lines and circles for label placement. These should still automatically be converted to line segments and circular arcs. We project all vertices of the island onto the computed line (/circle). The final label is the minimal line segment (/circular arc) that contains all projected vertices. As a consequence the computed label position best spans the width of the island group. We note that the actual label may have a different length, but any length label can trivially be computed from this.

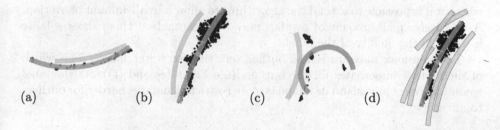

(a)   (b)   (c)   (d)

Fig. 9. Island groups with their original, manual label (grey) and the computed label (green). General labels under different points of measurement. (a) Closest point. (b) All points. (c) Closest point. (d) Candidate generation may use general labels to compute several possible high-quality label positions (manual example). (Color figure online)

4.2 Visual Inspection

General Labels. We observe that the computed labels capture the shape of the island group well when the island group forms a coherent and clear shape (see Fig. 9(a) and (b)). The label overlaps many islands and is less suitable for label placement, but is a good basis for label candidate generation. By slightly changing the radius and center of the arc we can generate many labels capturing the shape of the group (see Fig. 9(d)).

The min-max distance measure is outlier-sensitive. This is no problem in most groups, but outliers may cause unexpected results (see Fig. 9(c)).

In general, we notice that the effect of the point of measure is small for the min-max measure. When the group contains large islands, however, the effect becomes more prominent. In Fig. 10 the same island group is shown (Antilles) for the three different points of measure. For large islands the closest point, farthest point, and centroid may give significantly different results.

Fig. 10. Antilles islands with the manually placed label (grey) and the computed label (green) for different points of measure: (a) All points. (b) Centroid. (c) Closest point. (Color figure online)

Water Labels. Generally, water labels give good results (see Fig. 11(a), (b) and (c)). Often the label is located outside the island group following the general shape. When the label is placed inside the group, the same holds for the manual label (see Fig. 11(c) and (d)).

We observe that, as expected, preventing any overlap between the label shape and the islands may be overly restrictive. In Fig. 11(d) two tiny islands prevent the label from following a more natural shape. Future work may investigate whether it is possible to extend the algorithms to allow a small amount of overlap. A relatively small amount of overlap may be acceptable if this causes a large increase in the quality of the label.

The min-max measure is less outlier sensitive for water labels. The width of the group compensates for the outliers (see Fig. 11(e) and (f)). As the label cannot intersect any island its shape is more restricted and it is harder for outliers to affect it.

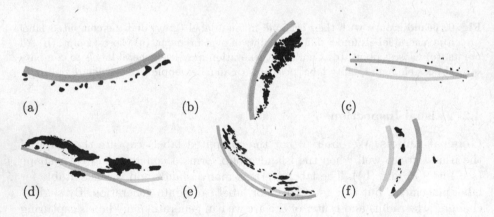

Fig. 11. Island groups with their original, manual label (grey) and the computed label (green). Water labels under different points of measurement. (a) All points. (b) Closest point. (c) Closest point. (d) Centroid. (e) Centroid. (f) Closest point. (Color figure online)

4.3   Comparison

To support our visual inspection we also perform a comparison in the line of the research outlined by Reimer et al. [12]. They suggest to analyze what criteria cartographers may subconsciously use when placing island labels, by comparing manually placed island labels from atlases with automatically generated labels according to geometric criteria. For example, suppose we consider the setting with water labels and min-max distance for circular labels, measured to the closest point of each island. If an analysis reveals that cartographers place island labels within a small percentage of what is possible in this setting, then it is likely that cartographers apply this criterion either explicitly or subconsciously. Note that the manually placed label and the automatically generated label may be very different even if cartographers realize nearly the same score on the measure.

For each measure and each setting we compute the label placement that minimizes the distance to the island group and the corresponding distance. We also compute the distance, according to this measure, from the manual label to the island group. For fair comparison we measure distance to the circle concentric to the manual label. The more correlated the distance of the manual label and optimal label position are, the more likely it is that the manual label was placed (subconsciously) according to the given measure. In Table 1 an overview of all distance measures (normalized to the minimal distance) is given for circular arc labels using the min-max distance. We make some observations.

First, the requirement that labels are strictly non-overlapping reduces our ability to optimize the given metrics. Consequently, the manual label for the Antilles is generally 'better' than the optimal water label. Second, the outlier sensitivity of the min-max distance measure is reflected in the 'quality' of the manual label for the Windward Islands, which appears to ignore a further removed island. Consequently, the manual label is a poor fit in the min-max

Table 1. Distance of the manual label to the island group for the different settings (normalized in comparison to the optimal label placement for that setting). For each combination of island group and label type (general or water), the distance of the label that minimizes its respective measure best is underlined. Note that the manual label may overlap the island group causing labels that are better than "optimal".

| | Circular-arc min-max label | | | | | |
| | General label | | | Water label | | |
	Centroid	All	Closest	Centroid	All	Closest
Aleut. Isl	3.39	_2.61_	9.53	1.04	0.97	1.18
Antilles	_1.42_	1.51	1.57	0.93	_0.84_	1.07
Carol. Isl	_1.71_	1.73	1.78	1.68	_1.66_	1.72
Dalm. Isl	_1.34_	1.40	1.65	_1.01_	1.02	1.02
Hebrides	2.54	_2.44_	3.03	_1.09_	1.11	1.10
Wind. Isl	_3.80_	3.81	6.35	_2.57_	2.35	2.87

distance measure. Finally, surprisingly, this preliminary test suggests that the closest point of each island may not be as important as the furthest point or centroid. We should take some care in interpreting these measures though, as the closest point measure is also most easily influenced by a small change in the label shape.

5   Discussion and Future Work

The time bounds we obtain are worst-case time bounds for optimal label placement. They show that optimal placement can be done at least this efficiently. The quadratic and cubic runtime bounds imply that we cannot expect interactive label placement for high-detail island groups. This need not be a problem as the geometry of the problem is likely only marginally affected by the exact details of the island shapes, and line simplification can be applied in preprocessing to reduce the input size. Alternatively, the algorithms could be adapted so that the cubic running time does not show up for realistic inputs.

We sketch two possible extensions of our algorithms and investigate the possibilities for future work.

Min-Sum Distance. In Sect. 2 we presented two algorithms for straight-line general and water labels using the *min-max* distance. Both can be extended to the *min-sum* distance as follows. An optimal straight-line label must have equally many islands placed to either side. Hence, for a fixed rotation, the optimal (general) solution can be found in $O(n \log k)$ time. The optimal water label is placed at the legal position closest to the above solution.

In both cases there always exists an optimal solution tangent to an island. Thus, in dual space it is located on an edge of the arrangement. We compute the arrangement in $O((n + k^2) \log k)$ time and detect the faces having an equal number of islands above and below it. We traverse the arrangement and update the required information in constant time per face, resulting in $O((n + k^2) \log k)$ time to find the optimal general and water label.

Non-zero-Height Labels. For general labels, using a non-zero-height does not change the solution. To place a water label of height h, we can offset all islands by $h/2$ and compute a zero-height label. The vertices of the islands in primal space, however, become circular arcs. Consequently, in dual space we have an arrangement of curves. We can still compute the arrangement, the centerline, and the closest legal segments in $O(nk + k^2 \log k)$ time. A similar approach works for the min-sum distance measure.

Future Work. We would like to lift the restriction that the full line or circle must be free for water labels; the part used by the text would be enough. However, the degrees of freedom in the problem increase, and it is unclear whether sufficiently efficient algorithms exist.

Alternatively, it would be interesting to see if we can relax the requirement that a water label cannot have any overlap with the islands. Allowing the label to overlap an outermost strip of fixed width for each island could easily be integrated in our approach. To achieve this, checks of intersection with the middle annulus circle could simply be done with a negatively offset version of the islands. When we instead would require that the label has a maximum amount of overlap in total over all islands, finding the optimal water label is non-trivial.

We note that labeling bodies of water and channels is similar to labeling island groups. In contrast to island group labeling though, we try to place the label in the middle of the body of water or channel. We may achieve this by maximizing the distance to the nearest points outside the body of water, while avoiding any islands. Hence, a max-min distance measure (maximizing the minimum distance) may be more applicable.

Acknowledgments. Thanks to Sarah Lohr, Susanne Heuser, and Andreas Reimer for providing us with the digitized maps.

References

1. Agarwal, P., Erickson, J.: Geometric range searching and its relatives. Contemp. Math. **223**, 1–56 (1999)
2. Agarwal, P., Sharir, M.: Efficient randomized algorithms for some geometric optimization problems. Discrete Comput. Geom. **16**(4), 317–337 (1996)
3. Agarwal, P., van Kreveld, M., Overmars, M.: Intersection queries in curved objects. J. Algorithms **15**(2), 229–266 (1993)
4. de Berg, M., Cheong, O., van Kreveld, M., Overmars, M.: Computational Geometry, 3rd edn. Springer, Berlin (2008)
5. Brewer, C.: Designing Better Maps: A Guide for Gis Users. ESRI Press, Redlands (2005)
6. Cheong, O., Everett, H., Glisse, M., Gudmundsson, J., Hornus, S., Lazard, S., Lee, M., Na, H.-S.: Farthest-polygon Voronoi diagrams. Comput. Geom. Theory Appl. **44**(4), 234–247 (2011)
7. Dehne, F., Maheshwari, A., Taylor, R.: A coarse grained parallel algorithm for Hausdorff Voronoi diagrams. In Proceedings of the International Conference on Parallel Processing, pp. 497–504 (2006)
8. Imhof, E.: Positioning names on maps. Am. Cartographer **2**(2), 128–144 (1975)
9. van Kreveld, M., Schlechter, T.: Automated label placement for groups of islands. In: Proceedings of the 22th International Cartographic Conference (2005)
10. Melkman, A.: On-line construction of the convex hull of a simple polyline. Inf. Process. Lett. **25**(1), 11–12 (1987)
11. Papadopoulou, E., Lee, D.: The Hausdorff Voronoi diagram of polygonal objects: a divide and conquer approach. Int. J. Comput. Geom. Appl. **14**(6), 421–452 (2004)
12. Reimer, A., van Goethem, A., Rylov, M., van Kreveld, M., Speckmann, B.: A formal approach to the automated labelling of groups of features. Cartography Geogr. Inf. Sci. (2015). http://dx.doi.org/10.1080/15230406.2015.1053986
13. Wolff, A., Strijk, T.: A map labeling bibliography (2009). http://i11www.iti. uni-karlsruhe.de/~awolff/map-labeling/bibliography/
14. Wood, C.: Descriptive and illustrated guide for type placement on small scale maps. Cartographic J. **37**(1), 5–18 (2000)

An Algorithmic Framework
for Labeling Road Maps

Benjamin Niedermann[1](✉) and Martin Nöllenburg[2]

[1] Karlsruhe Institute of Technology, Karlsruhe, Germany
benjamin.niedermann@kit.edu
[2] TU Wien, Vienna, Austria

Abstract. Given an unlabeled road map, we consider, from an algorithmic perspective, the cartographic problem of placing non-overlapping road labels embedded in the roads. We first decompose the road network into logically coherent road sections, i.e., parts of roads between two junctions. Based on this decomposition, we present and implement a new and versatile framework for placing labels in road maps such that the number of labeled road sections is maximized. In an experimental evaluation with road maps of 11 major cities we show that our proposed labeling algorithm is both fast in practice and that it reaches near-optimal solution quality, where optimal solutions are obtained by mixed-integer linear programming. In direct comparison, our algorithm consistently outperforms the standard OpenStreetMap renderer Mapnik.

1 Introduction

Due to the increasing amount of geographic data and its continual change, automatic approaches become more and more important in cartography. This particularly applies to the time-consuming and demanding task of label placement and much research has been done on its automation. Badly placed labels of features of interest can easily make maps unreadable [4]. Depending on the type of map feature, label placement is done differently. For *point features* (e.g., cities on small-scale maps) labels are typically placed closely to that feature, while for *line features* (e.g., roads, rivers) the name is either placed along or inside the feature. The latter approach is also used for *area features* (e.g., lakes). Regardless of the applied technique and feature type, labels should not overlap each other and clearly identify the features [8].

The cartographic label placement problem has also attracted the interest of researchers in computational geometry and has been thoroughly investigated from both the practical and theoretical perspective [13, Chapter 58.3.1], [14]. While algorithms for labeling point features get a lot of attention, much less work has been done on line features and area features. In this paper we address labeling line features, namely labeling the entire road network of a road map. We take an algorithmic, mathematical perspective on the underlying optimization problem and build on our recent theoretical results for labeling tree-shaped networks [3]. We apply the quality criteria for label placement in road maps elaborated by

© Springer International Publishing Switzerland 2016
J.A. Miller et al. (Eds.): GIScience 2016, LNCS 9927, pp. 308–322, 2016.
DOI: 10.1007/978-3-319-45738-3_20

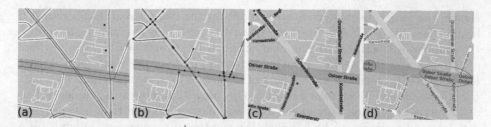

Fig. 1. The presented workflow. (a) The road network given by polylines (thin, blue segments). (b) Phase 1: a graph G is created whose embedding is the simplified road network; blue segments: road sections, red segments: junction edges. (c) Phase 2: creating the labeling using G. (d) A labeling produced by the OSM renderer Mapnik. The six labels of road *Osloer Straße* are enclosed by red ellipses. (Color figure online)

Chirié [2] based on interviews with cartographers. They include that (C1) labels are placed inside and parallel to the road shapes, (C2) every road section between two junctions should be clearly identified, and (C3) no two road labels may intersect. Similar criteria have been described in a classical paper by Imhof [4].

Variations of embedded labels have been considered in road maps before. Chirié [2] and Strijk et al. [11, Chapter 9] presented simple, local heuristics that place non-overlapping labels based on a discrete set of candidate positions – in contrast we consider the problem globally applying a continuous sliding model. Seibert and Unger [10] utilized the geometric properties of grid-based road networks and proved that it is NP-complete to decide whether at least one label can be placed for each road. For the same grid-based setting Neyer and Wagner [6] evaluated a practically efficient algorithm that is not applicable for general road networks.

Road labeling with embedded labels has also been considered for interactive and dynamic maps. Maass and Döllner [5] provided a heuristic for labeling interactive 3D road maps taking obstacles into account. Vaaraniemi et al. [12] presented a study on a force-based labeling algorithm for dynamic maps considering both point and line features. Schwartges et al. [9] investigated embedded labels in interactive maps allowing panning, zooming and rotation of the map. They evaluated a simple heuristic for maximizing the number of placed labels.

For labeling point features a typical objective is to maximize the number of non-overlapping placed labels, because every placed label enhances the map with further information. While this is mostly true for point features, maximizing the number of labels is not the right objective for label placement of roads since not every label that is placed necessarily contributes more information to the map. For example, consider the placed labels of the road *Osloer Straße* in Fig. 1(d). We can easily remove some of those labels without losing any information, because the map user can still identify the same road sections; see Fig. 1(c). In online map services, however, one often finds such redundant labels; see the full version of this paper [7] for two examples. Some roads may have unnecessarily many labels, which ma y in turn cause others to remain completely unlabeled.

Hence, the user cannot identify such roads on the map, a real disadvantage if headed for that road. Due to these observations we do not aim to maximize the number of labels, but the number of labeled *road sections*. For the purpose of this paper, a *road section* forms a connected piece of the road network that logically belongs together, e.g., a part of a road between two junctions or a part that distinguished by its color or width. Our algorithm, however, is independent of the actual definition of road sections; any partition of the road network into disjoint road sections can be handled. We say that a road section is *labeled* if a label (partly) covers it.

As the underlying model for maximizing labeled road sections we re-use the planar graph model that has been introduced in our theoretical companion paper [3]. In that paper we proved that labeling a maximum number of road sections is NP-hard, even for planar graphs and if no road consists of multiple branches. However, we presented a polynomial-time algorithm for the case that the road graph is a tree. While this result for trees is mostly of theoretic interest (road networks rarely form trees), we will show in this paper that our tree-based algorithm can be used successfully as the core of an efficient and practical road labeling algorithm that produces near-optimal solutions.

Contribution and Outline. We introduce a versatile algorithmic framework for placing non-overlapping labels in road networks maximizing the number of labeled road sections. We keep the algorithmic components easily exchangeable. In Sect. 2 we discuss and expand the model introduced in [3]. Afterwards, we present a workflow for labeling road networks in two phases; see Fig. 1.

Phase 1 (Sect. 3). We translate the given road network into a semantic representation (an abstract road graph) that identifies pieces of the road network that belong semantically together. To that end, we simplify the road network, e.g., we merge lanes closely running in parallel. By design this simplification maintains the overall geometry of the road network and only merges structures in the data that should not be labeled independently. Phase 1 is not part of the labeling optimization process.

Phase 2 (Sect. 4). Based on the abstract road graph, we create an actual labeling using one of three algorithms: a naive base-line algorithm, a heuristic extending our tree-based algorithm [3] and a mixed-integer linear programming (MILP) formulation.

As proof of concept we implemented the core of the framework only taking the most important cartographic criteria into account. However, with some engineering it can be easily enhanced to more complex models, e.g., enforcing minimum distances between labels, abbreviating road names, or using alternative definitions of road sections. In Sect. 5 we present a detailed evaluation of our framework on 11 sample city maps. Due to its availability and popularity in practice, we compare our results against the standard OpenStreetMap (OSM) renderer Mapnik as a representative of local heuristics; it uses a strategy similar to [2, 11]. We show that our tree-based algorithm is fast and yields near-optimal labelings that outperform Mapnik.

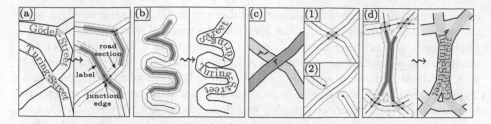

Fig. 2. Illustration of model and arising issues. (a) Sketch of a road network and its abstract road graph. (b) Labels are possibly curvy and have sharp bends making the text hardly legible. (c) ISSUE 2: two ways to represent bridges and tunnels in the abstract road graph. (d) ISSUE 4: the text representation of labels may overlap, while the curve representation in the abstract road graph does not. (Color figure online)

2   Semantic Representation of Road Networks

At any given scale, road networks are typically drawn as follows. Each road or road lane is represented as a thick, polygonal curve, i.e., a polygonal curve with non-zero width; see the background of Fig. 1(a). If two (or more) such curves intersect, they form junctions. If two or more lanes of the same road closely run in parallel they merge to one even thicker curve such that individual lanes become indistinguishable. We then want to place road labels inside those thick curves. More precisely, a *road label* can again be represented as a thick curve (the bounding shape of the road name) that is contained in and parallel to the thick curve representing its road; see Fig. 1(c).

For the purpose of this paper it is sufficient to use a simplified representation, which represents the road network and its labels as thin curves instead [3]. More precisely, a road network is modeled as a planar embedded *abstract road graph* whose edges correspond to the skeleton of the actual thick curves. In this model a label is again a thin curve of certain length that is contained in the skeleton. Following the cartographic quality criteria (C1)–(C3), we want to place labels, i.e., find sub-curves of the skeleton, such that (1) each label starts and ends on road sections, but not on junctions, (2) no two labels overlap, and (3) a maximum number of road sections are labeled. Requiring that labels end on road sections avoids ambiguous placement of labels in junctions where it is otherwise unclear how the road passes through it. Note that this does not forbid labels across junctions. From a labeling of the abstract road graph it is straight-forward to transform each label back into its *text representation* by placing the individual letters of each label along the thick curves; see Fig. 2(a).

Abstract Road Graph Model. We have introduced the abstract road graph in [3], but for the convenience of the reader we repeat it here, see also Fig. 1(b) and Fig. 2(a). A road network (in an abstract sense) is a planar geometric *graph* $G = (V, E)$, where each vertex $v \in V$ has a position in the plane and each edge $\{u, v\} \in E$ is represented by a polyline whose end points are u and v. Each edge further has a *road name*. A maximal connected subgraph of G consisting of edges with the same name forms a *road* R. The length of the name of R is denoted by $\lambda(R)$. Each edge $e \in E$ is either a *road section*, i.e., the part of a road in between two junctions, or a *junction edge*, which models road junctions. Formally, a *junction* is a maximal connected subgraph of G that only consists of junction edges. Typically, when two roads cross, a junction is a star with one center vertex and four outer vertices, but more complex junctions are possible. We require that no two road sections in G are incident to the same vertex and that vertices incident to road sections have at most degree 2. Thus, the road graph G decomposes into road sections, separated by junctions.

We say a point p lies on G, if there is an edge $e \in E$ whose polyline contains p. Hence, a polyline ℓ (in particular a single line segment) lies on G if each point of ℓ lies on G. Further, ℓ *covers* e, if there is a point of ℓ that lies on e. If each point of e is covered by ℓ, e is *completely covered*. The *geodesic distance* of two points on G is the length of the shortest polyline on G connecting both points.

A *label* of a road R is a simple open polyline ℓ on G that has length $\text{len}(\ell) = \lambda(R)$, ends on road sections of G, and whose segments only lie on edges of R. The start point of ℓ is denoted as the *head* $h(\ell)$ and the endpoint as the *tail* $t(\ell)$. Obviously, the edges that are covered by ℓ form a path $\mathcal{P}_\ell = (e_1, e_2, \cdots, e_{k-1}, e_k)$ such that e_1, and e_k are (partly) covered and $e_2, \ldots, e_{k-1}$ are completely covered by ℓ. If e_i is a road section (and not a junction edge), we say that e_i is *labeled* by ℓ. We restrict ourselves to *well-shaped* labels, i.e., labels that are not too curvy or do not contain broken type setting due to sharp bends; see Fig. 2(b). Similar to Schwartges et al. [9], we apply a local criterion to decide whether a label is well-shaped; see also [7]. A *labeling* $\mathcal{L}$ for a road network is a set of mutually non-overlapping, well-shaped labels, where two labels ℓ and ℓ' *overlap* if they intersect in a point that is not their respective head or tail.

Following the criteria (C1)–(C3), the problem MAXLABELEDROADS is to find a labeling $\mathcal{L}$ that labels a maximum number of road sections, i.e., no other labeling labels more road sections. In [3] we showed that MAXLABELEDROADS is NP-hard in general, but can be solved in $O(|V|^3)$ time if G is a tree.

Shortcomings for Real-world Road Networks. While the abstract road graph model allows theoretical insights, we cannot directly apply it to real-world road networks. Due to the following issues, we need to invest some effort in a pre-processing phase (see Sect. 3) to guarantee that the resulting labels in the text representation do not overlap, look nice and are embedded in the roads' shapes.

ISSUE 1: If lanes run closely in parallel, their drawings in the road network merge to one thick curve and individual lanes become indistinguishable. Hence, in our abstract model, such lanes should be aggregated to a single road section that represents the skeleton of the merged curve, and labels should be contained in it; see Fig. 1(c).

ISSUE 2: Real-world road networks are not planar, but edges may cross, namely at tunnels and bridges; see Fig. 2(c). To avoid overlaps between labels placed on those road sections, we either can model the intersection as a regular junction of two roads or we split one into two shorter road sections that do not

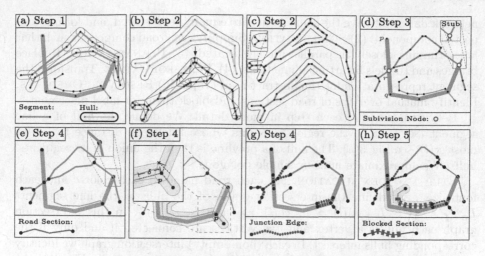

Fig. 3. Illustration of the steps applied in Phase 1. Segments of the same thickness and color have the same road name. For more details see the description of Phase 1. (Color figur online)

cross the other road section. In both cases the road graph becomes planar. For our prototype we use the first variant (also used by Mapnik), because more road sections can be labeled.

ISSUE 3: In real-world road networks some road sections are possibly so long that the label should be repeated after appropriate distances.

ISSUE 4: Labels have a certain font size so that when transforming an abstract label curve into its text representation, labels of different roads may overlap due to their road sections being too close; see Fig. 2(d).

3 Phase 1 – Construction of Abstract Road Graphs

The first phase of our framework consists of transforming the input road network data into an abstract road graph while resolving the four issues mentioned in Sect. 2. Typically, road networks are given as a set of polylines that describe the roads and road lanes. Individual polylines do not necessarily form semantic components such as road sections. So as a first step, we break all polylines down into individual line segments (whose union forms the road network). Let L be the set of all these line segments. We further require that each line segment $l \in L$ is annotated with its *road name* rn(l), the stroke width st(l) and the color co(l) that are used to draw l, and finally the *font size* fs(l) that shall be used to display the name. We say that two line segments $l, l' \in L$ are *equally represented* if st$(l) = $ st(l') and co$(l) = $ co(l'). We assume that fs$(l) < $ st(l) for any l; otherwise we set st$(l) := $ fs(l).

The workflow consists of the following five steps; see Fig. 3. (1) IDENTIFI-CATION. Identify single *road components*, i.e., sets of line segments in the road network data that have the same name, are equally represented, and form a connected component. (2) SIMPLIFICATION. Simplify each road component such that lanes running closely in parallel are aggregated. (3) PLANARIZATION. Replace bridges and tunnels by artificial junctions. (4) TRANSFORMATION. Transform the segment representation into an abstract road graph. (5) RESOLVING OVERLAPS. Identify mutual overlaps of road sections and block them for label placement.

Below we describe each step in more detail. We define the *hull* of a line segment $l \in L$ to be the region of points whose Euclidean distance to l is at most st(l); see Fig. 3(a). The hull of a polyline is then the union of its segments' hulls. We approximate hulls by simple polygons.

STEP 1 – IDENTIFICATION. For each road name n, each color c and each font size f we define the intersection graph of the hulls of the line segments $L_{n,c,f} = \{l \in L \mid \text{rn}(l) = n, \text{ co}(l) = c \text{ and fs}(l) = f\}$. In this intersection graph each hull is a vertex and two vertices are connected if and only if the corresponding hulls intersect. In each (non-empty) intersection graph we identify all connected components, which we call *road components*; e.g., in Fig. 3(a) the blue segments form a road component. Thus, based on L we obtain a set $\mathcal{C}$ of road components. By definition, each component $C \in \mathcal{C}$ has a unique name rn(C), stroke width st(C), color co(C) and font size fs(C).

STEP 2 – SIMPLIFICATION. For each road component $C \in \mathcal{C}$ we geometrically form the union of the according hulls. Thus, the result is a simple polygon P (possibly with holes) describing the contour of the road component; see Fig. 3(b), top. Following Bader and Weibel [1] we use the conforming Delaunay Triangulation of P to construct a *skeleton* as a linear representation of C. In this way, after some further simplifications (see Fig. 3(c)), we obtain a skeleton for each component C such that labels centered on the skeleton are guaranteed to be contained in P. This resolves ISSUE 1. We annotate each skeleton edge with the name, stroke width, color and font size of C. For more details see [7].

STEP 3 – PLANARIZATION. So far polylines describing the skeletons of different road components may intersect at other points than their end points, e.g., polylines representing bridges and tunnels may cross other polylines. As motivated in Sect. 2, we subdivide these polylines to resolve intersections; see Fig. 3(d). More precisely, if two line segments $\overline{pq}$ and $\overline{rs}$ of two polylines intersect at a point t, we replace them by the four segments $\overline{pt}$, $\overline{tq}$, $\overline{rt}$ and $\overline{ts}$. We do the intersection tests with a certain tolerance to identify T-crossings safely. However, this may yield short stubs that protrude junctions slightly; we remove those stubs. This resolves ISSUE 2 and yields a set of annotated polylines only intersecting in vertices.

STEP 4 – TRANSFORMATION. Next we create the abstract road graph from the polylines of the previous step. As a result of Step 3 we know that any two polylines intersect only in vertices. We first take the union of all polylines, identify vertices that are common to two or more polylines and mark these vertices as *junction seeds*. This induces already a planar graph $G = (V, E)$ with polyline edges whose vertices V are either junction seeds or have degree 1. It remains to partition the edges of G into road sections and junction edges. Initially, we mark all edges as road sections. We distinguish two types of junction seeds in G.

If a junction seed v has degree at least 3, only two of its incident edges e and e' belong to the same road R and all other incident edges belong to different roads (and have a different road type than R) then we do not create any junction edges at v, see Fig. 3(e), small box. Since R is the only road that may use the junction at v and it is visually clear that all other roads end at v we can safely treat v as an internal vertex of a road section of R. So we disconnect all incident edges of v except e and e' from v and let each of them end at its own slightly displaced copy of v. The edges e and e' are merged at v and the new edge remains a road section. This resolves the situation as desired.

For all other junction seeds we create junction edges as follows. Let v be a junction seed and let E_v be the set of edges incident to v. We intersect the hulls of all edges in E_v and project their intersection points onto the corresponding edges, see Fig. 3(f). For each edge $e \in E_v$ we determine the projection point p_e that is farthest away from v (in geodesic distance). If the distance between p_e and v exceeds a given threshold δ, we shift p_e to the point on e that has distance δ from v. Now we subdivide e at p_e and mark the edge $\{v, p_e\}$ as a junction edge; the other edge at p_e (if non-empty) remains a road section. The threshold δ ensures that roads running closely in parallel are not completely marked as junction edges. Figure 3(g) shows the resulting abstract road graph.

To resolve ISSUE 3 we subdivide road sections whose length exceeds a certain threshold (in our experiment 350 pixels) by inserting a very short junction edge.

STEP 5 – RESOLVING OVERLAPS. By Step 2 the hulls of edges that belong to the same road component do not overlap. However, if two sections of different roads run closely in parallel, their hulls (and hence their labels) may overlap. We identify overlaps of the hulls of non-incident edges in G and block the corresponding parts of the edge whose road is less important for placing labels; ties are broken arbitrarily. More complex approaches using road displacement could be applied, however, we have chosen a simple solution. By design hulls of incident edges may only overlap if both are junction edges; those overlaps are handled by the labeling algorithms; see Sect. 4. This resolves ISSUE 4.

4 Phase 2 – Label Placement in Road Graphs

In this section we present the four different methods for solving MAXLABELED-ROADS that we subsequently evaluate in our experiments in Sect. 5. Furthermore, we describe a technique for decomposing road graphs into several smaller, independent components that may speed up computations.

4.1 Labeling Methods

BASELINE. An obvious base-line heuristic to obtain lower bounds is to simply place a well-shaped label on each individual road section that is long enough to admit such a label without extending into any junctions. We use this approach to show that it is beneficial to position labels across junctions.

MAPNIK. Mapnik (http://mapnik.org) is a standard open source renderer for OpenStreetMap that includes an road labeling algorithm. The algorithm iteratively labels so-called *ways*, which are polylines describing line features in OpenStreetMap. Along each way it places labels with a certain spacing and locally ensures that labels do not intersect already placed labels of other ways. It does not use any semantic structure from the road network (e.g., road sections), but relies on how the contributors of OpenStreetMap modeled single ways. We may run the rendering algorithm and extract all placed labels from its output.

TREE. The tree-based heuristic makes use of our recently proposed algorithm that optimally solves MAXLABELEDROADS if G is a tree [3]. The basic idea for trees is that a placed label splits the tree into several independent sub-trees, which then are labeled recursively. Using dynamic programming we reuse already computed results so that the algorithm's complexity becomes polynomial, namely $O(|V|^5)$ running time and $O(|V|^2)$ space. Applying some further intricate modifications we improved this to $O(|V|^3)$ time and $O(|V|)$ space, and $O(|V|^2)$ time if each road in G is a path. We omit the details (see [3]) and use that algorithm as a *black box*. If G is a tree, our heuristic optimally labels G. Otherwise it computes a spanning tree T on G using Kruskal's algorithm and computes an optimal labeling for T. We construct T such that all road sections of G are contained in T. Since a road section is only incident to junction edges, this is always possible. In Sect. 5 we show that large parts of realistic road networks can actually be decomposed into paths and trees without losing optimality.

MILP. In order to provide upper bounds for the evaluation of our labeling algorithms, we implement a mixed-integer linear programming (MILP) model that solves MAXLABELEDROADS optimally on arbitrary abstract road graphs. The basic idea is to discretize all possible label positions and to restrict the space of feasible solutions to non-overlapping sets of labels; see also [7]. Although solving a MILP is generally NP-hard, we can apply specialized solvers to find optimal solutions for reasonably sized instances in acceptable times.

4.2 Decomposition of Road Networks

We may speed up both our heuristic TREE and the exact approach MILP by decomposing the road graph into smaller, independent components to be labeled separately, i.e., components whose individual optimal solutions compose to a conflict-free optimal solution of the initial road graph. Such a decomposition allows us to compute solutions in parallel with either of the above methods and it further decreases the total combinatorial complexity. The decomposition rules guarantee that the labelings of the components can always be merged without creating any label overlaps. We name this technique D&C and sketch the decomposition and composition steps; see also [7].

STEP 1 – DECOMPOSITION. For many road sections, e.g., long sections, of real-world road networks labels can be easily placed preserving the optimal labeling. We iterate through the edges of G and cut or remove some of them if one of the following rules applies. As a result the graph decomposes into independent connected components; see Fig. 4(a)–(d). Let e be the currently considered edge and let R be the road of e.

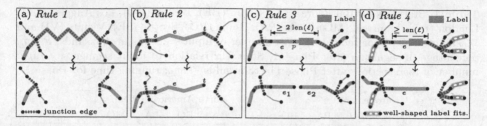

Fig. 4. Illustration of the rules. Edges of the same thickness and color belong to the same road. (Color figure online)

Rule 1. If e is a junction edge and it cannot be completely covered by a well-shaped label, i.e., e is not well-shaped, then remove e.

Rule 2. If e is a road section that ends at a junction that is not connected to any other road section of R, then detach e from that junction.

Rule 3. If e is a road section, a well-shaped label ℓ fits on e, and e is at least twice as long as ℓ, then cut e at its midpoint.

Rule 4. If e is a road section, a well-shaped label ℓ fits on e, and e is connected to a junction that is only connected to road sections of R that may completely contain a well-shaped label, then detach e from that junction.

On each edge we apply at most one rule. If we apply *Rule 3* or *Rule 4* on an edge e, we call e a *long-edge*. Afterwards, we determine all connected components of the remaining graph G', which are then independently labeled.

STEP 2 – LABEL PLACEMENT. For the constructed components we compute solutions in parallel with either of the above methods.

STEP 3 – COMPOSITION. Finally, we compose the labelings of the second step to one labeling. Due to the decomposition, no two labels of different components can overlap. If a long-edge e is not labeled, we place a label on it, which is possible by definition. We adapt the algorithms of Step 2 such that they do not count labeled road sections that were created by *Rule 3*, but we count the corresponding long-edge in this step.

5   Evaluation

We evaluate our framework and in particular the performance of our new tree-based labeling heuristic by conducting a set of experiments on the road networks of 11 North American and European cities; see Table 1. While the former ones are characterized by grid-shaped road networks, the latter ones rarely posses such regular geometric structures. Since the road networks in rural areas are much sparser than those of cities, we refrained from considering these networks and focused on the more complex city road networks. We extracted the abstract road graphs from the data provided by OpenStreetMap[1]. We applied the spherical Mercator projection ESPG:3857, which is also known as *Web Mercator* and used

---

[1] openstreetmap.org.

Table 1. Statistics for Baltimore (BA), Berlin (BE), Boston (BO), Los Angeles (LA), London (LO), Montreal (MO), Paris (PA), Rome (RO), Seattle (SE), Vienna (VI) and Washington (WA) for zoom 16. *OSM*: number of input segments in thousands. *Segm.*: percentage of segments after Phase 1, Step 3 in relation to input segments. *Graph*: number of road sections after Phase 1 in thousands. *Time:* running time for Phase 1.

	European cities					North American cities					
	BE	LO	PA	RO	VI	BA	BO	LA	MO	SE	WA
OSM	225.0	563.4	292.5	117.0	119.9	332.1	225.0	327.0	161.4	433.1	103.9
Segments	55	73	62	62	54	40	50	67	72	59	37
Graph	37.9	105.4	49.9	15.4	18.9	33.8	27.8	80.6	40.2	77.1	11.4
Time (sec.)	21	65	32	12	11	28	21	44	21	42	9

Table 2. *Speedup*: ratio of running times of two algorithms. *Quality:* ratio of the number of labeled road sections computed by two algorithms.

	Ratio	European cities					North American Cities						
		BE	LO	PA	RO	VI	BA	BO	LA	MO	SE	WA	Avg.
Speedup	$\frac{\text{MILP}}{\text{D\&C+MILP}}$	3.44	3.07	2.51	1.71	3.12	1.44	2.33	1.3	1.79	3.1	1.32	**2.29**
	$\frac{\text{TREE}}{\text{D\&C+TREE}}$	1.77	1.8	1.73	1.62	1.71	1.57	1.71	1.37	1.75	1.68	1.35	**1.64**
	$\frac{\text{D\&C+MILP}}{\text{D\&C+TREE}}$	2.82	2.32	3.33	2.54	2.74	6.84	3.06	21.59	6.36	5.32	10.59	**6.14**
Quality	$\frac{\text{D\&C+TREE}}{\text{TREE}}$	1.01	1.0	1.0	1.0	1.01	1.01	1.0	1.01	1.02	1.01	1.02	**1.01**
	$\frac{\text{D\&C+TREE}}{\text{MILP}}$	1.0	1.0	0.99	0.99	0.99	0.96	0.99	0.96	0.97	0.97	0.91	**0.97**
	$\frac{\text{Mapnik}}{\text{MILP}}$	0.74	0.85	0.83	0.91	0.76	0.71	0.8	0.62	0.61	0.8	0.68	**0.75**
	$\frac{\text{BaseLine}}{\text{MILP}}$	0.58	0.49	0.4	0.38	0.48	0.39	0.42	0.39	0.46	0.37	0.24	**0.42**
	$\frac{\text{D\&C+TREE}}{\text{Mapnik}}$	1.36	1.19	1.2	1.09	1.29	1.37	1.25	1.55	1.58	1.21	1.33	**1.31**

by several popular map-services. We considered the three scale factors 4.773, 2.387 and 1.193, which approximately correspond to the map scales 1:16000, 1:8000, 1:4000[2]. Further, they correspond to the *zoom levels* 15, 16 and 17, respectively, which are widely used by map services as OpenStreetMap. Those zoom levels show road networks in a size that already allows labeling single road sections, while the map is not yet so large that it becomes trivial to label the roads. We applied the standard drawing style for OpenStreetMap, which in particular includes the stroke width and color of roads as well as the font size of the labels. Further, this specifies for each zoom level the considered road categories; the higher the zoom level the more categories are taken into account.

Our implementation is written in C++ and compiled with GCC 4.8.4 using optimization level -O3. MILPs were solved by Gurobi[3] 6.0. The experiments were performed on a 4-core Intel Core i7-2600K CPU clocked at 3.4 GHz, with 32 GiB RAM. The D&C-approach labels single components in parallel. For computing the Delaunay triangulation we used the library Fade2d[4].

---

[2] wiki.openstreetmap.org/wiki/Zoom_levels.

[3] www.gurobi.com.

[4] www.geom.at.

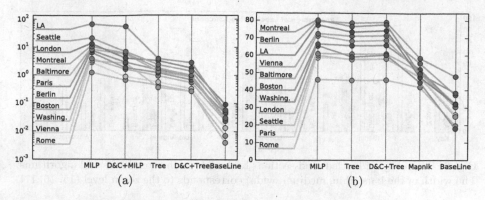

Fig. 5. (a) Running times in seconds of the algorithms (logarithmic scale). (b) Percentage of labeled road sections over all zoom levels broken into the different algorithms.

For each city and each zoom level we applied the algorithms BASELINE, TREE, D&C+TREE, MILP and D&C+MILP. We adapted the algorithm such that short road sections (shorter than the width of the letter W) are not counted, because they are rarely visible. Further, we let Mapnik (Version 3.0.9) render the same input. For each label we identified for each of its letters the closest road section r with the same name and counted it as labeled. Since Mapnik does not optimize the labeling by the same criteria as we do, we compensate this by also counting neighboring road sections as labeled if the junction in between them is not incident to any other road section. This accounts for those long road sections that we split artificially to resolve ISSUE 3.

The raw data of our experiments is made available on i11www.iti.kit.edu/roadlabeling. On this page we also provide interactive maps of the cities Berlin, London, Los Angeles and Washington, which present the computed labelings.

Phase 1. With a maximum of 67s (London, zoom 17) and 27s averaged over all instances, Phase 1 can be applied on large instances in reasonable time. During Phase 1 the number of segments is reduced to between 40% and 83% of the original instance (measured after Step 3, before creating junction edges); see also [7]. This clearly indicates that the procedure aggregates many lanes, since by design the approach does not change the overall geometry, but the simplification maintains the shape of the original network. This is also confirmed by the labelings; see Fig. 1(b)–(c) and interactive maps.

Phase 2, Running Time. We first consider the average running times over all zoom levels; see Fig. 5(a). We did not measure the running times of Mapnik, because its labeling procedure is strongly interwoven with the remaining rendering procedure, which prevents a fair comparison. As to be expected MILP is the slowest method (max. 126s, Los Angeles, ZL 15), while BASELINE is the fastest procedure (max. 0.17s). Combining MILP with D&C yields an average speedup of 2.29 over all instances and a max. speedup of 3.44; see Table 2.

The algorithm TREE needs less than 4.7s and its median is about 1.3s. Hence, despite its worst-case cubic asymptotic running time, it is fast in practice.

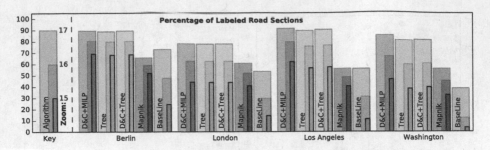

Fig. 6. Percentage of labeled road sections broken down in zoom levels and algorithms. The width of the bars (thin, medium, wide) corresponds to the zoom level (15, 16, 17).

Similar to MILP, it is further enhanced by combining it with D&C for a speedup of 1.64 with respect to TREE, and an average speedup of 6.14 with respect to D&C+MILP; see Table 2. In the latter case it even has a maximum speedup of about 21.6. Since decomposing and composing the labelings is done sequentially, the theoretically possible speed up using D&C is not achieved.

If we break down the running times into single zoom levels, we observe similar results; see also [7]. Since with increasing zoom level the instance size grows, for most of the algorithms also the running time increases. Only for North American cities and MILP do we observe that the running time for instances of smaller zoom levels are higher than for larger zoom levels.

Phase 2, Quality. First we analyze the average percentage of labeled road sections over the three zoom levels; see Fig. 5(b). As an upper bound, MILP, which provably solves MAXLABELEDROADS optimally, yields results from 46.2 % (Rome) to 80.3 % (Montreal). Considering zoom levels independently, we obtain a minimum of 27.5 % (Rome, ZL 15) and a maximum of 91.7 % (Montreal, ZL 17). We think that the wide span is attributed to the different structures of road networks and road names, e.g., Rome has a lot of short alleys and long road names. Hence, many road components are too short or convoluted to contain a single label. Abbreviating road names could help to overcome this problem.

The algorithm D&C+TREE yields marginally better results than TREE, but only 1 % on average, see Table 2. Comparing D&C+TREE with MILP we observe that D&C+TREE yields near-optimal results with respect to our road-section based model. On average it reaches 97 % of the optimal solution; see Table 2. While the quality ratio is only 91 % for Washington, more than half of the instances are labeled with a quality ratio of ≥99 %. For European cities the percentage of road sections that belong to components that are optimally solved by TREE (long edges, paths, and trees) is notably higher (88.3 %) than for North American cities (60.5 %). Nonetheless, we obtain similar percentages of labeled road sections for North American Cities. Hence, the heuristic computing a spanning tree of non-tree components is both fast and yields near-optimal results. The additional implementation effort of TREE is further justified by the observation that the naive way to place labels only on single road sections lags far behind;

only 42 % on average, 58 % as maximum and 24 % as minimum compared to the optimal solution. Mapnik achieves on average 75 % of the optimal solution and a maximum of 91 %. For more than the half of the instances Mapnik achieves at most 76 % of the optimal solution. So in direct comparison, D&C+TREE consistently outperforms Mapnik. Moreover, D&C+TREE has a better utilization of labels and achieves an average ratio of 1.61 labeled road sections per label, compared to Mapnik with a ratio of 1.37; see also [7].

With increasing zoom level the number of labeled road sections is increased, which is to be expected, since more road sections become long-edges; see Fig. 6 for four cities (similar results apply for the others). For each zoom level, we observe similar results as described before: TREE and D&C+TREE achieve near-optimal solutions and Mapnik labels considerably fewer road sections. However, for smaller zoom levels the gap between MILP and Mapnik shrinks.

From a visual perspective, labels lie on the skeleton of the road network, which is achieved by design; see Fig. 1(c) and the interactive maps. Instead of unnecessary repetition of labels, labels are only placed if they actually convey additional information. In particular, visual components are labeled, but not single lanes that are indistinguishable due to the zoom level.

6   Conclusion

We introduced a framework for labeling road maps based on an abstract road graph model that is combinatorial rather than geometric. We showed in our experimental evaluation that our proposed heuristic for decomposing the road graph into tree-shaped subgraphs and labeling those trees provably optimally is efficient and effective. It has running times in the range of seconds to one minute even for large road networks such as London with more than 100,000 road sections and achieves near-optimal quality ratios (on average 97 %) compared to upper bounds computed by the exact method MILP. Our algorithm clearly outperforms the labeling algorithm of the standard OSM renderer Mapnik, with an average improvement in the number of labeled road sections of 31 %. Interestingly, MILP is able to compute mathematically optimal solutions within a few minutes for all our test instances, even though it is slower by a factor of about 6 compared to the tree-based algorithm. So for practical purposes there is a trade-off between a final, but rather small improvement in quality at the cost of a significant and by the very nature of MILP unpredictable increase in running time. We only implemented essential cartographic criteria to evaluate the algorithmic core of our framework; further criteria (e.g., abbreviated names) and alternative definitions of road sections can be easily incorporated. The framework can be pipelined with labeling algorithms for other map features, e.g., after placing labels for point features, one may block all parts of the road network covered by a point label and label the remaining road network such that no labels overlap. While this allows the labeling of different types of features sequentially, constructing a labeling of all features in a single step remains an open problem.

Acknowledgment. We thank Andreas Gemsa for many interesting and inspiring discussions, and his help on the implementation.

References

1. Bader, M., Weibel, R.: Detecting and resolving size and proximity conflicts in the generalization of polygonal maps. In: International Cartographic Conference (ICC 1997), pp. 1525–1532 (1997)
2. Chirié, F.: Automated name placement with high cartographic quality: city street maps. Cartogr. Geogr. Inf. Sci. **27**(2), 101–110 (2000)
3. Gemsa, A., Niedermann, B., Nöllenburg, M.: Label placement in road maps. In: Paschos, V.T., Widmayer, P. (eds.) CIAC 2015. LNCS, vol. 9079, pp. 221–234. Springer, Heidelberg (2015)
4. Imhof, E.: Positioning names on maps. Am. Cartogr. **2**(2), 128–144 (1975)
5. Maass, S., Döllner, J.: Embedded labels for line features in interactive 3D virtual environments. In: Computer Graphics, Virtual Reality, Visualisation and Interaction (AFRIGRAPH 2003), pp. 53–59. ACM (2007)
6. Neyer, G., Wagner, F.: Labeling downtown. In: Bongiovanni, G., Petreschi, R., Gambosi, G. (eds.) CIAC 2000. LNCS, vol. 1767, pp. 113–124. Springer, Heidelberg (2000)
7. Niedermann, B., Nöllenburg, M.: An algorithmic framework for labeling road maps. CoRR. arXiv:1605.04265 (2016)
8. Reimer, A., Rylov, M.: Point-feature lettering of high cartographic quality: a multi-criteria model with practical implementation. In: EuroCG 2014 (2014)
9. Schwartges, N., Wolff, A., Haunert, J.-H.: Labeling streets in interactive maps using embedded labels. In: Advances in Geographic Information Systems (ACM-GIS 2014), pp. 517–520. ACM (2014)
10. Seibert, S., Unger, W.: The hardness of placing street names in a Manhattan type map. Theor. Compt. Sci. **285**, 89–99 (2002)
11. Strijk, T.: Geometric algorithms for cartographic label placement. Dissertation, Utrecht University (2001)
12. Vaaraniemi, M., Treib, M., Westermann, R.: Temporally coherent real-time labeling of dynamic scenes. In: Computing for Geospatial Research Applications (COM. Geo 2012), pp. 17:1–17:10. ACM (2012)
13. van Kreveld, M.: Geographic information systems. In: Handbook of Discrete and Computational Geometry, 2nd edn., Chap. 58, pp. 1293–1314. CRC Press (2010)
14. Wolff, A., Strijk, T.: The map labeling bibliography (2009). http://liinwww.ira.uka.de/bibliography/Theory/map.labeling.html

Measuring Cognitive Load for Map Tasks Through Pupil Diameter

Peter Kiefer[1(✉)], Ioannis Giannopoulos[1], Andrew Duchowski[2], and Martin Raubal[1]

[1] Institute of Cartography and Geoinformation, ETH Zürich,
Stefano-Franscini-Platz 5, 8093 Zürich, Switzerland
{pekiefer,igiannopoulos,mraubal}@ethz.ch
[2] School of Computing, Clemson University,
100 McAdams Hall, Clemson, SC, USA
duchowski@clemson.edu

Abstract. In this paper we use pupil diameter as an indicator for measuring cognitive load for six different tasks on common web maps. Two eye tracking data sets were collected for different basemaps (37 participants and 1,328 trials in total). We found significant differences in mean pupil diameter between tasks, indicating low cognitive load for *free exploration*, medium cognitive load for *search, polygon comparison, line following*, and high cognitive load for *route planning* and *focused search*. Pupil diameter also changed over time within trials which can be interpreted as an increase in cognitive load for *search* and *focused search*, and a decrease for *line following*. Such results can be used for the adaptation of maps and geovisualizations based on their users' cognitive load.

1 Introduction

The cognitive load users must cope with has been identified as a major criterion for the design of geographic visualizations and geographic human-computer interfaces [5,7,9]. This has become even more relevant in the age of mobile computing where geographic information is presented on constrained interfaces and under stressful, distractive and multi-tasking conditions [11,14].

The notion of cognitive load was introduced by Cognitive Load Theory (CLT) as a means of describing how the mental effort of learners is influenced by the design of learning material [23,30,31]. The geovisualization community has considered CLT mainly with respect to *extraneous cognitive load*, which denotes the cognitive load determined by the complexity of the information presentation, e.g., by the design of the map [5], or the number and type of animations [12]. It has been argued that high cognitive load may lead to less efficient and less effective map reading [20] and spatial orientation [28], as well as decreased spatial learning [21]. Recently, the cognitive load of experts and novices during a visual search task on a map was compared [24,25] and interpreted as differences in *germane cognitive load*, which 'reflects the effort that contributes to the construction of schemas' [32] in permanent memory.

© Springer International Publishing Switzerland 2016
J.A. Miller et al. (Eds.): GIScience 2016, LNCS 9927, pp. 323–337, 2016.
DOI: 10.1007/978-3-319-45738-3_21

This paper takes a different view: for two map designs and one level of expertise, it investigates the *intrinsic cognitive load* [31] of six different tasks people typically perform on maps. We hypothesize that certain tasks (e.g., route planning) are more demanding for the working memory than others (e.g., comparing the area of polygons), thus inducing a higher cognitive load. This hypothesis is investigated through a user study on two different basemaps (Google Maps™ and OpenStreetMap; 1,328 trials in total, taken from 37 participants).

We use the pupil diameter while performing these tasks as a measure for cognitive load, which has repeatedly been shown to be a reliable indicator [2,13, 15,19]. Significant within-subject differences in the mean pupil diameter between tasks were found which we interpret as evidence for differences in the intrinsic cognitive load these tasks evoke. More precisely, we conclude on low cognitive load for the task *free exploration*, medium cognitive load for *search*, *polygon comparison*, *line following*, and high cognitive load for *route planning* and *focused search*. A further analysis reveals changes of pupil diameter over time within trials, suggesting an increase in cognitive load for *search* and *focused search*, and a decrease for *line following*.

Our paper is the first using pupillometry for the analysis of eye tracking data recorded during map interaction, thus aiming to contribute to *"fundamental empirical research and state-of-the-art evaluation methods within [...] geographic information visualization and cognition"* [8]. Further, since pupil diameter can be measured in real-time, our results have the potential to be used for adaptive maps [27] that change based on the user's current cognitive load.

We proceed as follows: Sect. 2 provides background on cognitive load and how it can be measured with eye tracking. Section 3 introduces our method, including experimental design and stimulus selection for the eye tracking experiments. We report and discuss results in Sects. 4 and 5, before concluding the paper in Sect. 6.

2   Related Work

2.1   Cognitive Load

Cognitive load was introduced in the 1980 s as a theory of learning [30], targeted at an improvement of learning material. The theory suggests *'that total cognitive load is an amalgam of at least two quite separate factors: extraneous cognitive load which is artificial because it is imposed by instructional methods and intrinsic cognitive load over which instructors have no control'* [31, p. 307].

Sweller identifies element interactivity as the main reason for intrinsic cognitive load [31] (we return to this in Sect. 5). Extraneous cognitive load, on the other hand, is determined by the presentation of the material and can be influenced by the instructor. For instance, the design of a map can be either supportive or impedient for task solving [5]. The higher the intrinsic and/or the extraneous load, the less capacity remains in working memory for germane cognitive load – a third type of cognitive load which occurs during schema acquisition and automation [26,32].

Bunch and Lloyd distinguish subjective from objective ways of measuring cognitive load. While the first are based on interviews or questionnaires, the latter can be achieved by measuring the performance of participants in a secondary (parallel) task [5]. What they omit are ways of measuring cognitive load using physiological sensors, such as eye tracking, galvanic skin response or electroencephalogram [10]. The first of these – eye tracking – has been applied to map tasks in recent work [24,25]: experts and novices were found to have different average fixation durations and frequency which has been attributed to differences in germane cognitive load. In this paper, we focus on intrinsic cognitive load of different tasks and utilize a different eye tracking measure: the pupil diameter.

2.2 Cognitive Load and Pupil Diameter

We are not the first using eye tracking methodology in GIScience and Cartography. Much progress has been made in topics such as map interpretation, map interaction, spatial decision-making, and wayfinding (see [17], Sect. 2 for a comprehensive overview). In this paper, instead of analyzing *where* on a stimulus someone is looking, we focus on pupil diameter - a novel approach in GIScience and Cartography.

It has long been recognized that a relationship exists between cognitive load and pupil diameter [4]. Although Hess and Polt [13] demonstrated correlation between pupil dilation and problem difficulty, i.e., pupil size increases with problem difficulty, their early study was limited by several factors, including state of the technology available at the time. Their study observations were based on camera recordings of five participants' eyes, with a 16-mm Arriflex camera taking image samples at 2 frames per second. Given multiplication problems of different complexity to solve, the pupils of each participant typically showed a gradual increase in diameter, reaching a maximum dimension immediately before a response was given, then reverting to the previous control size.

Pupil dilation, in response to a given assignment meant to elicit mental activity, is referred to as Task-Evoked Pupillary Dilation (TEPD) or Task-Evoked Pupillary Response (TEPR) [1,3]. Using a television pupillometer sampling at 20 Hz, Ahern and Beatty [1] measured pupil diameter in a slightly updated replication of Hess and Polt's mental arithmetic experiment. In all correct responses to the assigned multiplication, pupillary responses showed a common pattern of dilation followed by a slight constriction after presentation of the multiplicand. A larger dilation was evoked by the multiplier; this increase in pupillary dilation was maintained during the problem-solving period. More difficult problems evoked larger pupillary dilations, reconfirming the relationship between problem difficulty and task-evoked activation.

Here, we test the *dilation reflex*, i.e., the relationship of pupil dilation to varying task demands, in the context of mentally processing geographic information. Instead of analog or digital cameras, we evaluate the utility of the pupil diameter as produced by a head-mounted eye tracker. Klingner et al. [18] review past uses of eye trackers for measuring TEPR. Confirming that an eye tracker can be used to measure cognitive load via measurement of pupil diameter, they suggest measurement following a 2 s delay after stimulus onset. While they advocate detailed timing and evaluation of short-term pupillary response, we adopt what Klingner et al. refer to as a coarse measurement of the time-aggregated style of data processing, i.e., an aggregated measurement of pupil diameter over a long period of time. Such coarse measurements have been successfully applied in previous studies, such as Hyönä et al.'s experiment on language tasks of different complexity [15, Experiment 1].

Marshall analyzes pupil diameter [22] suggesting that the dilation reflex undergoes oscillatory changes during different levels of cognitive load. They claim the measurement is reliable across hardware platforms and sampling rates [2]. Their approach relies on a sophisticated multiscale (wavelet) analysis of the pupil diameter frequency, e.g., effectively measuring pupillary hippus, or *pupil unrest* [29]. However, according to Beatty and Lucero-Wagoner [4], in addition to reflexive control of pupillary size, the tiny, cognitively related, fluctuations in pupillary diameter are visually insignificant and appear to serve no functional purpose whatsoever. Whether characterization of pupil unrest is a reliable measure of cognitive load appears debatable.

Here we intend to evaluate pupil diameter in state space instead of frequency space, a more straightforward and accessible method albeit potentially more susceptible to confounds stemming from the light reflex, or the pupil's response to light levels.

3   Method

Data collection was performed in two separate studies following the same design, setup, and procedure but differing in the basemap used: Google Maps^{TM} for the first study (**GMaps**), OpenStreetMap for the second (**OSM**). Both studies took place in 2013. The **GMaps** dataset has previously been used for a paper on activity recognition [16].

We are not studying map design here, i.e., we will not compare cognitive load of **GMaps** vs. **OSM**. The rationale for using two datasets is rather to get an indication on whether results generalize over at least two map designs.

3.1   Experimental Design

The study followed a within-subject design with one independent variable (*task*) and one dependent variable (*mean pupil diameter*, measured in millimeters). Six test conditions were considered for *task* (see also [16]):

T1 *free exploration*: exploring the map at free will. ("You have 20 s for exploring the map. You can look at whatever you want.")

T2 *search*: searching for a point of interest ("On the following map, please search for X", where X is given by its label.)

T3 *route planning*: planning the shortest route between two cities ("Do you see X and Y? Please, plan the shortest route from X to Y.")

Fig. 1. Hardware setup for the two studies.

T4 *focused search*: searching for the 3 closest points of interest of a certain type on a 'you are here'-map ("Do you see your position (the blue dot)? Please, search for the three closest Z", where Z is an object type.)

T5 *line following*: counting intersections while following a road with one's gaze ("Do you see X? Please, follow X from North to South and count the number of intersections", where X is a road name and cardinal directions were systematically varied)

T6 *polygon comparion*: comparing the area of two lakes ("Do you see X and Y? Please compare the areas of these two lakes and name the bigger one.")

3.2   Participants

Participants for each of the two experiments were recruited through a university mailing list. All were university students or already holding a university degree. None of them used maps in their profession (i.e., no cartographer, geographer, land planner etc.); therefore they can all be regarded as having the same level of expertise. A monetary compensation of 15 CHF (Swiss Francs) was offered.

GMaps: 19 participants took part; 2 were excluded from further analyses due to calibration errors. From the remaining 17 participants, 10 were female. The average age was 28 years (SD: 8.7). **OSM**: 20 participants (11 female) took part and none was excluded. The average age was 23.8 years (SD: 7.4).

3.3   Apparatus

Data were recorded using the SMI (v1.8) head-mounted eye tracking glasses (30 Hz)[1] and transmitted via a USB cable to a laptop. A chin rest was placed at a distance of 65 cm to the stimulus in order to guarantee that the viewer would look at the monitor along an axis perpendicular to the monitor plane. We used

---

[1] http://www.smivision.com/en.html.

two 24" widescreen LED monitors (1920 × 1200 pixels, Samsung *S24A850DW*). One monitor was used to display the stimulus, the other one for controlling the experiment (see Fig. 1). The experiment was controlled through our own software framework which chooses and presents a random set of stimuli, including instructions and previews, plus an (optional) re-calibration screen (refer to Sects. 3.4 and 3.5). Shutters and constant ceiling lights ensured the same lighting conditions in the room over all trials.

3.4   Procedure

Participants were introduced to the experiment. They were told they would have to solve simple tasks on maps. The eye tracker was mounted, and the participant was asked to rest her head on the chin rest. A three-point calibration was performed. Each participant had 36 trials on different stimuli (refer to Sect. 3.5), presented in randomized order, where no two successive trials were from the same task. Each trial consisted of three phases:

1. *Instruction phase*: the participant was presented a textual description of the task (in German) and could ask questions.
2. *Preview phase*: either a preview showing small parts of the stimulus (**T3**, **T4**, **T5**, **T6**), or a black dot in the center (**T1**, **T2**) was shown. The goal of this phase was to clearly separate the task to be analyzed from an orientation activity beforehand. For instance, start and destination points for the route planning tasks were shown here. At the end of the preview phase the participant was asked to fixate a certain point in order to provide equal start conditions for all participants.
3. *Task phase*: the stimulus was shown, and the eye movements recording was started. The recording was either ended as soon as the participant indicated with a move of her hand that she had solved the task, or after a maximum of 20 s.

The experimenter checked the calibration after each trial. In case the calibration had been lost, the previous trial was considered 'not valid' (excluded from later analyses) and a re-calibration was performed.

3.5   Stimuli

Since we are not investigating map design here we chose stimuli from standard web maps as used by people in their daily routines[2]. Two different web maps were used as sources for the stimuli: Google Maps[TM][3] for the **GMaps** study, and OpenStreetMap[4] for the **OSM** study.

In order to ensure that participants see the exact same map extents, stimuli were static images (screenshots) without the possibility of panning or zooming.

---

[2] Studies on standard web maps have become quite common recently, e.g. [6].
[3] Before the 2013 redesign (classic style); not available online any more (6 May 2016).
[4] http://www.openstreetmap.org/.

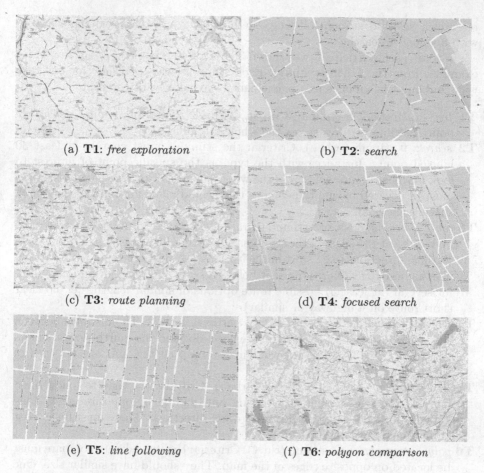

(a) **T1**: *free exploration* (b) **T2**: *search*

(c) **T3**: *route planning* (d) **T4**: *focused search*

(e) **T5**: *line following* (f) **T6**: *polygon comparison*

Fig. 2. Example stimuli (**GMaps** dataset). Zoom levels: 12 for (a, c, f), 18 for (b, d, e). (Color figure online)

Participants were supposed to be unfamiliar with the geographic area shown in the stimulus, but familiar with the language and cultural context to allow for reasonable search tasks. Since all participants were from Switzerland and native German speakers, we chose map extents from Germany and Austria. With a brief interview after the experiment we asserted they were indeed unfamiliar with the areas they had seen during the trials.

It is not possible to identify *the* representative instance of a certain task type, which implies that complexity within a task type generally varies (we return to this issue in Sect. 5). Stimuli were chosen in a way that all task instances for one task type were of a similar difficulty level. More specifically, easy tasks were avoided to ensure a certain task duration which would allow us to collect a sufficient amount of data. Selection criteria are detailed in the following. One

researcher selected stimuli following these criteria and discussed the selection
with a second researcher.

In total, each participant was shown 36 out of 40 stimuli (see Fig. 2 (a–f) for
examples from the **GMaps** study). Each stimulus was used only for one task
type and only shown once to a participant to avoid a learning effect:

T1 *free exploration*: 6 stimuli (3 urban, 3 rural). Criterion: similar density of
point and line features across the whole stimulus.

T2 *search*: 9 stimuli (urban). Criteria: the stimulus must contain at least 30
labeled points. Instances with the type as the specific point of interest to
look for must be present across the whole stimulus No large part of the map
must be covered by empty polygons that would allow for limiting the search
space, such as an ocean.

T3 *route planning*: 6 out of 8 stimuli (rural). Criteria: start and destination must
be located at the edges of the stimulus. One stimulus for each pair of opposite
cardinal directions (e.g., start in the North-East, destination in South-West).
The highest road priority present between start and destination must allow
for several (at least 5) possible route options of similar length (i.e., no clear
short route on a highway or similar).

T4 *focused search*: 5 stimuli (urban). Criteria: as for **T2**. The distance between
the third closest point to the 'you are here'-dot and the fourth closest should
be similar. One stimulus with dot in the map center, and one for each of
North/South/East/West.

T5 *line following*: 6 out of 8 stimuli (urban). Criteria: the road to follow must
traverse the whole stimulus, starting and ending at opposite edges. One
stimulus for each pair of opposite cardinal directions. There must be at least
10 intersections along the road.

T6 *polygon comparion*: 4 stimuli (rural). Criteria: the two lakes to compare must
be located on opposite edges of the map. They should have similar size. One
stimulus for each pair of opposite cardinal directions.

Stimulus selection criteria were the same for both studies (**GMaps** and **OSM**),
therefore 80 stimuli were used in total.

The luminance was measured at the distance of the participant's eyes to
the stimulus (accumulated local luminance) for each map stimulus. The results
showed that the luminance was constant throughout the whole experiment, with
a constant *lux* value of *270* (measured with testo 540, ISO 9001:2008). This
ensures changes in pupil diameter are not caused by different color, hue, or
contrast profile of the individual stimuli.

4   Results

As described in Sect. 3.4, some trials were considered 'invalid' due to calibration
issues. The number of valid trials (out of 1,404 recorded) used for the analysis
was 1,328 (**T1**: 222; **T2**: 332; **T3**: 220; **T4**: 185; **T5**: 221; **T6**: 148). The average
trial duration was 15.27 s (SD=5.55 s).

The eye tracker recorded for each gaze (at 30 Hz) the pupil diameter in millimeters which will be used as the basis for the following analyses.

4.1 Differences in Mean Pupil Diameter Between Tasks

The mean pupil diameter was calculated for every single task (aggregated trials) performed by each participant and was used as input for within-subjects analyses [15]. A Friedman test revealed that there were statistically significant differences between the measured mean pupil diameter for the six map tasks, $\chi^2(5) = 89.649$, $p < .001$. Post-hoc analyses with the Wilcoxon signed-rank test were performed, revealing statistically significant differences between several map tasks (see Table 1(a)). Median (IQR) pupil diameters for tasks **T1** to **T6** were 2.53, 2.63, 2.68, 2.66, 2.64 and 2.61, respectively. Minimum and maximum pupil diameters for tasks **T1** to **T6** were (1.85, 3.09), (1.96, 3.47), (1.97, 3.59), (2.03, 3.65), (2.00, 3.55), (1.89, 3.44) (all in millimeters).

Figure 3(a) illustrates an ordering between the tasks, based on the above results. An example is illustrated in Fig. 3(b), showing the results obtained from a single user.

Analyses were also performed on the two different map services separately. A Friedman test revealed that there was a statistically significant difference between the measured mean pupil diameter for the six map tasks in each of the two map cases, GMaps and OSM, $\chi^2(5) = 46.681$, $p < .001$ and $\chi^2(5) = 63.629$, $p < .001$, respectively.

Post-hoc analyses with the Wilcoxon signed-rank test revealed statistically significant differences between several map tasks (see Table 1(b) for **GMaps** and Table 1(c) for **OSM**). Median (IQR) pupil diameters for task **T1** to **T6** for **GMaps** were 2.64, 2.78, 2.90, 2.83, 2.81 and 2.95, respectively. Median (IQR) pupil diameters for task **T1** to **T6** for **OSM** were 2.36, 2.49, 2.54, 2.53, 2.48 and 2.45, respectively.

4.2 Change in Pupil Diameter Within Trials

To evaluate the change in pupil diameter within each task, we follow to a certain extent Klingner et al. [18]. That is, Klingner et al. compute the change in pupil diameter (in mm), presumably with respect to a baseline signal. It is not clear, however, how large a temporal window was used over which the baseline was measured. They note that stimulus onset (spoken multiplicand) occurred 5 s after measurement began. From the data reported, it appears that the baseline measurement occurred over the first 2 s. Klingner et al. note that a smoothing filter was used to smooth the pupil diameter data.

We follow Klingner et al. [18] by computing our within-trial pupil change with respect to a baseline signal, captured over a variable-length temporal window (0.5, 1.0, 1.5, and 2.0 s). Prior to our computation, following Klingner et al., we also apply a Butterworth filter to smooth the raw pupil diameter data (see Fig. 5). We use a 2nd degree Butterworth filter set to 1/30 half-cycles per sample (the point at which the gain drops to $1/\sqrt{2}$ of the passband). Smoothing of the

Table 1. Differences in avg. pupil diameter within participants between tasks. Read the tables as follows: avg. pupil diameter for *task in line* is significantly smaller than for *task in column*.

(a) All trials (both datasets combined).

	T1		T2		T3		T4		T5		T6	
	Z	p	Z	p	Z	p	Z	p	Z	p	Z	p
T1	-		-5.288	<.001	-5.303	<.001	-5.137	<.001	-5.273	<.001	-4.956	<.001
T2	-		-		-3.432	<.001	-4.247	<.001	-		-	
T3	-		-		-		-		-		-	
T4	-		-		-		-		-		-	
T5	-		-		-2.663	<.01	-3.251	<.001	-		-	
T6	-		-		-2.663	<.01	-2.467	<.05	-		-	

(b) **GMaps** dataset.

	T1		T2		T3		T4		T5		T6	
	Z	p	Z	p	Z	p	Z	p	Z	p	Z	p
T1	-		-3.621	<.001	-3.621	<.001	-3.621	<.001	-3.621	<.001	-3.574	<.001
T2	-		-		-		-3.574	<.001	-		-2.817	<.01
T3	-		-		-		-		-		-	
T4	-		-		-		-		-		-	
T5	-		-		-		-2.627	<.01	-		-2.533	<.05
T6	-		-		-		-		-		-	

(c) **OSM** dataset.

	T1		T2		T3		T4		T5		T6	
	Z	p	Z	p	Z	p	Z	p	Z	p	Z	p
T1	-		-3.883	<.001	-3.920	<.001	-3.659	<.001	-3.845	<.001	-3.211	<.001
T2	-		-		-2.912	<.005	-2.539	<.05	-		-	
T3	-		-		-		-		-		-	
T4	-		-		-		-		-		-	
T5	-		-		-2.576	<.05	-2.240	<.05	-		-	
T6	-		-3.845	<.001	-3.920	<.001	-3.509	<.001	-4.247	<.001	-	

pupil diameter effectively denoises the signal by removing the high frequency component, attributable to high frequency pupil diameter oscillation known as *pupil unrest* or *hippus* [29].

For each of the temporal windows, we used a univariate type-III repeated-measures ANOVA assuming a 2×6 mixed design where the independent variables were map type (between-subjects at two levels: **GMaps**, **OSM**) and task (within-subjects at 6 levels; see Sect. 3.1). The dependent variable was mean pupil change computed as the mean of the pupil diameter difference from the mean diameter over the baseline time window, averaged over 20 s.

For a 0.5 s baseline (see Fig. 4(a)), the effect of task was significant ($F(5, 175) = 14.64, p < 0.01$) but the map type was not ($F(1, 35) = 1.38, p = 0.25$, n.s.). The mean pupil difference was smallest during task **T5** ($M = -0.07$), and differed significantly from each of the tasks **T2**, **T4**, and **T6** ($p < 0.01$).

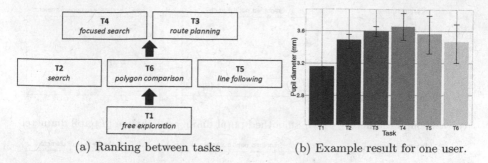

(a) Ranking between tasks.   (b) Example result for one user.

Fig. 3. Figure 3(a) ranks the tasks from significantly smaller (bottom) to significantly bigger (top) mean pupil diameter based on the results illustrated in Table 1(a). Figure 3(b) exemplifies the results obtained from a single user.

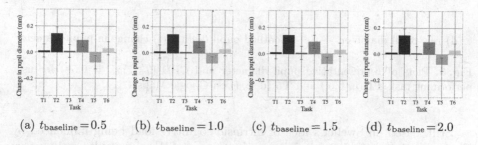

(a) $t_{baseline} = 0.5$ (b) $t_{baseline} = 1.0$ (c) $t_{baseline} = 1.5$ (d) $t_{baseline} = 2.0$

Fig. 4. Change in Pupil Diameter (CPD) with different baseline windows.

Significant differences between mean pupil difference (at the $p < 0.01$ level) were also observed between tasks **T1** and **T2**, **T2** and **T3**, and **T2** and **T6**. Similar results were observed at larger baseline windows of 1.0–2.0 s (see Figs. 4(b)–(d)).

5   Discussion

Although pupil diameter is a well-known indicator for cognitive load [2,13,15,19], it is also influenced by other factors, most importantly luminance (which was controlled for by the study setup) and fatigue. A potential effect of fatigue would apply to all tasks which were shown in a randomized order, therefore it is safe to assume the observed effect has been caused by differences in cognitive load.

Figure 3(a) summarizes our main results: we hypothesized differences in cognitive load between 6 tasks, and we indeed were able to group them into 3 classes of significantly different mean pupil diameter, suggesting differences in cognitive load. Setting up more detailed hypotheses from the beginning would have been speculation since, to our knowledge, no complete and heuristically proven cognitive model for these 6 map tasks exists yet.

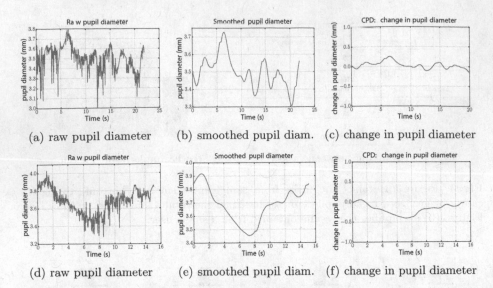

(a) raw pupil diameter   (b) smoothed pupil diam.   (c) change in pupil diameter

(d) raw pupil diameter   (e) smoothed pupil diam.   (f) change in pupil diameter

Fig. 5. Plots of representative pupil diameter and CPD ($T_{baseline} = 0.5\,s$) for user performing task **T1** (a, b, c) and **T5** (d, e, f) on Google Maps^{TM}.

Still, based on Sweller's idea of intrinsic cognitive load being influenced by the interactivity[5] of elements relevant for the task [31], our results make sense: it is no surprise that *free exploration* (**T1**) has the lowest cognitive load since nothing needs to be kept in working memory. *Polygon comparison* (**T6**), with medium cognitive load, can be solved by regarding the interaction of two map elements (the two lakes). During *line following* (**T5**), the participant at any moment needs to keep in working memory the road, the previous and current intersection, and a counter. *Search* (**T2**), another task with medium cognitive load, can be solved by keeping in memory all point objects that have been looked at already and their positions. *Focused search* (**T4**) is similar to **T2**, but with the additional requirement to estimate distances to the blue dot. Finally, a high cognitive load for *route planning* (**T3**) is reasonable since it requires a large number of map elements and their interaction to be considered.

The temporal within-trial analyses (Sect. 4.2) added further insights: they indicate that the cognitive load of tasks **T1**, **T3**, and **T6** remained on the level it was at the start of the task. For instance, *free exploration* does neither have higher or lower cognitive load in later phases of the task than at the beginning (see Fig. 5 (a,b,c) for an example). Cognitive load of the two search tasks (**T2**, **T4**) seems to increase, which is plausible since the number of visited points that needs to be kept in working memory increases as well. The decrease of cognitive load for **T5** (refer to Fig. 5 (d,e,f)) is more difficult to interpret: peripheral vision

---

[5] A potential definition of element interactivity here would be the number of elements whose relation needs to be kept in working memory to solve a task successfully without having to keep the relations to or between any other elements in memory.

might play a role here. The next intersection(s) relevant for the counting is/are most likely already perceived in the periphery in later phases of the task, which is not true when the stimulus 'pops up' at the start of the task.

Is it possible that we are observing changes in extraneous or germane, instead of intrinsic cognitive load [26,31]? Germane cognitive load would occur if the participants learned schemata for the tasks. Our tasks are common, so it is unlikely participants created new schemata for, say, *route planning*. Extraneous cognitive load would be an issue if the design of the basemap were specifically supportive or obstructive for some tasks. We approach this question by comparing the overall results (Table 1(a)) with the basemap-specific results (Tables 1(b, c)). The differences between **OSM** and the overall results are small; instead of being in the same 'medium cognitive load' class, **T6** in **OSM** causes significantly less cognitive load than **T2** and **T5** (which makes sense w.r.t. the number of interacting elements). In **GMaps**, on the other hand, there are larger differences: **T6** is now on the same (high) level as **T4**, while **T3** is only significantly higher than **T1**, but not than any other task. This result might be interpreted as Google Maps^{TM} being more supportive for *route planning* (**T3**), but less supportive for *polygon comparison* (**T6**) than OpenStreetMap.

As described in Sect. 3.5, the stimuli were selected by two human raters following a set of criteria with the aim of identifying '*common*' cases for each task (neither too easy nor too difficult). The results are thus generalizable to tasks that are close to the introduced selection criteria, but probably do not apply to *all* potential instances of a task type, such as route planning with start and destination being directly connected by one street segment. Also, performing the same task on a different scale (e.g., route planning in a city) might lead to different results. Concerning generalizability over maps, the presented ranking (see Fig. 3(a)) is based on the two tested popular map services. We do not claim that the presented results will hold independent of any map service.

Though we controlled for familiarity with the geographic areas, we did not control for familiarity with the map design. It can be assumed that participants were more familiar with **GMaps** than with **OSM**, potentially leading to lower cognitive load for **GMaps**. A comparison of cognitive load between map types, however, was not the aim of this study.

We did not include a short delay before task onset (unlike, e.g., Klingner et al. [18]). Instead, task onset began as soon as the stimulus appeared, and our analyses relied on either coarse (aggregated) pupil diameter (Sect. 4.1, similar to [15]) or within-trial changes (Sect. 4.2). Determination of the baseline temporal window for the latter is difficult. In our case it appears that cognitive demand begins fairly quickly. This gives credence to the use of a short temporal window. On the other hand, the longer the temporal baseline window, the less change in pupil diameter, on average, can be expected.

6   Conclusion

This paper is the first using pupil diameter as a measure for cognitive load while solving map tasks. We applied this measure to two datasets collected through studies on different web maps. We were able to group 6 map tasks into 3 classes of significantly different mean pupil diameter which we interpreted as differences in cognitive load: low (*free exploration*), medium (*search, polygon comparison, line following*), and high (*focused search, route planning*).

These results may motivate pupillometry to be used for future studies on cognitive load in GIScience research, such as during wayfinding [17]. It would further be interesting to investigate the correlation between the number of interacting elements on a map and cognitive load more systematically (refer to Sect. 5, [31]). Future gaze-contingent map interfaces may use our method to recognize cognitive load in real-time and adapt accordingly.

Acknowledgement. Supported by the Swiss National Science Foundation (grant no. 200021-162886).

References

1. Ahern, S., Beatty, J.: Pupillary responses during information processing vary with Scholastic Aptitude Test Scores. Science **205**(4412), 1289–1292 (1979)
2. Bartels, M., Marshall, S.P.: Measuring cognitive workload across different eye tracking hardware platforms. In: ETRA 2012: Proceedings of the 2008 Symposium on Eye Tracking Research and Applications, Santa Barbara, CA. ACM (2012)
3. Beatty, J.: Task-evoked pupillary responses, processing load, and the structure of processing resources. Psychol. Bull. **91**(2), 276–292 (1982)
4. Beatty, J., Lucero-Wagoner, B.: The pupillary system. In: Cacioppo, J.T., Tassinary, L.G., Bernston, G.G. (eds.) Handbook of Psychophysiology, 2nd edn, pp. 142–162. Cambridge University Press, Cambridge (2000)
5. Bunch, R.L., Lloyd, R.E.: The cognitive load of geographic information. Prof. Geogr. **58**(2), 209–220 (2006)
6. Coltekin, A., Lokka, I.E., Boér, A.: The utilization of publicly available map types by non-experts - a choice experiment. In: Proceedings of the 27th International Cartographic Conference (ICC2015), Rio de Janeiro, Brazil, pp. 23–28 (2015)
7. Fabrikant, S.I., Goldsberry, K.: Thematic relevance and perceptual salience of dynamic geovisualization displays. In: Proceedings 22th ICA/ACI International Cartographic Conference, Coruna (2005)
8. Fabrikant, S.I., Lobben, A.: Introduction: cognitive issues in geographic information visualization. Cartogr.: Int. J. Geogr. Inf. Geovisualization **44**(3), 139–143 (2009)
9. Giannopoulos, I., Kiefer, P., Raubal, M., Richter, K.-F., Thrash, T.: Wayfinding decision situations: a conceptual model and evaluation. In: Duckham, M., Pebesma, E., Stewart, K., Frank, A.U. (eds.) GIScience 2014. LNCS, vol. 8728, pp. 221–234. Springer, Heidelberg (2014)
10. Haapalainen, E., Kim, S., Forlizzi, J.F., Dey, A.K.: Psycho-physiological measures for assessing cognitive load. In: Proceedings of the 12th ACM international conference on Ubiquitous computing, pp. 301–310. ACM (2010)
11. Harrison, R., Flood, D., Duce, D.: Usability of mobile applications: literature review and rationale for a new usability model. J. Interact. Sci. **1**(1), 1–16 (2013)
12. Harrower, M.: The cognitive limits of animated maps. Cartogr.: Int. J. Geogr. Inf. Geovisualization **42**(4), 349–357 (2007)

13. Hess, E.H., Polt, J.M.: Pupil size in relation to mental activity during simple problem-solving. Science **143**(3611), 1190–1192 (1964)
14. Hirtle, S.C., Raubal, M.: Many to many mobile maps. In: Raubal, M., Mark, D., Frank, A. (eds.) Cognitive and Linguistic Aspects of Geographic Space - New Perspectives on Geographic Information Research, pp. 141–157. Springer, Heidelberg (2013)
15. Hyönä, J., Tommola, J., Alaja, A.M.: Pupil dilation as a measure of processing load in simultaneous interpretation and other language tasks. Q. J. Exp. Psychol. **48**(3), 598–612 (1995)
16. Kiefer, P., Giannopoulos, I., Raubal, M.: Using eye movements to recognize activities on cartographic maps. In: Proceedings of the 21st SIGSPATIAL International Conference on Advances in Geographic Information Systems, pp. 488–491. ACM, New York (2013)
17. Kiefer, P., Giannopoulos, I., Raubal, M.: Where am I? Investigating map matching during self-localization with mobile eye tracking in an urban environment. Trans. GIS **18**(5), 660–686 (2014)
18. Klingner, J., Kumar, R., Hanrahan, P.: Measuring the task-evoked pupillary response with a remote eye tracker. In: Proceedings of the 2008 Symposium on Eye Tracking Research and Applications, pp. 69–72. ACM (2008)
19. Kruger, J.L., Hefer, E., Matthew, G.: Measuring the impact of subtitles on cognitive load: eye tracking and dynamic audiovisual texts. In: Proceedings of the 2013 Conference on Eye Tracking South Africa, pp. 62–66. ACM (2013)
20. Lloyd, R.E., Bunch, R.L.: Explaining map-reading performance efficiency: gender, memory, and geographic information. Cartogr. Geogr. Inf. Sci. **35**(3), 171–202 (2008)
21. Lloyd, R.E., Bunch, R.L.: Learning geographic information from a map and text: learning environment and individual differences. Cartogr.: Int. J. Geogr. Inf. Geovisualization **45**(3), 169–184 (2010)
22. Marshall, S.P.: Method and Apparatus for Eye Tracking Monitoring Pupil Dilation to Evaluate Cognitive Activity. US Patent No. 6,090,051, 18 July 2000
23. Mayer, R.E., Moreno, R.: Nine ways to reduce cognitive load in multimedia learning. Educ. Psychol. **38**(1), 43–52 (2003)
24. Ooms, K., De Maeyer, P., Fack, V.: Study of the attentive behavior of novice and expert map users using eye tracking. Cartogr. Geogr. Inf. Sci. **41**(1), 37–54 (2014)
25. Ooms, K., De Maeyer, P., Fack, V., Van Assche, E., Witlox, F.: Interpreting maps through the eyes of expert and novice users. Int. J. Geogr. Inf. Sci. **26**(10), 1773–1788 (2012)
26. Paas, F., Renkl, A., Sweller, J.: Cognitive load theory and instructional design: recent developments. Educ. Psychol. **38**(1), 1–4 (2003)
27. Reichenbacher, T.: Adaptive concepts for a mobile cartography. J. Geogr. Sci. **11**(1), 43–53 (2001)
28. Rossano, M.J., Moak, J.: Spatial representations acquired from computer models: cognitive load, orientation specificity and the acquisition of survey knowledge. Br. J. Psychol. **89**(3), 481–497 (1998)
29. Stark, L., Campbell, F.W., Atwood, J.: Pupil unrest: an example of noise in a biological servomechanism. Nature **182**(4639), 857–858 (1958)
30. Sweller, J.: Cognitive load during problem solving: effects on learning. Cogn. Sci. **12**(2), 257–285 (1988)
31. Sweller, J.: Cognitive load theory, learning difficulty, and instructional design. Learn. Instr. **4**(4), 295–312 (1994)
32. Sweller, J., Van Merrienboer, J.J., Paas, F.G.: Cognitive architecture and instructional design. Educ. Psychol. Rev. **10**(3), 251–296 (1998)

Author Index

Printed in the United States
By Bookmasters